Transformers

Principles and Applications

Second Edition

Kenneth L. Gebert

Kenneth R. Edwards

 AMERICAN TECHNICAL PUBLISHERS, INC.
HOMEWOOD, ILLINOIS 60430

Basis for cover illustrations by
courtesy of McGraw-Edison, Power
Systems Division; and Distribution
Transformer Department, General
Electric Company.

Basis for title page illustration
by courtesy of Distribution
Transformer Department, General
Electric Company.

Preface to the Second Edition

To give the apprentice and journeyman, along with the newcomer to the field, a necessary background for understanding transformers, this Second Edition of *Transformers* has been completely revised, updated and presented in a simplified, easy-to-read format. The authors have extended every effort to provide numerous up-to-date photos to complement the latest information on transformer application.

As a necessary background the Second Edition of *Transformers* presents basic electrical and magnetic principles as applied to transformers. The interesting details of building the core, winding the coils and assembling them on the core, and connecting them to the external leads or terminals are then presented. The new edition discusses both the core and shell-types of construction as well as some of the newer methods used in building transformers.

Chapter 2 explains the principles of operation and classifies and subclassifies transformers. Various vector diagrams are provided. Chapter 3 covers cooling methods, including the NEMA and ANSI industry standards for transformer insulating materials that stipulate the maximum temperature rise. Further chapters include information on tap changers (automatic and manual types), transformer connections, autotransformers, and reactors.

Other new and important material added to the Second Edition of *Transformers* is a chapter on "Transformer Maintenance," including an "Inspection Check List for Transformers and Regulators." It is very important to inspect and maintain a transformer not only to prevent costly failures and interruption of service but also to prolong the life of the transformer. This final chapter also includes the latest safety rules that must be followed when working with transformers.

Because of the wide and thorough coverage presented by the authors, *Transformers* promises to be a useful and valuable tool for apprentices and tradesmen working all levels in the industry.

The Publishers

Contents

Acknowledgments

The authors wish to extend their sincere appreciation to the companies and associations who have supplied the illustrations and technical data that appear throughout this book. The revision of this book would have been impossible without this assistance. Special appreciation is extended to:

Mr. Bert Chase, Director of Safety and Training, Florida Power and Light, Miami, Florida.

Mr. William Eldridge, General Supervisor of Customer Services, Northern States Power Company, Minneapolis Division.

Mr. Samuel Fletcher, Vice President, Virginia Electric Power, Richmond, Virginia.

Mr. Gary G. Forcey and Mr. Dave Tucker, Marketing Communications Account Representative, Westinghouse Electric Corporation, Pittsburgh, Pennsylvania.

Mr. Marvin Gellman and Mr. Ed Herbert of General Electric News Bureau, Washington, D. C.

Mr. Earl Kammerer, Electrical Engineer, Supervisor of Electrical Inspections, City of Minneapolis.

Mr. E. D. Olson, District Engineer, Westinghouse Electric Corporation, Minneapolis, Minnesota.

Mr. William Semmer, Manager of Construction Sales, Agency and Distributor Sales Division, Minneapolis District, General Electric Company.

Mr. Ernie Thompson, Sales Engineer, McGraw-Edison Power Systems Division, Minneapolis, Minnesota.

Mr. C. V. Walker, Manager of Marketing Communications, McGraw-Edison Power Systems Division, Canonsburg, Pennsylvania.

Mr. John Webb, Chairman, Joint Apprenticeship and Training Program, Southwestern Line Constructors, Arlington, Texas.

NEMA (National Electrical Manufacturers Association).

ANSI (American National Standards Institute).

NFPA (National Fire Protection Association).

Mr. Edwards wishes to thank his wife, Marie, who compiled parts of the manuscript and read the original course version.
Mr. Gebert wishes to thank his wife, Marion, who with much patience and understanding typed the manuscript.

Transformer Construction

Introduction

Use of Transformers

If an inventory were taken of the electrical appliances in the home or of the modern electrical equipment in industry that has been added in the past ten years, it would be found that the use of electrical appliances and equipment has greatly increased. The demand for electrical power has more than doubled in the past ten years and it is predicted that the demand will double again in less than ten years. This increase along with the trend of people and industry to migrate to the suburbs has created the problem of transmitting larger amounts of electrical power over longer distances than ever before.

Fortunately, there is at our disposal a "Transformer" which is a piece of electrical supply equipment that makes it possible to efficiently transmit large amounts of electrical power over long distances. In construction a transformer is an efficient, relatively simple, quiet device that has no moving parts thus limiting maintenance costs. Although it is simple in construction, it is somewhat complex in operation.

Increase of Voltage. The transformer is used in the transmission and distribution of power to reduce the conductor size in line construction. Electric power, being the product of voltage and current, may result from multiplying a very large current by a very small voltage; that is, in transmitting any given power, if the voltage factor is increased, the current factor will be decreased. Electrical voltage (volts) is the force which moves electrons through conductors to result in current (amps) flow. As the cross-section or size of the line conductors is determined by the amount of current it will carry, it may be smaller in size if the current is decreased.

It has been found that the size of the conductor necessary for transmitting a given amount of power a given distance depends inversely upon the *square* of the voltage, that is, if the voltage between transmission lines is doubled, a conductor only one-fourth as large is necessary to transmit the same power over the same distance with the same line loss. If the voltage is increased three times, the size of the line conductor is only one-ninth.

In the transmission of electrical power the function of the transformer is to raise or "step-up" the voltage at the generator to a much higher value, thus reducing the size of the transmission conductor. It is obvious that the installation and maintenance cost will be less for a smaller conductor.

Alternating current generators (alternators) normally produce voltages of up to 22,000 volts which must in turn be stepped-up to a higher voltage for economical transmission to the load or to the place where the power will be used. To transmit industrial loads for short distances or for low power demand areas such as residential the transmission lines usually operate at voltages of 12,470; 13,200; 13,800 or 14,400. To transmit large amounts of power over a considerable distance very high voltages are necessary. Experimental work has progressed to the point where engineers have found that voltages of 750,000 volts can be successfully and economically used over relatively short distances. For longer distances an even higher voltage is used. The transmission lines which carry power from Boulder Dam to the west coast operate at a voltage of about 500,000 volts. It may be of interest that even at this extra high voltage, special design is used for the transmission lines to reduce their weight. They are constructed of interlocking sections of hollow aluminum tubes which permits them to be strung between widely-separated towers.

Reduction of Voltage. While it is desirable to resort to very high voltages for transmission purposes, such voltages are undesirable when electrical power is distributed among buildings or within densely populated areas. It is the function of a transformer to lower or step-down the voltage at the load or the building where the power is to be used to a value that is safe and economical for use inside a residential, commercial or industrial building. For residential wiring, the normal voltage is 115/230 volts, whereas the voltage in a commercial or industrial building is usually 120/208 volts for appliances and incandescent lighting and 277/480 volts for fluorescent light and power. In large commercial buildings and industrial plants the modern trend is to use a 12,470 to 24,000 volt distribution system. Power at this voltage is delivered to substations within the building where transformers step down the voltage to the level needed.

Definition of Transformer

It will be well at this point to define a transformer. A transformer is a device to which electrical power is supplied at a definite voltage and from which electrical power is obtained at a different voltage. The transformer basically is a voltage-changing device. It does not change one form of power into another, as does a generator or a motor. Although one factor of power, the voltage, may be increased, that does not mean that something is being created by the transformer. Every *increase in voltage* means a corresponding *decrease in current*.

Having defined a transformer with reference to its function, its mechanical construction should next be considered. A transformer consists of two electrical circuits interlinked by a magnetic circuit. This should appeal to Odd Fellows who recognize the insignia of the "three links." One of the two electrical circuits is designated as the *primary*, and the other as the *secondary*, while the link connecting the two is the *core*. The *primary* circuit receives the energy and is referred to as the *input* whereas the *secondary* circuit discharges the energy and is referred to as the *output*. A transformer is a very efficient device and if we exclude the core and copper losses (usually about 1%) the input power will equal the output power.

Theory of Transformer Operation—Elementary Principles

Relation Between Electric Current and Magnetic Field

It was first discovered about 1819 that when an electric current was allowed to flow in a conducting wire of copper, a magnetic field always existed in the space around the conductor. This was the beginning of electrical engineering, and in the consideration of transformers it is very important to have a very thorough understanding of the fixed relation existing between the direction

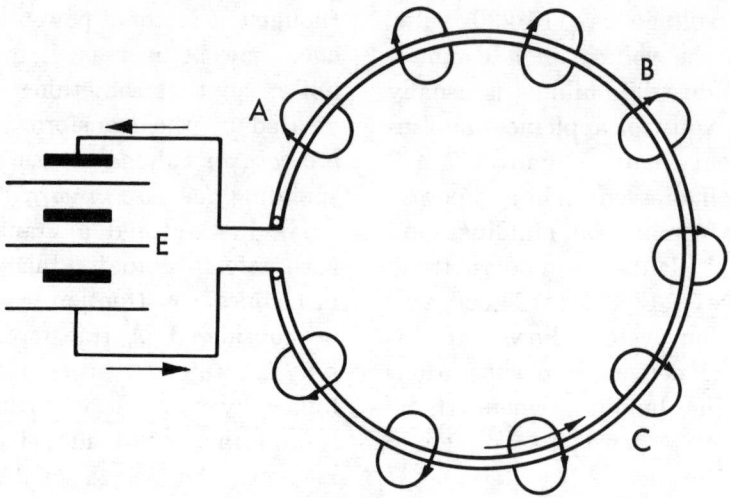

Fig. 1-1. Magnetic lines of force around a single conductor coil. (Magnetic lines of force are coming up through center of coil.)

of the current in a conductor and the magnitude and the direction of the resulting magnetic field. This relation is shown graphically in Fig. 1-1. A circular conductor *ABC* is connected with a battery of three cells (*E*) connected together in series and sending a current around the circular conductor in a counter-clockwise direction, as denoted by the large arrow near *C*. The magnetic lines of force always surround current-carrying conductors in the form of complete curves, which are circular in the case of a straight wire removed some distance from other current-carrying wires. In the illustration, the surrounding magnetic lines are shown in circles having directions indicated by the arrow-heads.

Right-Hand Screw Rule. There is a definite relation between the direction of the current and the direction of the magnetic flux. This leads us to what is generally referred to as the *right-hand screw rule* which is illustrated in Fig. 1-2 and is stated as follows: *The direction of advance of a right-hand screw corresponds with the direction of current, while the direction in which the screw is turned corresponds with the direction of the magnetic field surrounding the wire.* In Fig. 1-3 the screwdriver is replaced by a piece of wire with the current flowing from left to right. You will notice that when the right hand is used with the thumb pointing in the direction of the current, the fingers will point in the direction

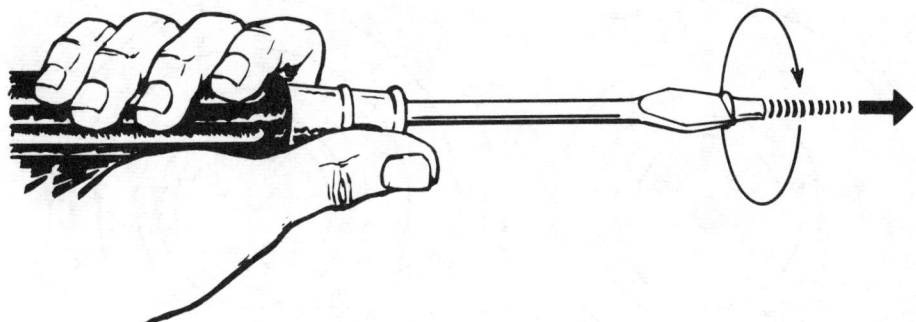

Fig. 1-2. Right-hand screw rule.

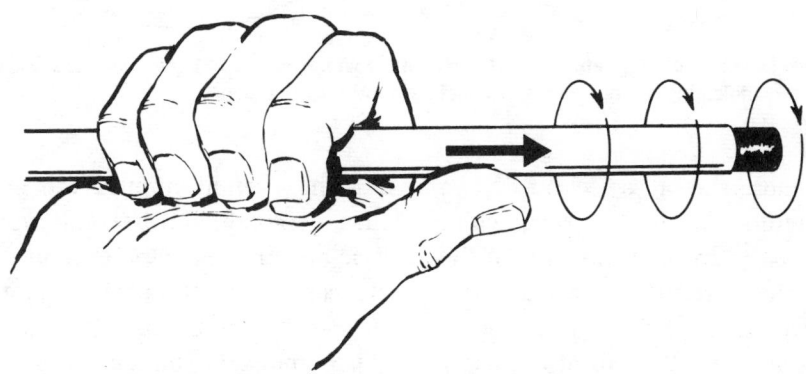

Fig. 1-3. Right-hand rule (current-flux relation).

of the magnetic flux around the wire.

The relation between the current and magnetic field is further emphasized in Fig. 1-4. The inner circle denotes the cross-section of a conductor arranged perpendicular to the page, in which the current is flowing down or away from the observer represented by the tail of an arrow. The magnetic field produced by the current consists of circular magnetic lines surrounding the conductor and having a clockwise direction. The two inner circles in Fig.

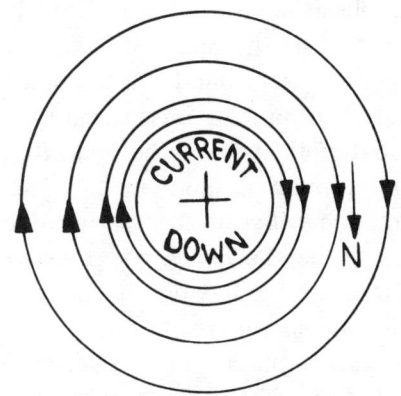

Fig. 1-4. Direction of magnetic lines of force with relation to current (one conductor). Note the arrow indicating the north pole in the magnetic field around the conductor.

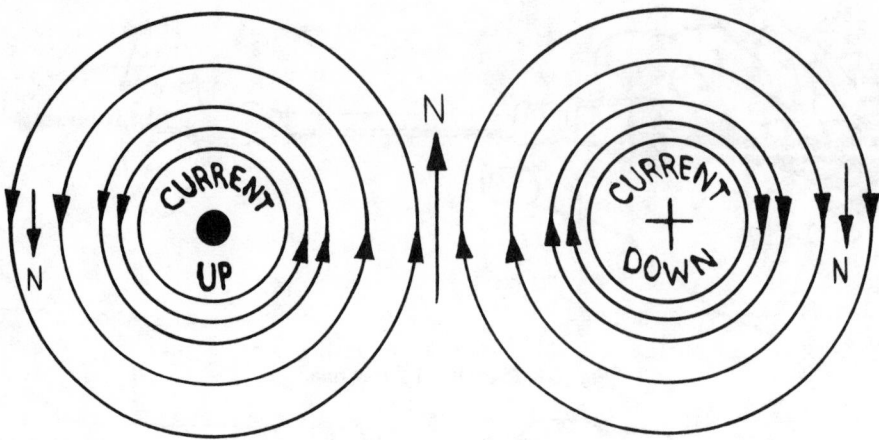

Fig. 1-5. Direction of magnetic lines of force with relation to current (two conductors). Note the arrows indicating the north magnetic field around the conductor.

1-5 denote the cross-section of two conductors arranged perpendicular to the page. In the conductor to the right the current is flowing down or away from the observer and is represented by the tail of an arrow while in the conductor to the left the current is flowing up or toward the observer and is represented by the head or the point of an arrow. When the right-hand rule is applied to each conductor in this figure, it is determined that the flux from each conductor flows in the same direction between the two conductors. Thus the fluxes add together to increase the flux density between the two conductors.

Another phase of the relation is that fact that a current flowing around a coil in a counterclockwise direction produces a north magnetic pole at the top of the coil. In Fig.

1-1 the current around the single turn of conductor is counterclockwise and the magnetic lines of force are coming up through the loop or turn. A north or *N* pole is one *from which* magnetic lines are pointing, while a south or *S* pole is one *into which* magnetic lines are pointing.

Magnetic Lines of Force

To establish magnetic lines of force in a magnetic circuit, a *magnetomotive force* must be applied. This force may come from the magnetic field around a permanent magnet or from the flow of an electric current through a wire or a coil of wire. Its unit of measurement is known as the *gilbert*. The lines of force always form closed loops and they travel in a path known as the *magnetic circuit*. The magnetic field is represented by lines of force and the total number of lines of force is called the *magnetic flux* whose unit is the *maxwell*. One *maxwell* is equal to one magnetic line of force. Whereas the magnetomotive force produces the magnetic flux, the *magnetizing force* is the magnetomotive force that produces the flux per unit length of the magnetic circuit. Its unit of measure is the *oersted* which is the magnetomotive force per unit length. The magnetizing force in *oersteds* is found by dividing the magnetomotive force in *gilberts* by the length in centimeters. Whereas in the foregoing the maxwell is the unit of measurement of the total number of lines of force in the magnetic field, the *flux density* is the measure of the lines of force per unit area taken at right angles to the direction of flux. Its unit of measure is the *gauss*, whose value is one maxwell per square centimeter.

The direction of the lines of force may be found by employing a small compass needle, the *N* end of which will always point in the direction of the lines, see Figs. 1-4 and 1-5. Fig. 1-6 shows the presence of magnetic lines of force around a conductor which is carrying an electric current. Concentric rings are formed by the iron filings around the conductor. Notice that the amount of flux in each path decreases as the length of the path increases, so that the flux density is stronger next to

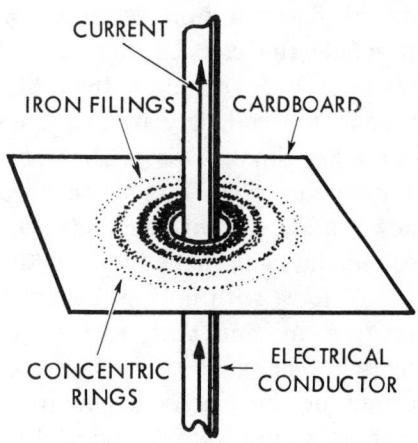

Fig. 1-6. Magnetic field around a current-carrying conductor.

the wire and gets weaker as it moves away from the wire.

Magnetic Conductivity, or Permeability

It is very important to note that different media allow magnetic lines of force to pass through them with different degrees of ease. For example, air is not a very good conductor for magnetic lines, neither is copper nor zinc. Iron, however, is an extremely good conductor for magnetic fields, or lines of force, and is therefore employed in place of air or other substances whenever possible. The advantages of employing iron instead of air as a medium for conducting magnetic lines may be appreciated better when it is stated that from experiment, with a given coil of wire consisting of a definite number of turns in which a certain current flows, if iron replaces the air inside the coil, the number of magnetic lines of force threading through the coil may be increased over 2000 times. This is the same as assuming that the single turn shown in Fig. 1-1 with a certain current in the loop might produce 200 lines of force with no iron near the arrangement, and that, without in any way changing the amount of current or the electrical circuit, if a soft-iron core were introduced so as to pass through the circular loop, the number of magnetic lines might be increased to 200 × 2000, or 400,-000 lines. The advantage of iron as a magnetic conductor is very evident.

The measure of the ease with which certain substances conduct magnetic lines of force is called magnetic permeability, or simply permeability. The Greek letter μ is often used to denote permeability. While iron is found to be the best conductor of magnetic lines of force, the various grades or kinds of iron differ greatly in magnetic conducting ability. The softer the iron the greater its conductivity, or its permeability; therefore it is advisable to employ pure soft iron in the cores of such devices as transformers.

Retentivity

One important characteristic of iron needs consideration in connection with its use in magnetic circuits, namely retentivity. Whenever the current is removed from such a circuit, Fig. 1-1, the magnetic lines all disappear or fade away. Should there be an iron core in the loop when the current is stopped, a considerable number of magnetic lines will remain in the iron, and the harder the iron, the greater the number of these retained lines. If very hard steel is used as a core instead of soft iron, many more lines will be retained. The steel will have become a so-called permanent magnet. The characteristic of retaining magnetic lines after the magnetizing cause has been removed is termed retentivity. Since high values of retentivity are very undesirable in transformers, only soft iron should be used.

Magnetic Induction

Steel, like all things, is composed of very small particles called molecules. Each molecule is regarded as a tiny magnet and the poles of these molecular magnets are distributed at random in an unmagnetized steel bar. Unmagnetized steel may be magnetized to a small degree if it is stroked by a permanent magnet, being careful that the same pole of the permanent magnet is used to stroke the unmagnetized steel in the same direction each time. When the unmagnetized steel bar is stroked with the pole of a permanent magnet some of the molecular magnets respond by lining up end-to-end thus having a combined field that makes the steel bar a weak permanent magnet. This is called *magnetic induction.* Whenever lines of force are produced, or set up, by an electric current, whether in air, steel, or iron, the magnetization is called *magnetic induction* or simply induction. When the current in circular loop, Fig. 1-1, is 1 ampere, a certain number of magnetic lines are set up in the air. Since the unit for amperes, or electric current is not the same as the unit for magnetic lines, the relation between the number of amperes of current and the number of lines must be expressed symbolically. In order to produce a magnetic field of a given number of lines, a current of a certain number of amperes must exist in the circular loop. This unit is known as the *ampere-turns* and is generally defined as *a unit of magnetomotive force which is obtained by multiplying the current in amperes by the number of turns in a coil.* Thus the magnetic lines of force are directly proportional to the ampere-turns of the coil.

An observation should be made at

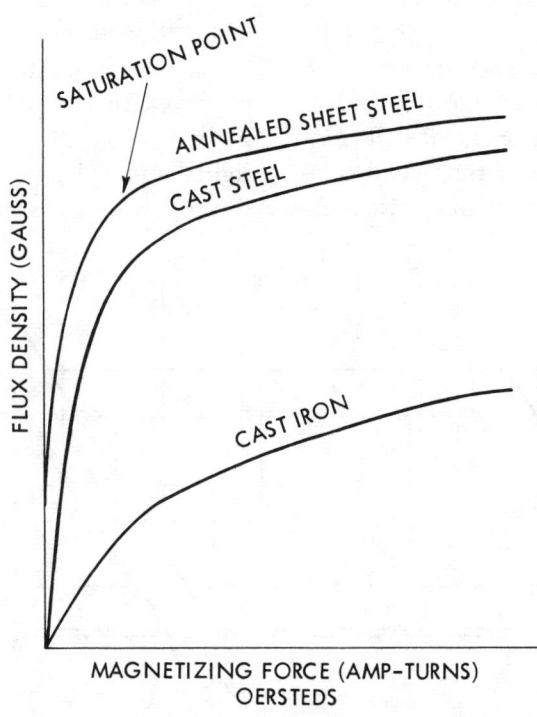

Fig. 1-7. Saturation curve.

this time in reference to what is called, the *saturation point* of a magnetic circuit. The relationship between the *flux density* and the magnetizing force (ampere-turns) for a magnetic material is shown in Fig. 1-7. When the point on the curve marked *saturation point* is reached, it is no longer practical to increase the current or turns of a coil as the magnetic circuit is said to be *saturated*.

Electromotive Force of Induction

Induced Current

In the foregoing paragraph there has been some discussion of magnetic induction, that is, the setting up of magnetic lines of force, or so-called magnetic flux, by an electric current. The reverse of this relation is of equal importance; namely, given the proper conditions, a changing magnetic field will produce an electric current. The relations of a changing magnetic field and a resulting induced current may best be explained by referring to Fig. 1-8. Let *AB* denote a conductor that may be moved back and forth along the rectangular circuit denoted by *CDEG* through the lines of force of a magnetic field having a direction downward, as indicated by the arrows. Suppose the slider is moved to the right, as indicated by the arrow *M*; according to the right-hand rule, which is given in the following paragraph, an electric

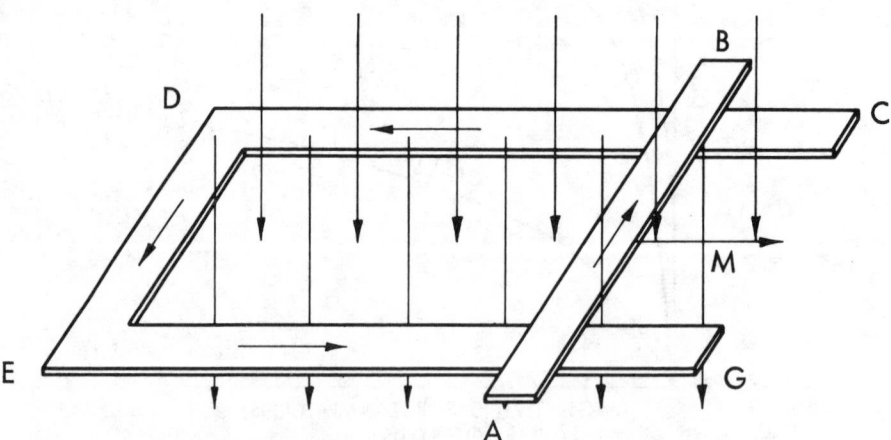

Fig. 1-8. Current induced by magnetic lines of force.

voltage having a direction from *A* to *B* will be induced in the moving slider *AB*.

Right-Hand Rule

The so-called right-hand rule should be very thoroughly understood by anyone who wishes to become proficient in the construction or operation of transformers. The rule is as follows: *Hold the right hand so that the thumb and the forefinger are perpendicular to each other and so that the second finger is perpendicular to both the thumb and the forefinger. Next place the hand in a magnetic field so that the forefinger points in the direction of the magnetic lines and the thumb points in the direction in which an inductor is moved across the lines of force. The second finger will then point in the direction of the induced voltage.*

This rule will become evident if Fig. 1-9 is studied carefully. The magnetic lines issue from the *N* pole, as shown by the forefinger. The lines are produced by a current from a battery of three cells, *E*, passing around the pole in a counterclockwise direction. If the moving inductor *AB* cuts the lines in the direction indicated by the thumb, the induced voltage will be from *A* to *B*, as indicated by the second finger.

Results of Moving Slider to Right

Returning to a consideration of Fig. 1-8, it should now be observed that an induced voltage from *A* to *B* in the moving inductor will cause a current to flow around the closed circuit *ABDE* in a counterclockwise direction. Further observation will show that a motion of *AB* to the right in effect causes more and more lines to be included by, or introduced into, the closed circuit *ABDE*. The next logical step in the process of our study is to obtain

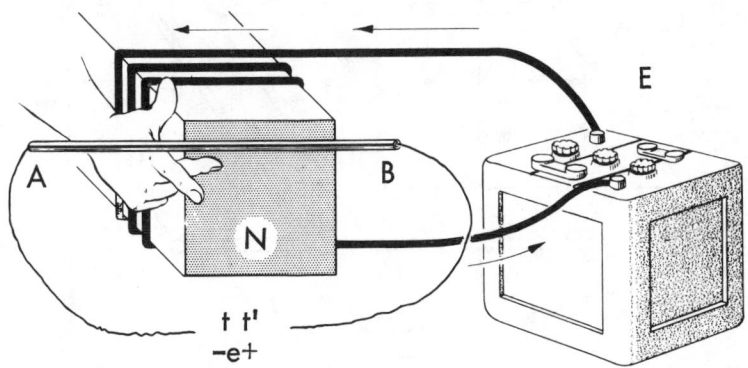

Fig. 1-9. Graphic representation of right-hand rule for induced currents.

some information as to the amount of pressure induced and the amount of the resulting current.

Relative Motion Between Circuit and Magnetic Field Necessary

It should be very carefully noted that an electrical voltage can only be *induced* while there is a *relative motion between a wire, or a circuit, and a magnetic field*. If the magnetic field that threads through a circuit is steady in value and the circuit is stationary, then there can be no induced voltage. A direct current that does not vary in its amount from one second to the next cannot therefore be employed in connection with transformers to effect changes, or transformations, in voltages. It should also be noted that if the circuit through which the number of magnetic lines is changing is an open one, no current can flow in the circuit, although there is an induced voltage. This is illustrated in Fig. 1-10, which shows the same arrangement as does Fig. 1-8 except that there is a cut at *tt'* in the portion of the circuit between *D* and *E*. In such a case when *AB* is moved to the right, a voltage is induced, as previously described and computed. The voltage exists between the portions of the circuit *t* and *t'* and according to the right-hand rule *t'* will be the positive, or +, terminal and *t* will be the negative, or −, terminal. If, while *AB* is moving and the voltage is being induced, *t* and *t'* are brought together, thus closing the circuit, the induced current can flow.

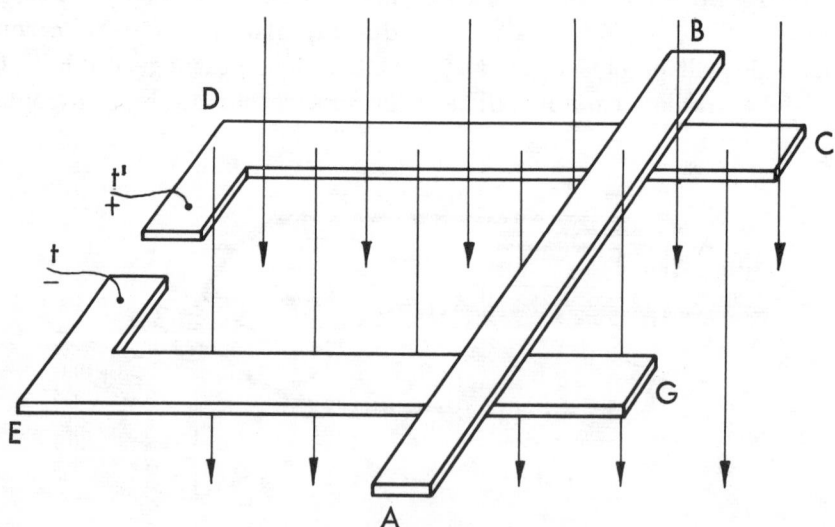

Fig. 1-10. Effect of broken circuit on induced current.

Left-Hand Rule

The opposite of the right-hand rule is quite naturally the left-hand rule and in transformer design and calculation it is quite as important as the right-hand rule. The left-hand rule serves to bring out the relation between the current in a wire and the field existing in the space about the wire and is as follows: *Hold the left hand so that the thumb, the forefinger, and the second finger are perpendicular to one another. Next place the hand in a magnetic field so that the forefinger points in the direction of the lines of force and the second finger points in the direction of the current in a conducting wire placed in the field perpendicular to the lines. Then the thumb will point in the direction in which the wire will tend to move.*

This rule may be better understood by referring to Fig. 1-11. The movable conductor is denoted by *AB;* the direction of the current is denoted by the arrows; and the direction of the magnetic lines of force is from the *N* end of the magnet. The conductor *AB* will tend to move to the left. In Fig. 1-12 *AB* denotes a conductor that can slide along the parallel conductors *CD* and *FG;* the uniform magnetic field has a direction downward, as shown by the vertical arrows; and a battery of cells at *E* can send a current through the circuit *FABD* in a counterclockwise direction. Under such conditions the sliding conductor *AB* will tend to move to the left, as indicated by the arrow *M*, and according to the left-hand rule this motion removes lines of force from the circuit.

Fig. 1-11. Graphic representation of left-hand rule for induced currents.

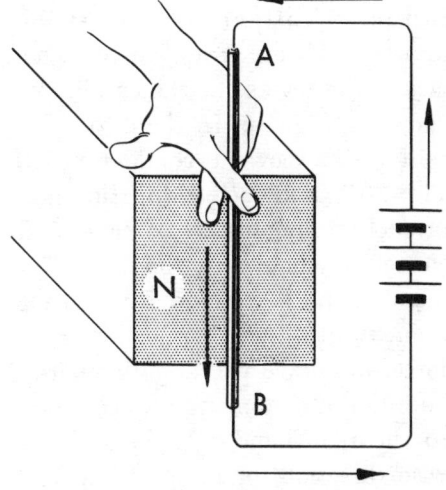

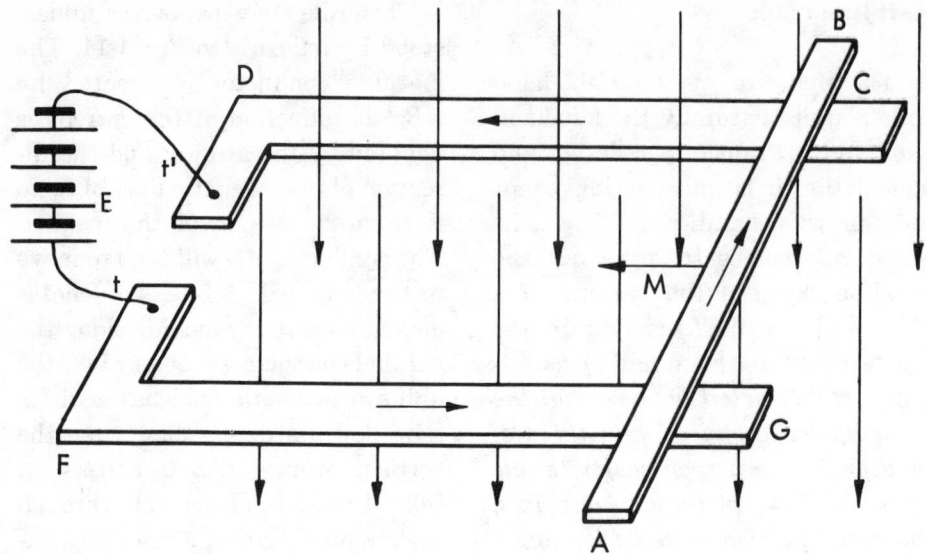

Fig. 1-12. Graphic representation of effect of motion on induced currents.

Counter Electromotive Force

According to the right-hand rule applying to a conductor moving across a magnetic field, whenever the number of lines threading through a circuit is decreased, the induced voltage is clockwise around such an arrangement. This is just the reverse of the conditions shown in Fig. 1-8. If, then, the current in the movable conductor *AB*, Fig. 1-12, is in such a direction and of such an amount as to cause *AB* to actually move to the left (Remember the Left-Hand Rule), thus *reducing* the number of lines of force, a voltage will simultaneously be induced in the circuit, according to the right-hand rule, in a clockwise direction, or directly opposite to that of the applied voltage of battery *E*.

This clockwise induced voltage due to the motion of *AB* sends a current around the circuit in a clockwise direction. The *resultant* current in the circuit must therefore be the difference between the current produced by the battery and that resulting from the induced voltage. Whenever the conductor *AB* stops moving, there is no induced voltage and hence no induced current, and the current in the circuit is simply one whose value may be expressed by Ohm's law:

$$I = \frac{E}{R}$$

in which *I* denotes the current, in amperes, in the circuit, *E* denotes the voltage, in volts, applied to the circuit, and R denotes the resistance, in ohms, of the complete circuit. Whenever the conductor AB is moving to the left, the current in the circuit is less than when the conductor is stationary. When the conductor is in motion, the current may be expressed by

$$I' = \frac{E-e}{R}$$

in which I' denotes the current in the circuit, E the applied voltage, and R the total resistance as before, but e denotes the induced voltage which, according to the right-hand rule, is directly opposite in direction to the applied battery voltage. Because of its opposite direction this voltage is called the *counter electromotive force;* and, being produced by induction or relative motion between an inductor and a magnetic field, it is called the electromotive force of induction.

If the direction of the current around the circuit $ABDEF$ is reversed, the movable portion AB will move to the *right*, according to the left-hand rule. In this case the number of lines of force threading through the circuit will be *increased* if AB moves, and the resulting induced voltage will be counterclockwise, according to the right-hand rule, or again in a direction exactly opposite to that of the applied voltage from the battery. Introducing lines of force into a circuit will produce an induced voltage equal in

value to, but opposite in direction to, that produced by removing the same number of lines in the same time, the *direction* of the lines being the same in both cases.

Rate of Change in Magnetic Flux

If the conductor AB, Fig. 1-12, moves to the left at a certain definite rate, the number of lines threading through the circuit is reduced in exact proportion to that rate. If the distance moved by the conductor is, say, 10 feet per second (120 inches), its velocity would be expressed as so many feet or inches per second. If the length of AB happens to be, say, 10 inches, and its velocity is 120 inches per second, it will sweep over an area equal to 120 times 10, which is 1200 square inches. If H, which is the number of magnetic lines per square inch passing downward through the circuit, happens to be 20,000, then the number of lines of force traversed by the moving inductor during one second of its motion will be 1200 times 20,000, or 24,000,000.

Thus the total area swept over by the moving portion of any circuit, such as that in Fig. 1-12, multiplied by the magnetic density, or the number of lines of force per unit area, will give the number of lines introduced into or removed from the circuit. The induced voltage in any circuit due to the rate of change in the magnetic lines of force induced by it, can be obtained by dividing the number of lines of force cut in one second by 100,000,000. According to Lenz's law, it is necessary for one hundred million magnetic lines of force (sometimes expressed at 10^8) to be cut in one second in order to produce one volt.

Thus the voltage produced by the above conductor will be 24,000,000 divided by 100,000,000 equals

$$\frac{24}{100}$$

or .24, which is about one-quarter of a volt.

Rate of Change in Current

It is very important now to realize that if the conductor *AB*, Fig. 1-12, should remain stationary and if 24,000,000 lines should be uni-formly removed each second from threading through the circuit, the voltage induced in the circuit would be exactly the same as though *AB* were moved to the left at the rate of 120 inches per second, if its length between the portions of the circuit upon which it slides were 10 inches and if the magnetic density were 20,000 lines of force per square inch. *Induced voltage always depends upon the relative motion of any circuit and a magnetic field threading through the circuit.* The question may be raised as to how the number of lines threading through the circuit may be removed. One obvious method is the reduction of the amount of current used to produce the field.

Suppose that a variable resist-

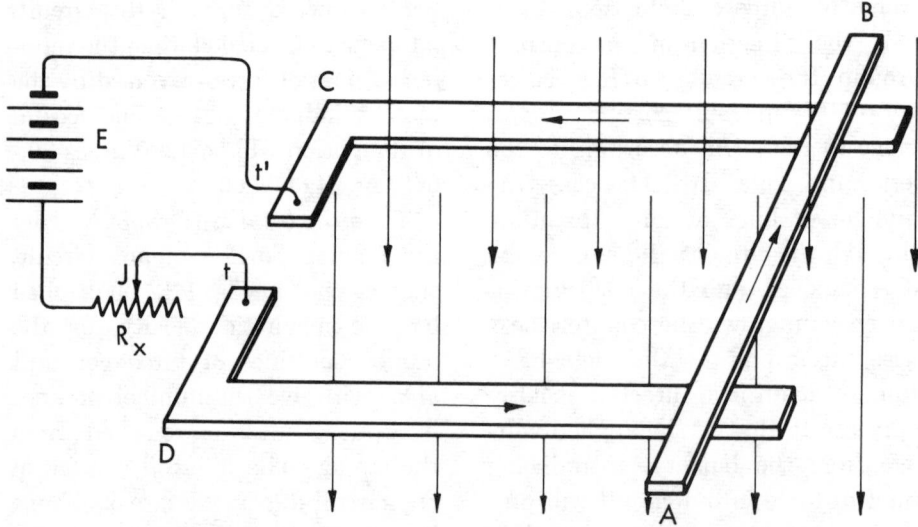

Fig. 1-13. Effect of change in current on magnetic lines of force.

ance, or rheostat, is connected into the circuit as indicated in Fig. 1-13 at R. By moving a variable contact represented by J to the right, the resistance at R_x is increased, the current in the circuit is decreased, and therefore the number of lines is reduced. The quicker the contact is moved, the greater the rate of change in the current and the greater the rate of change in the number of lines. Should the circuit be suddenly opened, thus suddenly stopping the flow of the current, the number of lines would be reduced to zero *very* suddenly—provided there was no iron in or near the circuit.

Operation of Transformer

It will now be in order to assemble the separate electrical and magnetic circuits which have been explained, to form an idealized transformer, Fig. 1-14. In this figure NS denotes a round bundle of soft-iron wires forming the core of a coil of three turns which has a current sent around it in a clockwise direction from a battery of cells denoted by E. This will cause the upper end of the core to be an S pole; the magnetic lines will have a downward direction as indicated. Suppose a circuit denoted by $ABCD$ is arranged as shown so that a certain number of magnetic lines thread downward through it. So long as the parts of the arrangement are stationary and the battery cur-

rent does not change in amount, *there will be no current* in the circuit $ABCD$. Should the battery circuit be opened, as indicated in Fig. 1-15, and the magnetizing current in the coil surrounding the iron core becomes zero, the number of lines threading downward through the circuit $ABCD$ will decrease rapidly and *a current will be induced* in circuit $ABCD$ in a clockwise direction according to the rules explained. The induced current in $ABCD$ is in the same direction as the current in the coil surrounding the core that produced the magnetic lines that threaded $ABCD$. The coil P surrounding the core NS may be designated as the *primary*, and the circuit $ABCD$ as the *secondary*, of a transformer in which the two electrical circuits are connected together, or interlinked, by the magnetic circuit consisting of the lines of force threading through the iron core, the primary, and the secondary. If the battery circuit is closed by joining t and t', the current in P will increase in value, thus *increasing* the number of lines threading through the secondary $ABCD$, and a *counterclockwise* current will be produced in the closed secondary circuit denoted by $ABCD$.

Magnetic Leakage

To one at all familiar with the form of the magnetic field produced by a coil of wire surrounding an

17

iron core, it will be evident that not all the magnetic lines produced by the magnetizing force of the ampere turns of the primary will pass through the secondary. The actual shape of the magnetic field produced by the magnetizing force of the primary coil is indicated in

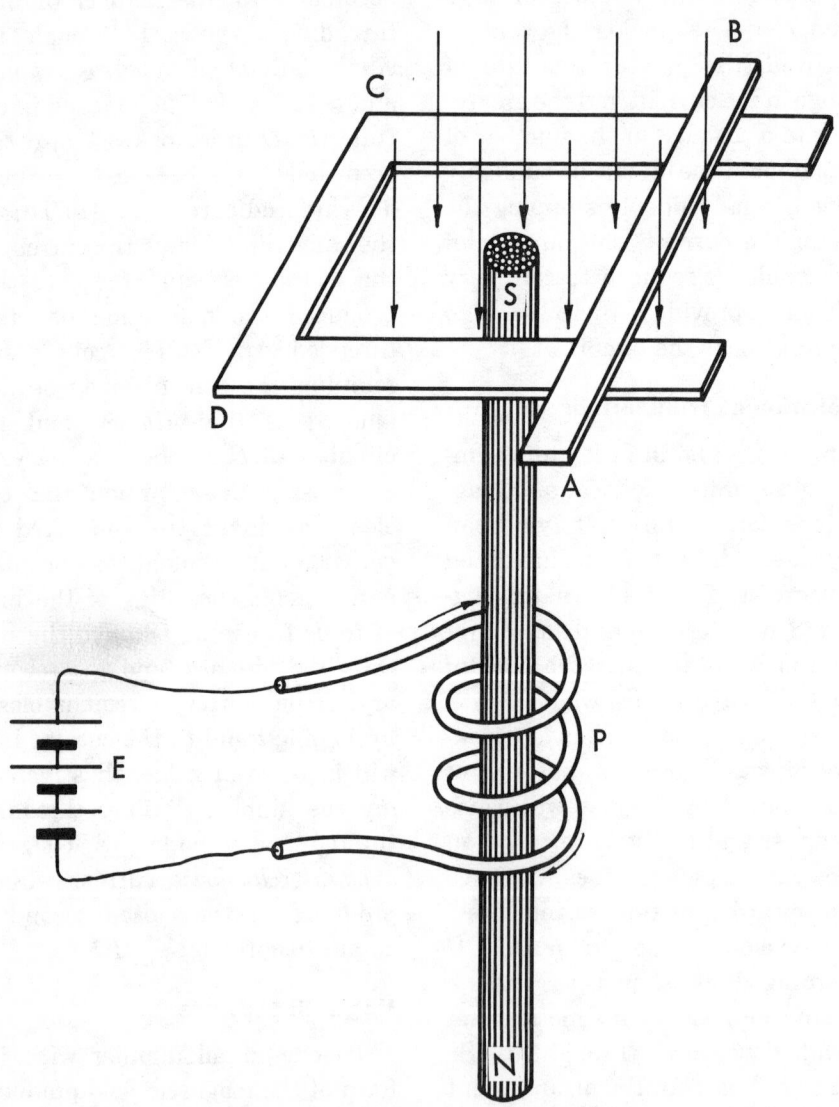

Fig. 1-14. Parts of simple transformer.

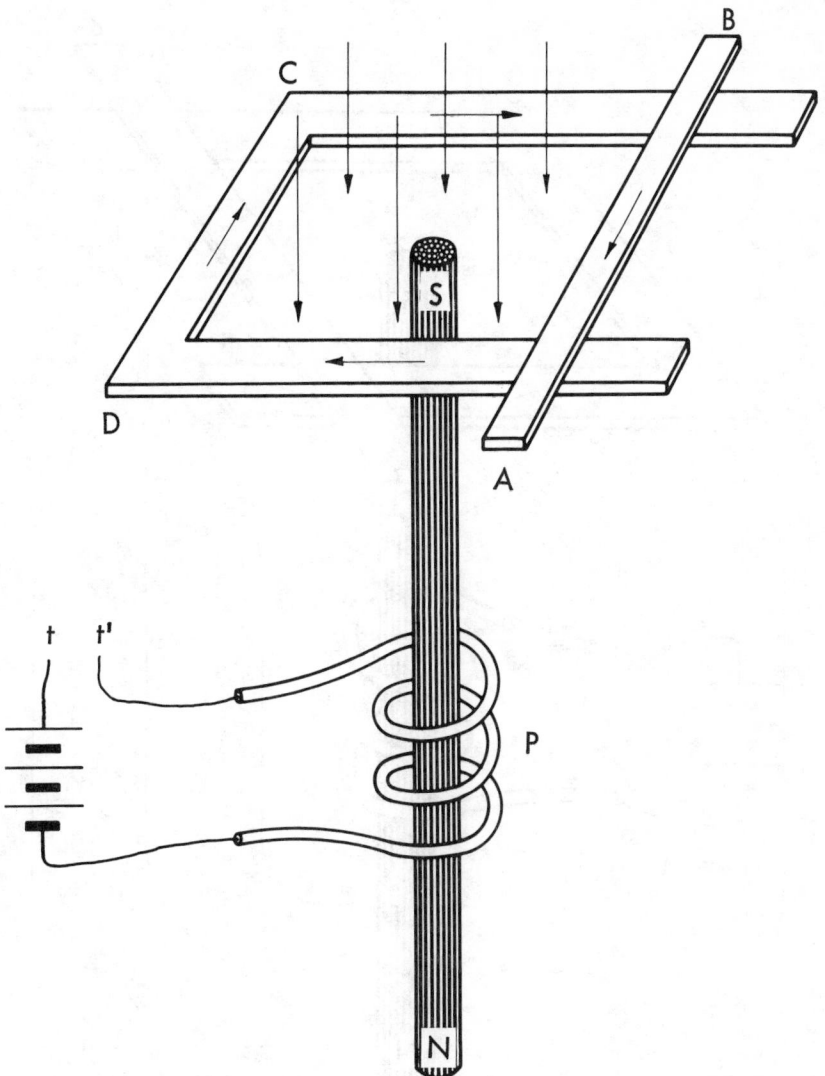

Fig. 1-15. Effect of opening battery circuit of simple transformer.

Fig. 1-16. Those magnetic lines produced by the primary coil which do not pass through the secondary are called *leakage lines*. The greater the number of leakage lines, the less the useful effect realized. It is advisable, therefore, to arrange both the primary and the secondary so as to have as many of the magnetic lines as possible thread through both

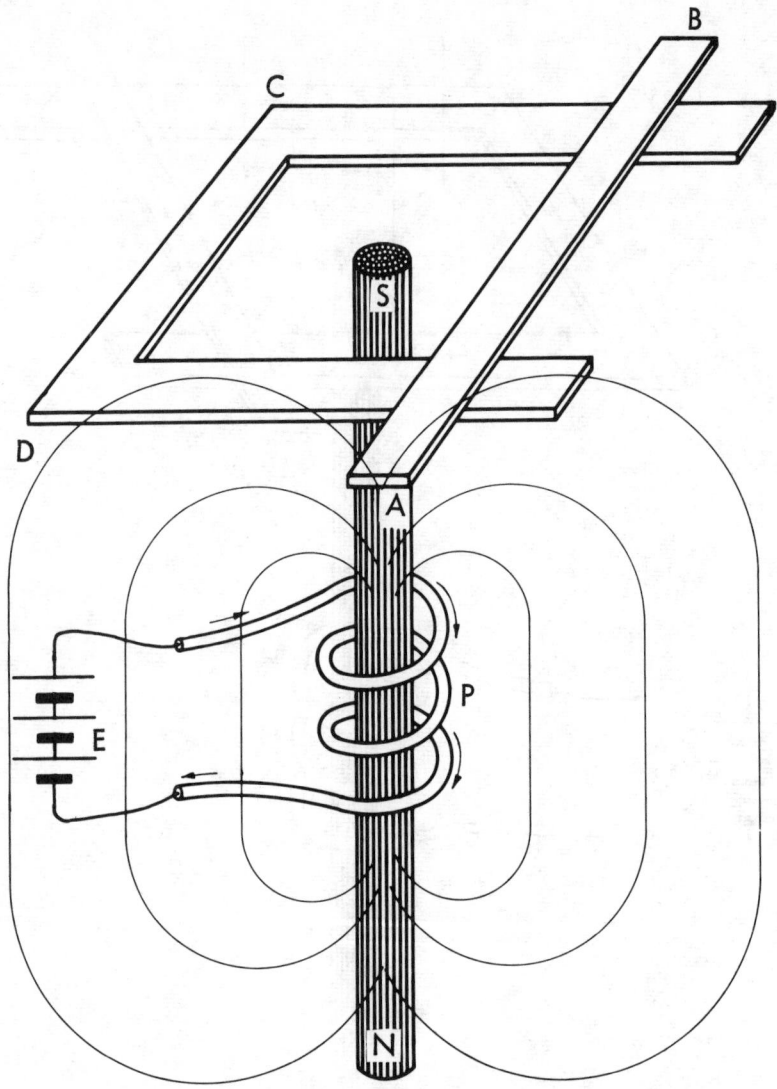

Fig. 1-16. Lines of force produced by primary coil of transformer.

primary and secondary. This may be accomplished by making the magnetic circuit, or core, a *complete*, or *closed*, circuit of iron, circular in shape, and winding the primary on this in close turns *PP*, Fig. 1-17. The secondary may be wound over the primary as shown at *S*. So far as magnetic efficiency is concerned, the form of transformer shown in Fig.

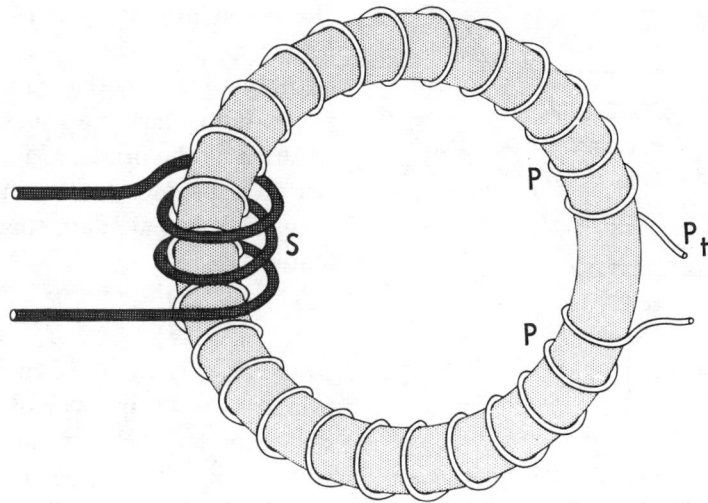

Fig. 1-17. Principal parts of closed-core transformer.

1-17 is the best that can be employed. There are no "joints" to interrupt the passage of the magnetic lines, and the path of the magnetic circuit is circular. The disadvantage of this form is the increased labor required in its building. Each turn of both primary and secondary must be threaded through the center hole by hand, so, if there are many turns to be wound on the core, the labor is considerable.

Open- and Closed-Core Transformers

A transformer having a magnetic circuit consisting partly of air and partly of iron is termed an open-core transformer, Figs. 1-14 and 1-15. One application where the characteristics of the open-core transformer are desirable is in the operation of certain kinds of wireless apparatus. For most classes of power service, transformers having magnetic circuits made up completely of iron are employed. These are called closed-core transformers.

Effect of Number of Turns in Primary

The effect of increasing the number of turns in the primary windings is to increase the possible magnetizing force due to the primary ampere turns; that is, to increase the amount of the magnetic flux with a given primary current.

Effects of Number of Turns in Secondary

The amount of the induced voltage in a secondary consisting of a

21

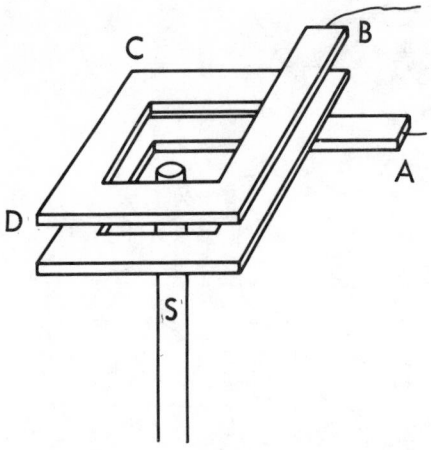

Fig. 1-18. Effect of additional turns on secondary terminal voltage.

single turn has been explained and computed. It will be interesting to note that if a second circuit, or turn, Fig. 1-18, is arranged connected in series with the first turn, or circuit, Figs. 1-14, 1-15, and 1-16, then the total induced voltage of the secondary will be twice what it was with the single turn. In fact, in respect to its single turn form, the secondary terminal voltage will always depend upon the number of turns wound to constitute the secondary coils. This fact of secondary voltage increase holds true for any arrangement of turns.

Transformer Ratio (Turns Ratio)

The ratio between the voltage and the number of turns on the primary and secondary windings of a transformer is called the *turns ratio*. The relation of the voltage produced in the secondary winding of a transformer to the primary will depend entirely upon whether the secondary winding has the same number of turns as the primary, a less number of turns, or a greater number of turns. Mathematically, this may be stated

$$\frac{N_p}{N_s} = \frac{E_p}{E_s}$$

where N_p = number of turns in the primary, N_s = number of turns in the secondary, E_p = the primary voltage, E_s = the secondary voltage. If the secondary has a greater number of turns than the primary, the secondary voltage will be higher. If the secondary has a less number of turns than the primary, the secondary voltage will be lower. If the secondary has one-half as many turns as the primary, the secondary voltage will be approximately one-half that of the primary. This later arrangement would be referred to as a step-down transformer, because the primary voltage is stepped down to the secondary voltage, which is lower than the primary. In a step-up transformer the secondary winding has a greater number of turns than the primary winding and, therefore, the secondary voltage is higher than the primary voltage.

Unless otherwise specified, the transformer ratio is the turns ratio. In practice, it is customary to specify the ratio of transformation by writing the *primary number first*.

The primary winding is the winding *which received the energy*, and it is not always the high-voltage winding.

Strict adherence to this policy eliminates the necessity of adding the terms "step-up" and "step-down" to a transformer. For example, when the transformer ratio is noted as being a 10:1 ratio, this obviously is a step-down transformer, and a ratio of 1:10 would be a step-up transformer. Although the terms *step-up* and *step-down* refer to the

value voltage, it should be obvious at this point that as the turns ratios vary in the same proportion, the transformer ratio may also be called the turns ratio.

The inverse ratio of the voltage applies to the current. Thus, mathematically:

$$\frac{N_p}{N_s} = \frac{I_s}{I_p} \text{ or } \frac{E_p}{E_s} = \frac{I_s}{I_p}$$

where $N =$ the number of turns
$E =$ voltage
$I =$ current.

Losses In Transformer

Primary and Secondary Losses

The energy losses in a transformer may be classified as the *primary loss*, the *secondary loss*, and the *core loss*. The primary loss is a true I^2R loss and is expressed in watts. This loss is caused by the resistance in the primary winding. The power loss in the secondary is also an I^2R loss, caused by the resistance in the secondary windings. When there is no load on the transformer, the transformer is not supplying current, and the power loss in the secondary is zero due to a very small current.

Core Loss

Eddy-Current Loss. The loss in the iron core is a power loss and is

expressed in watts. The core loss may be divided into two parts, one called *eddy-current* loss, the other *hysteresis* loss. A so-called eddy current in any iron core is a true electric current induced in the iron by the changing magnetic flux, just as if the iron were the desired conductor. Induced currents in the iron core are undesirable, since such currents require power input to the device and simply convert the electrical input into heat that is not only unavailable for useful electrical output but tends to heat the primary and the secondary windings, thus increasing the resistance and therefore the power loss in both coils.

The eddy-current loss may be reduced if the iron core is constructed

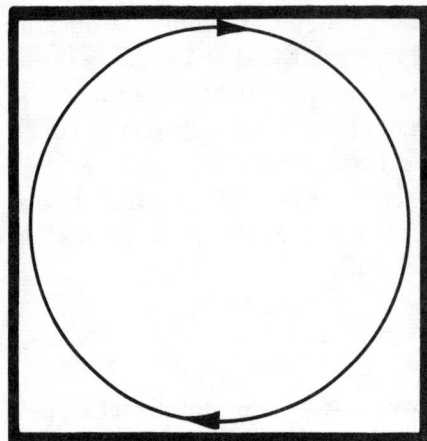

Fig. 1-19. Solid core eddy current.

of thin sheets of soft iron, called laminations, which are insulated from each other.

If the core were made of a solid piece of iron the current would circulate in the core as indicated in Fig. 1-19. If the core were divided

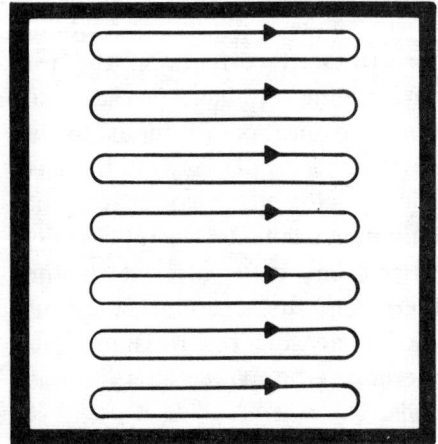

Fig. 1-20. Laminated core eddy current.

into two parts, then there would only be half the current in each part, and the loss in each core would be only ¼ as much as the loss in the solid iron core. The core in Fig. 1-20 is divided into seven parts so that only ⅐ of the current is present in each part, and the loss in this core would be ¹⁄₄₉ of the solid core.

Hysteresis Loss. "Hysteresis" is a Greek meaning *to lag*. Hysteresis in an iron core means that the magnetic flux, or lines of force, lag behind the magnetizing force that causes them. The nature of the hysteresis effect may be understood by considering that whenever a new piece of iron is subjected to a magnetizing force and the magnetizing force is removed, a portion of the magnetic flux remains in the iron. In order to remove this residual magnetism, another magnetizing force must be applied to the iron in a direction *opposite* to that of the initial magnetizing force. In other words, energy has to be supplied to demagnetize the iron. In the operation of the transformer the primary current is constantly changing not only in value but also in direction, so that the magnetizing force of the primary is first in one direction and then in the opposite direction. For each single change in magnetization from one direction to the opposite one a certain amount of power or energy is demanded. The

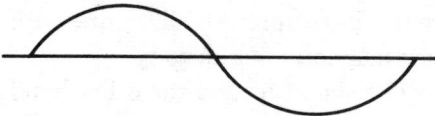

Fig. 1-21. Variation of current during one hertz.

peat a similar cycle of values. One complete cycle of changes in value is represented in Fig. 1-21; if there were 60 such changes in each second, the frequency of the current would be 60 hertz.*

Since magnetization affects the molecular constitution of the iron of the core, changing the position of the molecules from pointing in a certain direction to pointing in the opposite direction, it has practically the same effect as the rapid hammering of the iron, which would produce heat.

An important point to remember is that the copper loss of the transformer varies as the square of the load current; whereas, the core loss depends on the terminal voltage and the frequency of the supply. The core loss is constant from no load to full load because the factors affecting it are also constant, i.e., frequency, voltage, etc.

total amount of energy thus required per second will evidently depend upon the number of reversals that occur each second as well as upon the quality of the iron; a soft iron retains few lines of force and hence requires less applied energy to demagnetize it. Fig. 1-21 will serve to represent the simplest form of an alternating voltage, current, or magnetic flux. The current, or the magnetic flux, increases from zero to a certain positive, or +, maximum, decreases to zero, increases in the opposite direction to a negative, or −, maximum, and finally returns to zero, ready to re-

Types of Transformers

Transformers are classified, according to their relative disposition of iron and copper, into three basic types. These are the core, the shell and the cross or *H* type which is also called the distributed core type.

There has been much discussion as to the relative merits of each type of transformer, some manufacturers going so far as to claim ex-

clusive advantages for one or the other. The fact is that no general conclusion can be drawn as to which type is better, for each possesses

---

*In the past, the term *cps* (cycles per second) or just *cycles* has been used to designate the frequency of an alternating circuit. The term *hertz* is now being used in lieu of cps or cycles, so instead of referring to the frequency as 60 cycles, we are now using the term 60 hertz.

inherent characteristics which specially adapt it to certain conditions. A brief comparison of these characistics will aid in evaluating which type is normally used for specific conditions of service, size, voltage and the like. The core type has relatively a lighter core of smaller sectional area but a greater length of magnetic circuit, while the copper is relatively heavier, containing more turns, although a shorter mean length. The core type is more easily wound as cylindrical formed coils, and the coils are more accessible and expose more surface to radiation. The core type with its relatively large winding space, is better adapted for high voltages which require many turns and heavy insulation. Since it handles smaller currents, it requires smaller wire with low magnetic flux density.

The shell type, on the other hand is particularly suited for transformers of moderate voltage, requiring few turns and little insulation, large currents, and low frequency with corresponding magnetic flux.

Core Type

Fig. 1-22 shows a complete single-phase core built up of thin sheet steel strips on stampings. The two upright portions of the core upon which the coils are placed as shown in Figs. 1-23 and 1-24 are called the *legs* of the core. The short horizontal parts of the core that do not have windings on them are called the yokes. These yokes serve to complete the magnetic circuit and

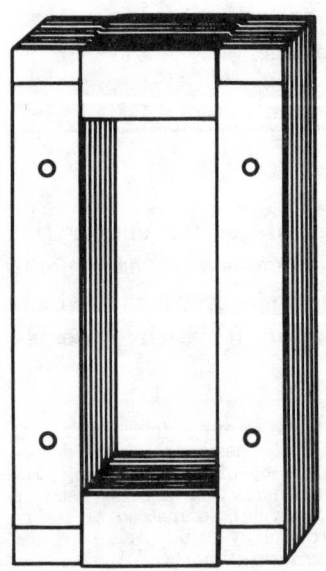

Fig. 1-22. Standard form of laminated transformer core.

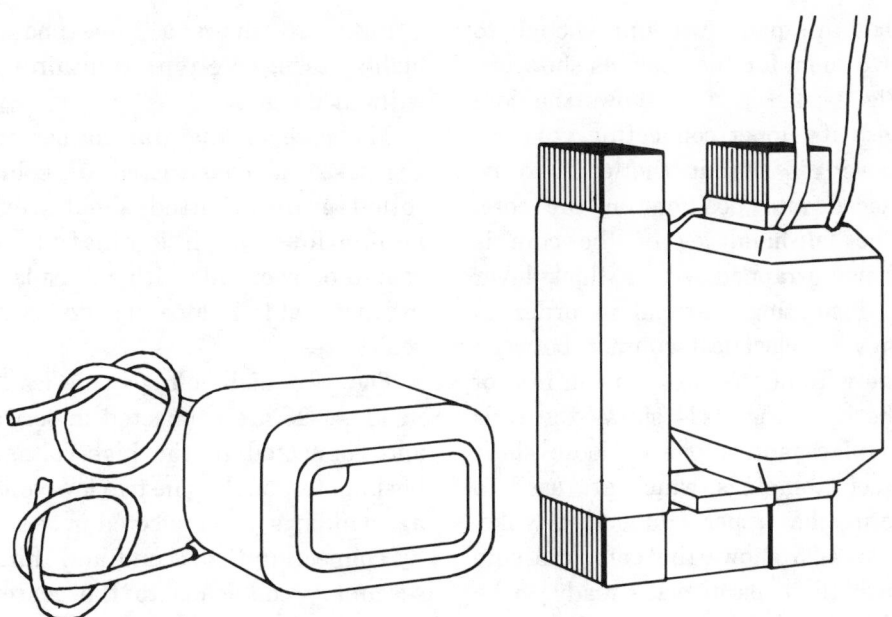

Fig. 1-23. Core with upper yoke removed, showing one coil in position.

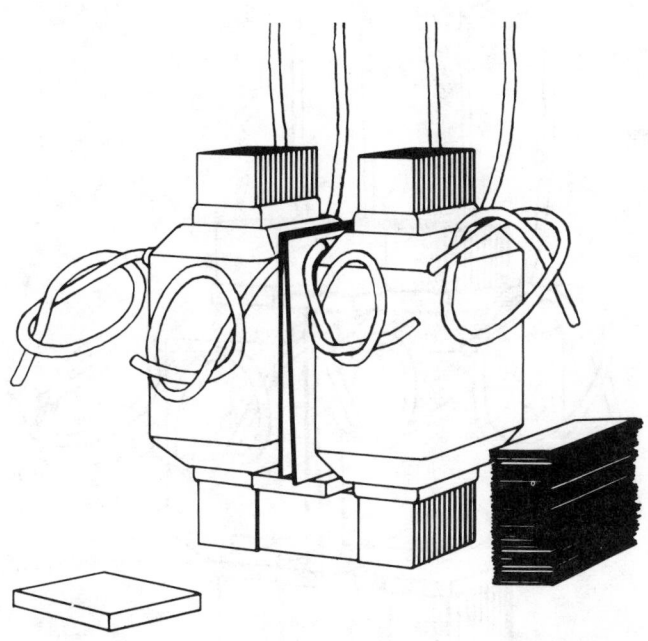

Fig. 1-24. Both transformer coils in position, showing laminated strips which form the upper yoke of core.

they are made just long enough to give room for the coils, as shown in Fig. 1-24. Fig. 1-23 shows the core with its upper connecting yoke removed to permit the coils to be placed into position on the core. The left-hand leg of the core is shown wrapped with a thick layer of insulating material in order to prevent electrical contact between the wire of the coil and the iron of the core. Fig. 1-24 shows the coils in place and a pile of loose sheet steel stampings, which are used to form the upper connecting yoke. Fig. 1-25 shows the complete core with the coils in place ready to be placed in a transformer tank or case.

Fig. 1-26 shows a three-phase, high-voltage, core-type transformer without its case.

The core of the transformer in Fig. 1-25 is constructed of cold-rolled grain-oriented sheet-steel laminations. Each lamination is coated on each side with an insulating material to reduce the eddy current losses.

The two high-voltage coil leads in Fig. 1-25 are connected in series and connected to the high-voltage bushing terminals. The two low-voltage windings are connected in series by connecting the second and third secondary coil leads to the center low-voltage terminal. The two out-

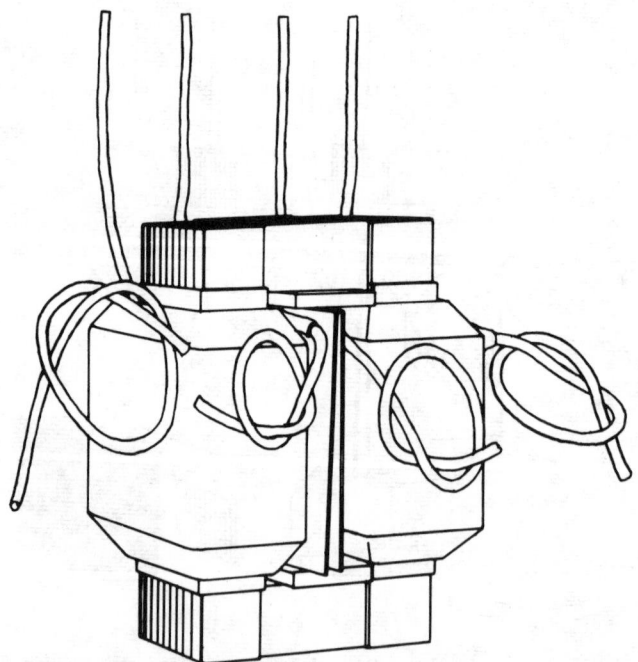

Fig. 1-25. Core-type of transformer assembled with case removed.

Fig. 1-26. Core-type transformer (core-and-coil assembly of a medium-type power transformer). (Westinghouse Electric Corp.)

side low-voltage leads are connected to the two outside low-voltage terminals in a manner similar to Fig. 1-40. This connection is used to form an Edison three-wire single phase system having voltages between the middle terminal and either one of the outside terminals of 120 volts and 240 volts between the two outside low-voltage terminals.

Normally on the small, single-phase transformers, the low-voltage winding is next to the core, and the high-voltage winding is placed over the low-voltage winding with a layer of insulating material between them. By placing the windings in such a manner the insulation required for the high-voltage winding is cut to a minimum. The sheet steel laminations and the coil are held firmly in place by angle or channel iron braces and bolts called core clamp brackets. The complete transformer is mounted in a welded, pressed-steel tank in a manner very similar to the one shown in Figs. 1-38 and 1-40. The mechanical construction of the transformer will vary depending upon the voltage and size. In higher voltage transformers, it is customary to use a tall case and immerse the coils and the terminal blocks in oil.

Distributed Core Type (Type *H* Core)

A modified form of the standard types of transformer cores is shown in Fig. 1-27. It is called the type *H* or the distributed core type. The large rectangular type of core, Fig. 1-22, is divided into four smaller cores with the magnetic circuit of each core in parallel as shown in Fig. 1-27. The core is cruciform in shape when viewed from the top. In this type of core the leakage flux is very small as the coil windings are wound on a center leg and surrounded by the four outside legs of the core structure. There are two high-voltage coils and two low-volt-

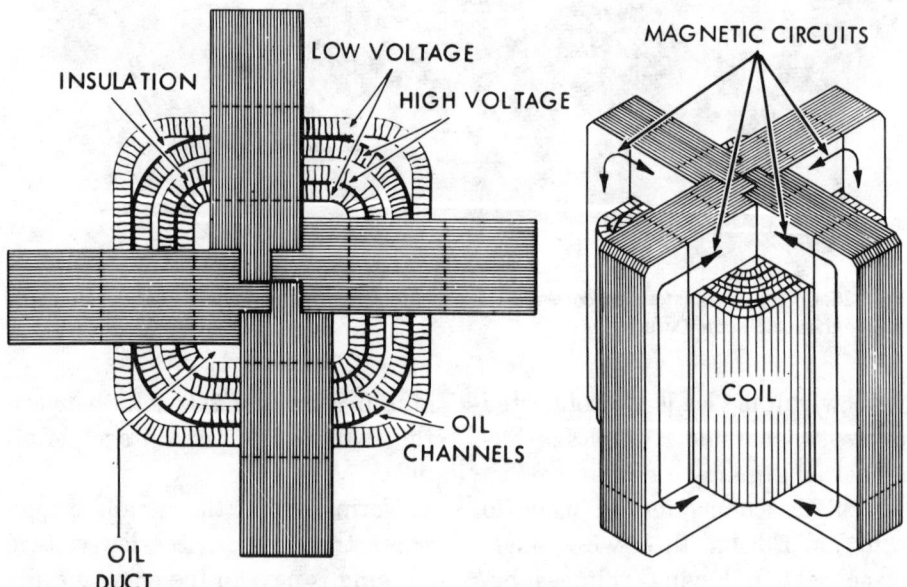

Fig. 1-27. Type H Transformer: Plan view (at left) and isometric projection of the assembled core and coils, showing four magnetic circuits in parallel (at right).

age coils. The coil arrangement in Fig. 1-27 is such that the high-voltage windings are placed between the two low-voltage windings. This coil arrangement keeps the insulation requirements at a minimum, with high-voltage insulation required only between the high-voltage and low-voltage windings. This method has been used to a large extent in the past in the construction of distribution transformers.

The purpose of the space between the primary and secondary coils is to allow a circulation of oil upward from the bottom of the case toward the top. This aids in conducting the heat produced by the energy losses in the transformer winding to the oil and to the outer case where it is radiated to the outside air.

Oil is a better heat-conducting medium than air. It carries heat from a transformer to the containing case much better than air, so that a transformer in oil shows a much lower temperature. Oil preserves the insulation, keeping it soft and pliable; it also prevents oxidation in the air. Consequently, its use is advantageous in producing proper conditions to maintain a uniform core loss and a superior insulation. Therefore, oil is in itself a very good insulator, having the valuable property common to all liquid insulators. It is not permanently damaged by a puncture caused by lightning, for the resistance of the oil is only momentarily broken down as the oil immediately flows into the break and seals the insulation. *Askarel* and *Transil* are the names of two special transformer oils that have a high insulating value and dielectric strength.

Shell-Type Core

In the shell-type of core the leakage flux is cut to a minimum because the coil windings surround the center core, and the outer core surrounds the coil. Fig. 1-28 shows the basic construction of a single-phase shell-type transformer. The core is constructed of thin sheet steel strips or stampings. In large shell-type power transformers the coils, both

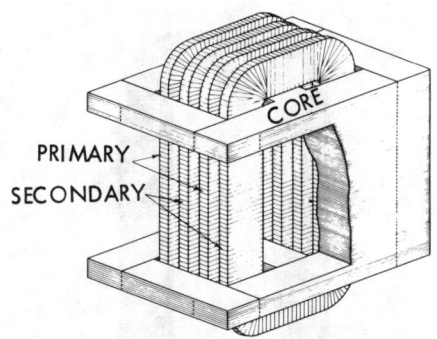

Fig. 1-28. Cutaway view showing construction of a shell-type transformer.

primary and secondary, are usually wound in pancake form on formers as shown in Fig. 1-29. A flat rectangular copper strip is used with one turn per layer, in many layers, Fig.

31

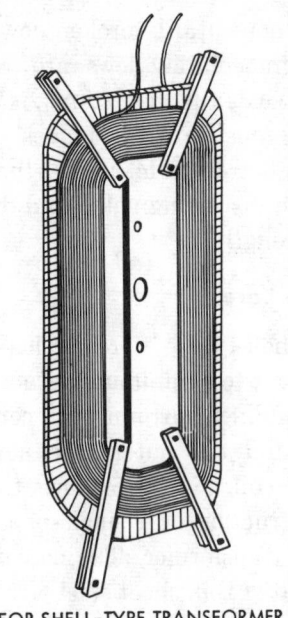

FOR SHELL-TYPE TRANSFORMER

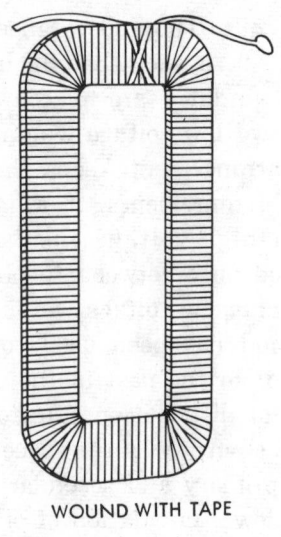

WOUND WITH TAPE

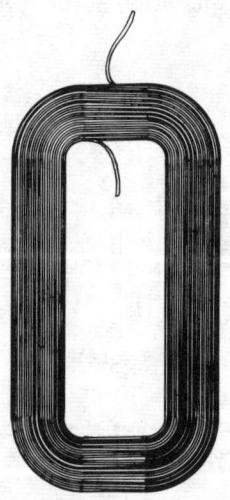

SHOWING USE OF INSULATED FLAT STRIPS

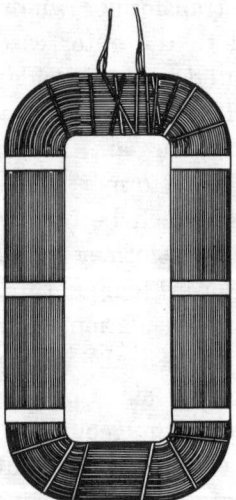

METHOD OF BINDING UP PANCAKE STRIP COILS

Fig. 1-29. Pancake coils.

1-29 *left*. Each flat rectangular copper strip is specially insulated and a special insulation between turns is used according to the other conditions, Fig. 1-29, *bottom left*.

The thin pancake coils are then

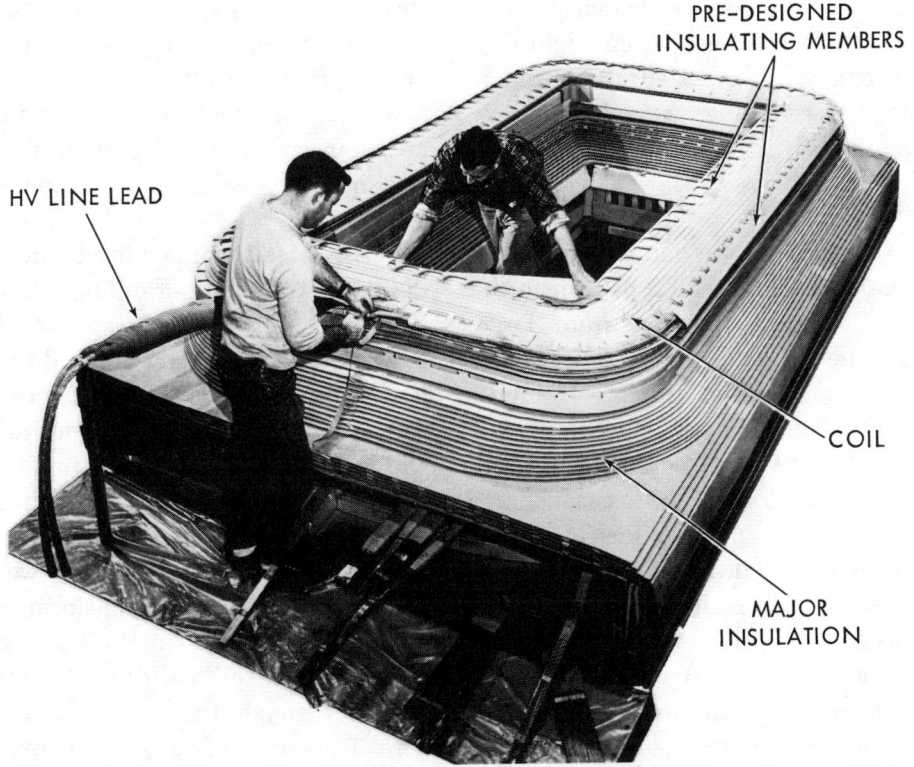

PRE-DESIGNED
INSULATING MEMBERS

HV LINE LEAD

COIL

MAJOR
INSULATION

Fig. 1-30. Partial phase assembly of a pancake coil for an EHV (Extra High Voltage) transformer. (This phase coil will be installed in a shell form fit core.) (Westinghouse Electric Corporation)

treated with an insulating compound and wound with a number of layers of tape according to the voltage for which they are designed, Fig. 1-29, *top right*, each layer being given several coats of insulating varnish baked on in ovens. The coils are then assembled into groups of two or more sections, Fig. 1-28, and the groups into complete windings, the primary and the secondary being intermixed or sandwiched in order to reduce magnetic leakage.

Suitable insulating barriers are interposed between the various groups. Fig. 1-30 shows a large pancake coil under construction.

In high-voltage transformers used for transmission of power, the insulation is heavily reinforced for a considerable length of the conductor nearest to the terminal leads. This is an important precaution, as the extra dielectric strength of the insulation of these end turns is a safeguard against breakdowns which might otherwise occur due to the excessive voltages from lightning discharges or other surges to which transmission lines are unfortunately subjected.

Wound Cores

The sheet steel laminations used in the core and shell type transformers were usually cut from large sheets, about 10 feet long and 3 feet wide which the steel rolling mills could make easily. These sheets were usually rolled, while red hot, down to the desired thickness. This rolling process improved the magnetic qualities of the sheet in the direction of its length. Thus, it became desirable to cut the laminations from the large sheet so that the magnetic flux or lines of force in the transformer core could coincide with the lengthwise direction of the large sheet. It also was desirable to keep the number of joints or pieces of steel in the transformer core to as small a number as possible. This caused "L" shaped punched laminations to be used extensively for constructing transformer cores.

It was not as easy to cut or punch these "L" shaped pieces or laminations from a sheet of steel as was the case with the simple straight strips. It required a careful layout of these "L" shaped strips on the large sheet to keep the waste or scrap material to a minimum. Thus

the increase in efficiency of a transformer due to the improved magnetic qualities of the iron in the cores was obtained only by an increase in cost of construction. This was not a very efficient improvement.

These steel manufacturers next developed a method of rolling, while cold, long strips of thin steel that have still better magnetic qualities lengthwise. These long strips of steel can be cut easily to any desired width. It is more convenient to handle the coil or roll of strip steel, than to handle the large flat sheets.

The spiral core, Fig. 1-31, was first used for current transformers where the primary consisted of one single insulated conductor that passed through the center of the core. The secondary was wound over the spiral core, usually by hand. The path of the magnetic lines of force is shown by the dotted lines

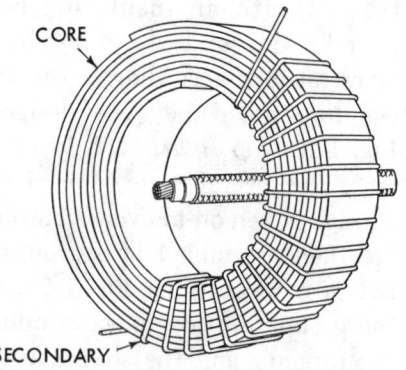

Fig. 1-31. The spiral core used in a current transformer.

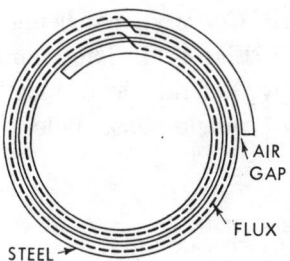

Fig. 1-32. The path of magnetic lines of force in a spiral core.

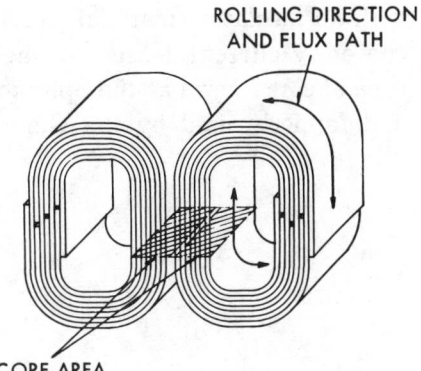

Fig. 1-33. Wound core. (Allis Chalmers)

in Fig. 1-32. The most difficult part with the spiral core is the placing of the core in the center of the coil, Fig. 1-43. The annealed laminated core has been squared and clamped in shape before it was annealed. This annealing process improves the magnetic qualities of the core and holds the turns of strip sheet steel closer together.

Manufacturers have developed the type of wound core they feel is best suited for their specific transformer design. Some manufacturers have developed a special core which consists of a long strip of silicon steel, wound in a spiral around the windings. This new type of core has several advantages in that the magnetic circuit path is relatively short and has a large cross-section, the flux path is always along the grain of the steel and flux leakage is at a minimum.

Fig. 1-33 shows the basic construction of a wound core. This core is spirally wound from a continuous strip of cold rolled steel which is cut

at every other turn. These cuts permit assembling the core around a pre-wound coil which passes through both openings. The cuts are made at random positions so that they do not fall in line, which permits the flux to flow between the layers of steel and not across the butt gap in every other turn. Each sheet must be coated on each side

Fig. 1-34. Old manual assembly line method of assembling core section of a G.E. 10-kVA Spirakore transformer. View shows how individual parts of core, each representing complete double turn around coil, form continuous spiral with butt joint every second turn. (General Electric Co.)

with an insulating material to limit the eddy current losses. A shell-type of core known as the Spirakore and formerly used by the General Electric Co. is shown being assembled in Fig. 1-34. The wound core used by General Electric Co. for the 25 kVA, single-phase, pole-type dis-

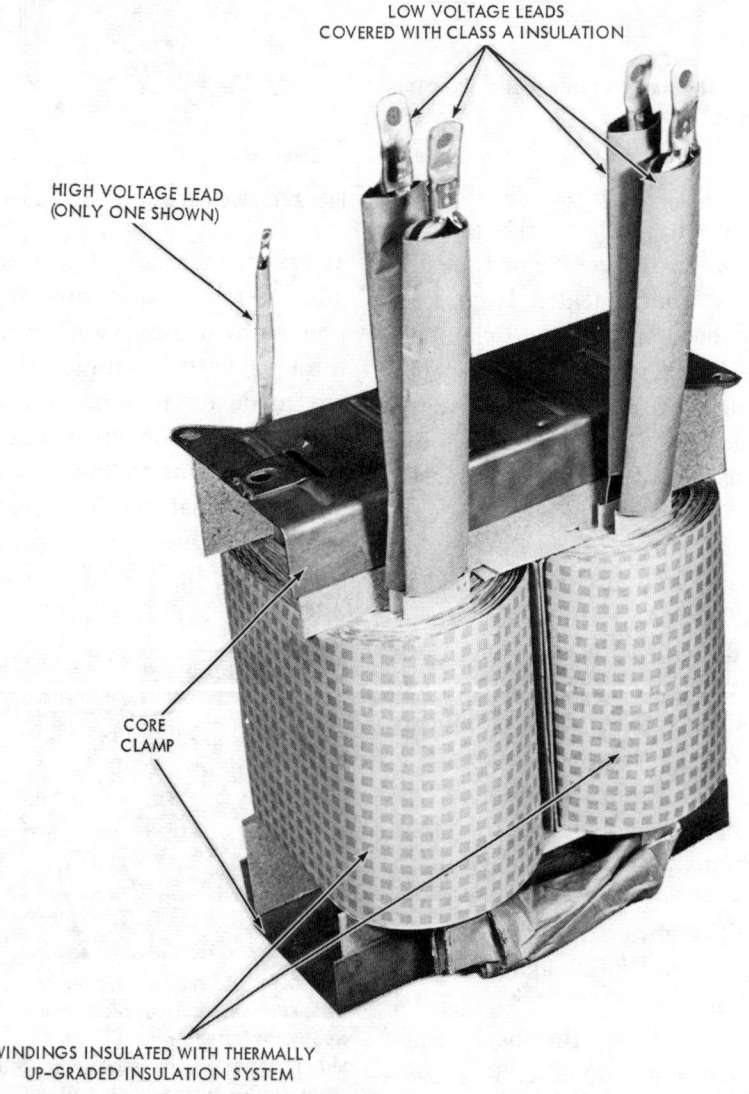

LOW VOLTAGE LEADS
COVERED WITH CLASS A INSULATION

HIGH VOLTAGE LEAD
(ONLY ONE SHOWN)

CORE
CLAMP

WINDINGS INSULATED WITH THERMALLY
UP-GRADED INSULATION SYSTEM

Fig. 1-35. Typical shell-type core-and-coil structure constructed of cold-rolled, grain-oriented steel. (Distribution Transformer Dept., General Electric Co.)

tribution transformer in Fig. 1-35 is constructed of strips of cold-rolled, grain-oriented steel. Fig. 1-36 shows the core clamp which holds the core in a permanent position. The transformers in Figs. 1-35 and 1-36 are ready to be installed in the pressure-tested, all-welded, oil-filled

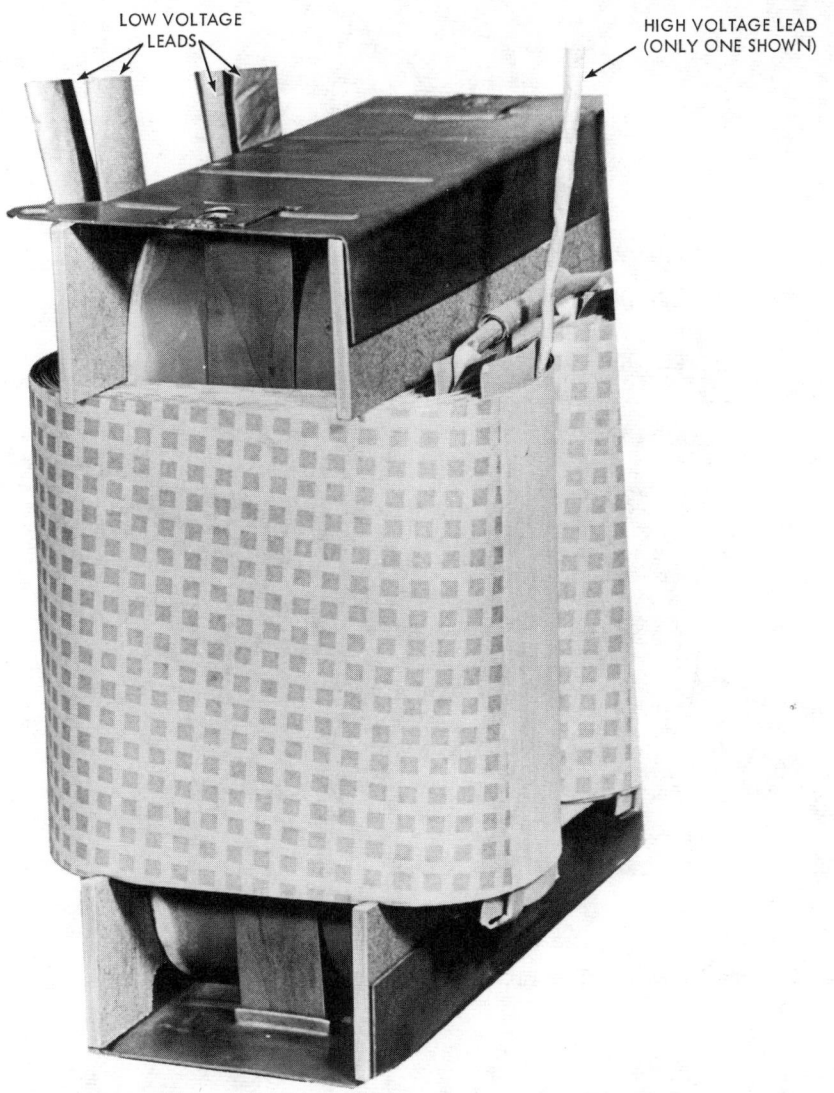

LOW VOLTAGE LEADS

HIGH VOLTAGE LEAD (ONLY ONE SHOWN)

Fig. 1-36. Core clamp. Permanently welded, without straps, it cradles the core and minimizes stress. It is rigidly built to withstand any operating conditions. (Distribution Transformer Dept., General Electric Co.)

Fig. 1-37. Transformer and lineman. (Distribution Transformer Dept., General Electric Co.)

tank shown in Figs. 1-37 and 1-38. Fig. 1-39 shows the external features of the same transformer. Fig. 1-40 shows a similar pole-type transformer manufactured by McGraw-Edison, Power Systems Division.

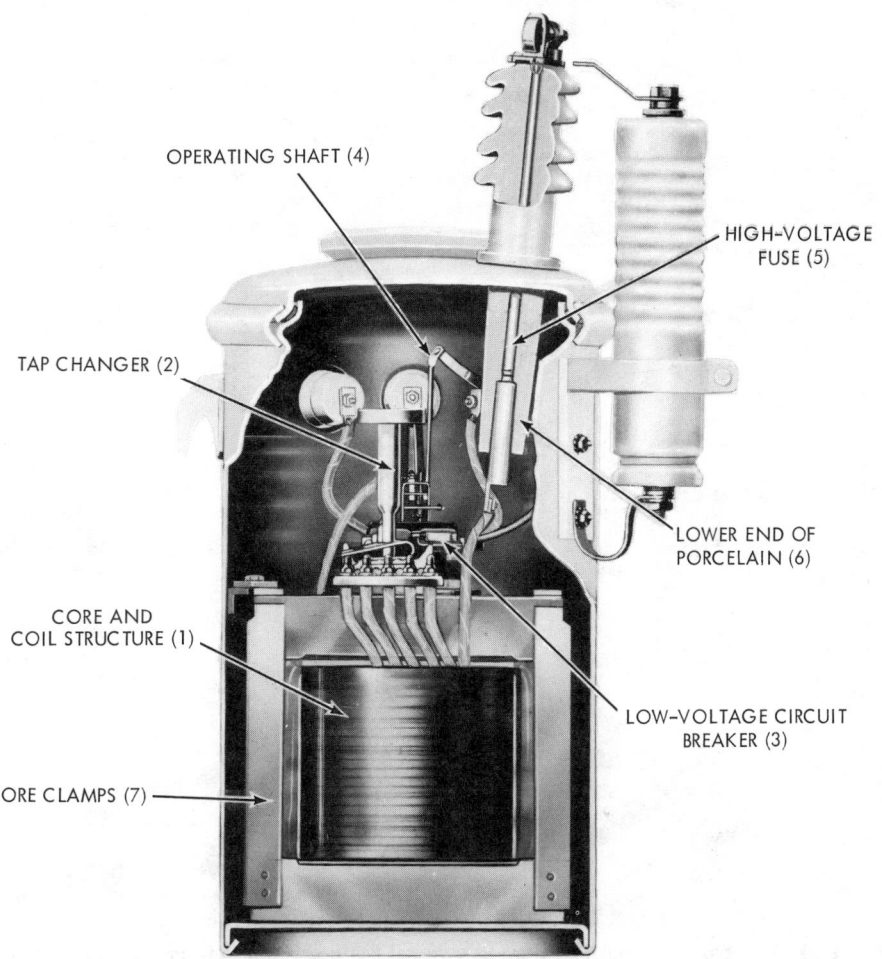

OPERATING SHAFT (4)

HIGH-VOLTAGE
FUSE (5)

TAP CHANGER (2)

CORE AND
COIL STRUCTURE (1)

LOWER END OF
PORCELAIN (6)

CORE CLAMPS (7)

LOW-VOLTAGE CIRCUIT
BREAKER (3)

Fig. 1-38. Cutaway of pole transformer.

(1) **CORE AND COIL STRUCTURE** is held in a close-fitting, steel-core cradle, permanently centered in the tank.

(2) **TAP CHANGER** is provided for voltage selection when the transformer is de-energized. The handle and dial are above the oil level.

(3) **LOW-VOLTAGE CIRCUIT BREAKER**, located below the oil level, is tripped by the deflection of bimetallic elements in series with the low-voltage leads on self-protected units.

(4) **OPERATING SHAFT** for the low-voltage circuit breaker is brought out above the oil level through a sealed bearing gland.

(5) **HIGH-VOLTAGE FUSE** on self-protected units disconnects the transformer from the line in the event of an internal fault.

(6) **LOWER END OF PORCELAIN** is below the oil level to eliminate the possibility of internal flashover of the bushing.

(7) **CORE CLAMPS**, made of steel channels, brace the core and minimize mechanical stress.

(Distribution Transformer Dept., General Electric Co.)

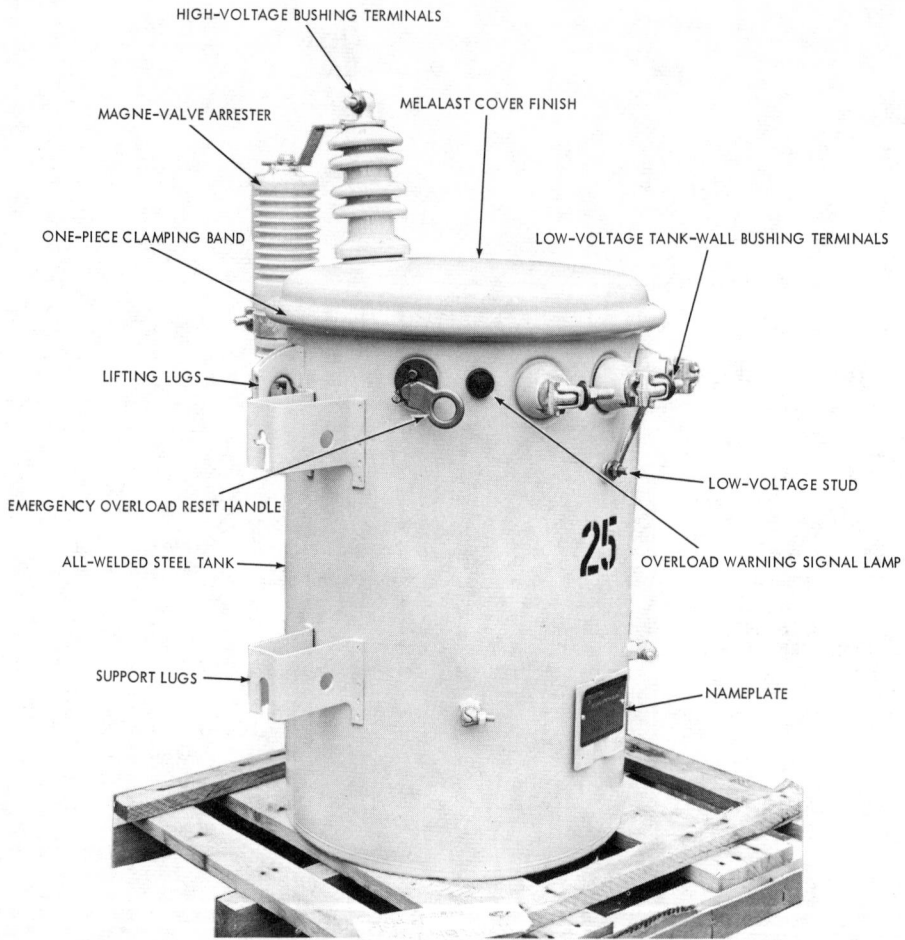

HIGH-VOLTAGE BUSHING TERMINALS

MAGNE-VALVE ARRESTER

MELALAST COVER FINISH

ONE-PIECE CLAMPING BAND

LOW-VOLTAGE TANK-WALL BUSHING TERMINALS

LIFTING LUGS

LOW-VOLTAGE STUD

EMERGENCY OVERLOAD RESET HANDLE

ALL-WELDED STEEL TANK

OVERLOAD WARNING SIGNAL LAMP

SUPPORT LUGS

NAMEPLATE

Fig. 1-39. Pole type transformer showing external features. Tank grounding provision is a welded stud on other side of tank (not shown). (Distribution Transformer Dept., General Electric Co.)

The wound core used by Westinghouse for the 3-phase 75-500 kVA, 18 kV class, pad mount transformer shown in Fig. 1-41 is known as their "Wescor." The step-lapped core joints result in a lower exciting current and a reduction of joint losses and consequently in a lighter transformer with a reduction of noise. The progressive wound coil that is used with this core is shown in Fig. 1-42. Fig. 1-43 (bottom) shows the complete core and coil assembly ready to be installed in the tank.

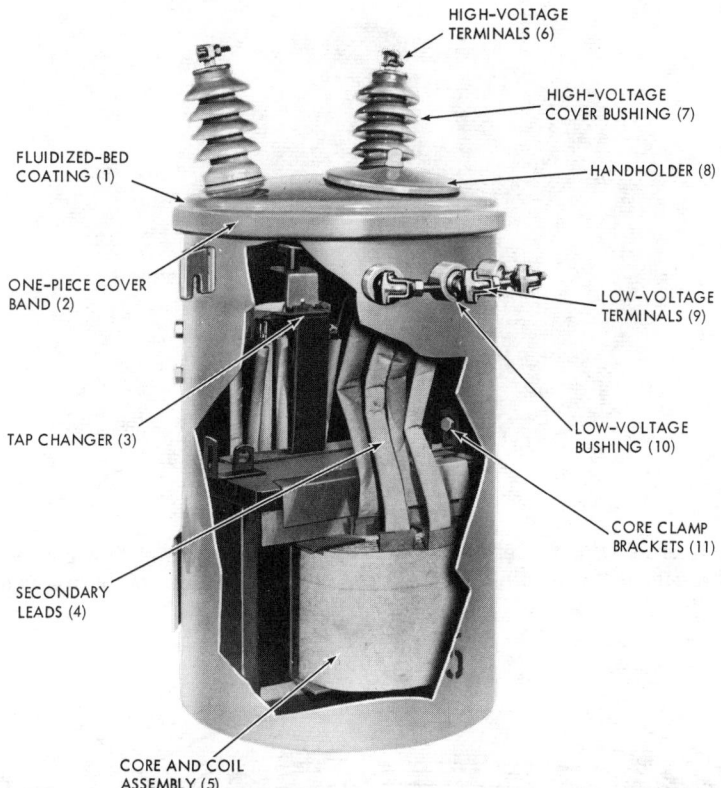

HIGH-VOLTAGE
TERMINALS (6)

HIGH-VOLTAGE
COVER BUSHING (7)

FLUIDIZED-BED
COATING (1)

HANDHOLDER (8)

ONE-PIECE COVER
BAND (2)

LOW-VOLTAGE
TERMINALS (9)

TAP CHANGER (3)

LOW-VOLTAGE
BUSHING (10)

CORE CLAMP
BRACKETS (11)

SECONDARY
LEADS (4)

CORE AND COIL
ASSEMBLY (5)

Fig. 1-40. Cutaway view of a typical conventional transformer with two cover bushings.

(1) **FLUIDIZED-BED COATING** completely insulated cover, handhole cover, and band closure; eliminates outages caused by birds or animals contacting high-voltage terminals. Cover is externally grounded to the tank.

(2) **ONE-PIECE COVER BAND** securely clamps cover to tank and provides optimum gasket compression.

(3) **TAP CHANGER,** snap action, heavy-duty contacts with fiberglass reinforced body.

(4) **SECONDARY LEADS** individually insulated and clearly marked for connections.

(5) **CORE-AND-COIL ASSEMBLY,** simple, compact shell-type core and balanced windings provide for high overload capacity, efficient, low-loss operation, and high short-circuit strength.

(McGraw-Edison, Power Systems Division)

(6) **HIGH-VOLTAGE TERMINALS** accept solid or stranded conductors; also equivalent aluminum conductors.

(7) **HIGH-VOLTAGE COVER BUSHINGS** are strong, wet-process porcelain; are gasketed and internally clamped to provide a moistureproof and oilproof seal.

(8) **HANDHOLE COVER** provides ready access to tank on cover bushing units with tap changers, and also on no-tap cover bushing units above 50 kVA.

(9) **LOW-VOLTAGE TERMINALS** are eyebolt connector or spade type.

(10) **LOW-VOLTAGE BUSHINGS,** two-piece porcelain securely keyed and sealed with nitrile gaskets.

(11) **CORE CLAMP BRACKETS** securely mount core/coil assembly to tank.

Fig. 1-41. Wescor® wound core. The Wescor® cores combine the highest quality of Hipersil steel with a patented method of forming a series of step-lapped core joints. (Westinghouse Electric Corp.)

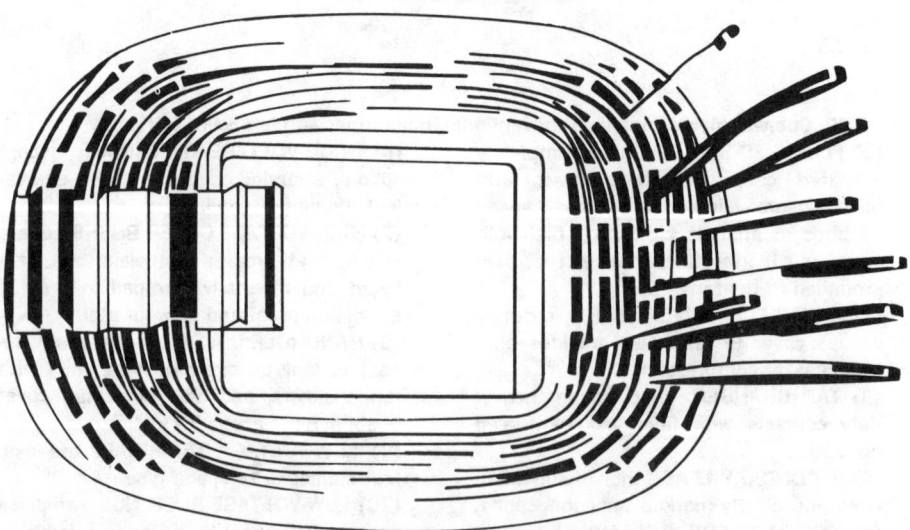

Fig. 1-42. Progressive wound coils. In a progressively wound coil the high-voltage winding is wound directly on top of the inner low-voltage section, and then the outer part of the low-voltage is wound directly on top of the high-voltage winding for tighter, stronger mechanical construction. This progressive winding provides a compact assembly with greater inherent mechanical strength against short-circuit forces. (Westinghouse Electric Corp.)

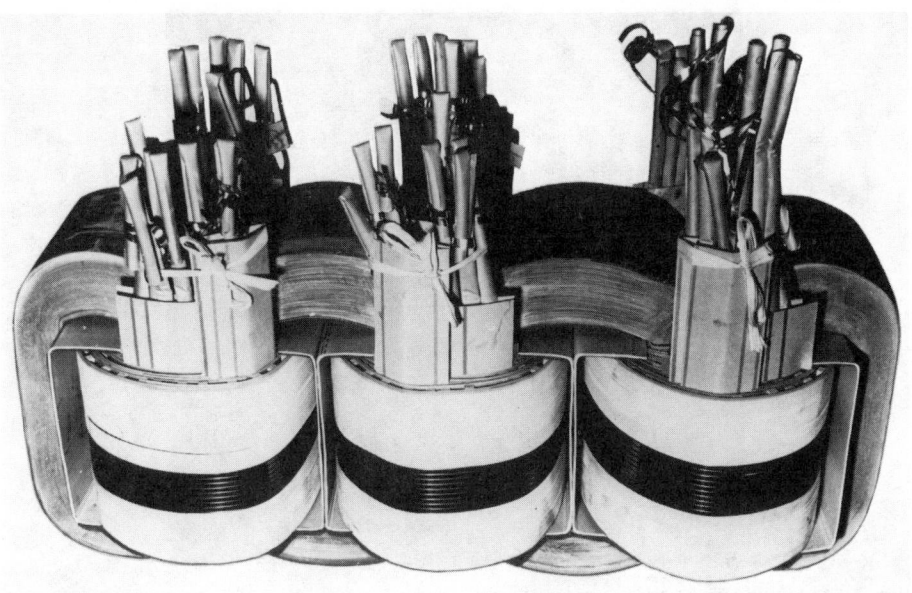

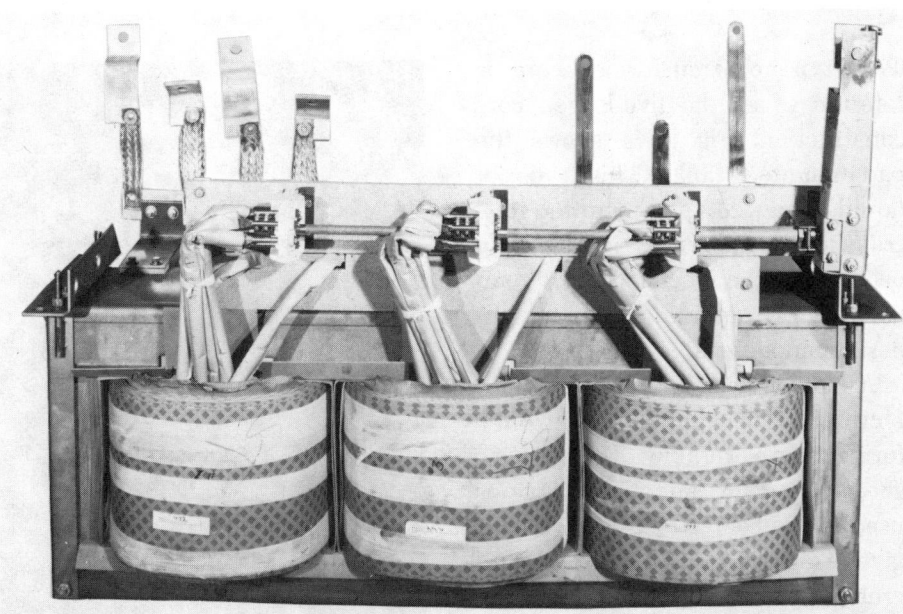

Fig. 1-43. Partially assembled core and coil (top) and fully assembled (bottom) showing wound five-legged construction used for commercial subsurface transformer (CST). (Distribution Transformer Dept., General Electric Co.)

Fig. 1-44. Liquid-immersed commercial pad-mounted distribution transformer. (Distribution Transformer Dept., General Electric Co.)

This type of transformer core is referred to as the five-legged core construction. Fig. 1-44 shows the pad-mounted tank. This type of liquid-immersed, pad-mounted distribution transformer is normally used when the electric power company feeds a building from an underground service.

A method used by another manufacturer is to wind the core on a form using several widths of strip sheet, Fig. 1-45. The widest strip is used in the center of the core. This gives a core with modified or stepped cruciform or cross-shaped section instead of a square cross-section. Then a loose-fitting cylinder of insulating material shown at *A*, Fig.

Fig. 1-45. Winding the strip steel core for a wound-core transformer. (McGraw-Edison Power Systems Division)

Fig. 1-46. Winding the coils on a wound-core transformer core. (McGraw-Edison Power Systems Division)

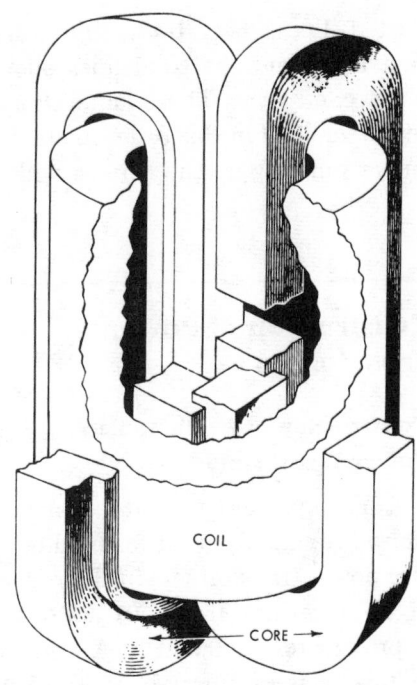

Fig. 1-47. How a cruciform-center core is constructed in large "Spirakore" transformers. (General Electric Co.)

1-46, is wound around and over the straight sides of the core shown at *B*, Fig. 1-46. The cylinder of insulating material is rotated or turned on the core by a specially built machine. In this way the primary and secondary coils are wound on the core. A completely wound coil is shown on the rear straight portion of the core at *C*, Fig. 1-46.

In the larger size distribution and power transformers, four cores may be used with the cores arranged in the center of the coil to give a cruciform or cross-shaped section shown in Fig. 1-47. This enables a tall coil of small diameter to be used, which reduces the diameter of the transformer core, and it also reduces the size of the transformer case. The cores are wound with strip sheet steel of two different widths.

Another method of taking advantage of the magnetic properties of a piece of strip sheet steel is to cut the ends of the laminations at an angle of 45°, Fig. 1-48, instead of with square ends as in Fig. 1-24. The

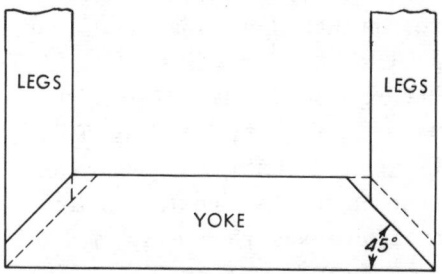

Fig. 1-48. A method of cutting and assembling the laminations on large transformers.

dotted lines, Fig. 1-48, show the first layer and the solid lines show the second layer. These laminations are assembled in this order in stacking or building up the core. A wider strip of steel is used for the yoke than is used for the legs, in order that angle iron bars can clamp and hold the laminations of the yoke together and in place.

Construction of Power Transformers*

Construction of Medium-Size Power Transformers

Core-form design and construction is commonly used for medium-size power transformers in the range of kVA ratings from 501 kVA to approximately 100,000 kVA and in voltage classes through 1050 kVA BIL (or 345 kV on the high side), Figs. 1-49 and 1-50. (BIL represents the Basic Insulation Level of a transformer.)

Insulation Systems. Temperature coordinated insulation systems for core-form transformers consist of a time-tested combination of oil ducts and specially developed insulating tapes, paper and boards that offer a high continuous kVA loading capability with a high life expectancy. TherMEcel is the name applied by McGraw-Edison Company to both the special treatment and to the insulating materials that have been chemically modified

to resist aging, while in transformer insulating oil, when subjected to high temperatures. This type of insulation has exceptional mechanical strength and thermal stability which is necessary so that transformers can effectively withstand shocks in-

Fig. 1-49. Typical LTC core-form transformer substation installation. (McGraw-Edison, Power Systems Division)

---

*Information in this section available through the courtesy of McGraw-Edison Power Systems Division.

HIGH VOLTAGE BUSHINGS   LIQUID LEVEL GAGE

LIFTING HOOK

PRESSURE GAGE
TEMPERATURE GAGE

COOLING RADIATORS

JACK LUG

PULLING HOLE

NAMEPLATE

SKI-RUNNER   DRAIN VALVE

Fig. 1-50. Typical transformer, rated 3750 kVA at 55C rise, self-cooled with provision for fans to increase the capacity to 5200 kVA at 65C rise. (McGraw-Edison, Power Systems Division)

curred during movement or short circuits. Depending upon the requirements of a particular design, a specified number of layers of specially treated tape is applied to the winding conductors. Specially designed wrapping machines, combined with strict quality control assure excellent insulation integrity, Fig. 1-51.

Parts made from sheet insulation, such as pressboard barriers, are precision-cut to specified dimensions and carefully shaped, when needed, with a heated forming press. Through careful control of the forming process, full dielectric strength is retained, Fig. 1-52.

Assemblies of barriers, collars and

Fig. 1-51. One bank of three machines wrapping TherMEcel-treated paper insulation on winding conductor. (McGraw-Edison, Power Systems Division)

Fig. 1-52. Portion of the insulation department in which major insulation parts for core-form transformers are prepared. (McGraw-Edison, Power Systems Division)

washers are used for major phase-to-phase and phase-to-ground insulation. Barrier collars are interleaved with barriers and spaced with insulating blocks in a way that results in overall coordination of the insulation and provides passages for free oil circulation.

Coils. There are several basic designs used for core-form windings. These include layer, helical and pancake types. The required voltage and current ratings normally dictate the particular design utilized to meet each transformer specification.

Layer windings of one or more layers, wound in the fashion of a spool of thread, are used where currents and voltages are low. Axial oil ducts between layers are established where needed for effective cooling without hot spots, Fig. 1-53.

Helical windings are most com-mon for low-voltage and high-current ratings. In this type of winding there are several conductors in parallel. Cooling ducts are established with locked-in radial spacers located between each multiple conductor turn to permit the flow of oil adjacent to each conductor. The individual radial ducts branch from main axial ducts, so that cooling oil flows freely throughout the entire winding, Fig. 1-54.

Pancake windings, continuously wound, are most common for high-voltage applications. These windings consist of multiplicity of coil sections, usually of single-conductor turns wound one on top of the other. Cross-overs between sections throughout the continuously wound coils are alternately on the inside and the outside of the winding. Radial spacers, locked to the axial

Fig. 1-53. Layer windings are used where currents and voltage are low. Axial oil ducts establish effective cooling between layers. (McGraw-Edison, Power Systems Division)

Fig. 1-54. Helical coils being wound with several conductors in parallel. Parallel conductors are carefully transposed to minimize circulating currents and stray losses. Cooling ducts permit the flow of oil adjacent to each conductor and throughout the winding. (McGraw-Edison, Power Systems Division)

spacers, separate pancake coil sections and establish oil ducts branching from the main axial duct system inside the winding. Efficient, uniform cooling results from the free flow of oil adjacent to each conductor, Figs. 1-55 and 1-56.

Full advantage is taken of the inherent characteristics of each type of winding when determining the de-

Fig. 1-55. Pancake coils being wound on tubular mold with pre-fixed axial spacers. Radial spacers, pre-treated with heat-reactive adhesive, are placed between coil sections. Coils are wound without brazed joints between pancake sections. (McGraw-Edison, Power Systems Division)

Fig. 1-56. Winding department for producing core-form power transformer coils. (McGraw-Edison, Power Systems Division)

sign for a particular transformer. Insulation systems for windings are designed to produce the best dielectric stress distribution, yet provide a liberal level of insulation to meet specified requirements. To secure the highest inherent level of strength, windings of appropriate type are precision-wound on round molds. The entire assembly is subjected to a bonding and curing process that assures a tightly wound and packed coil assembly that will not move during transportation or under short-circuit stress in service.

The circular, concentrically wound and assembled type of winding inherently provides high strength to withstand short circuits. Forces on the high-voltage winding, located on the outside, put the conductors in tension which they are best able to resist. The low-voltage winding is compressed against a strong insulating tube that fits over the core. Longitudinal forces are minimized by balancing the high- and low-voltage windings.

Because of carefully designed balance, there is a minimum tendency

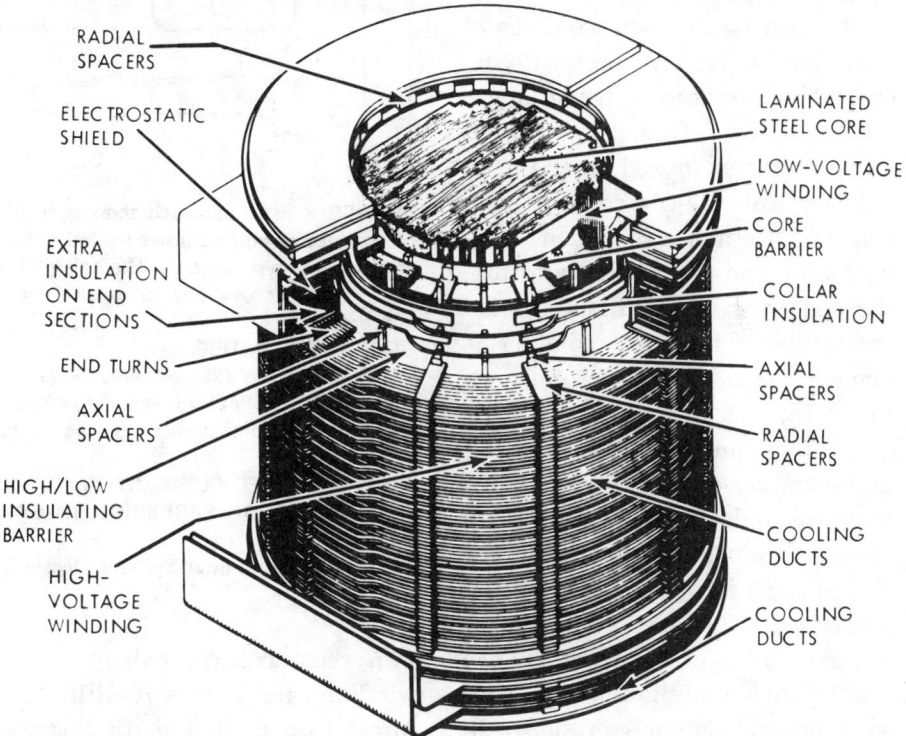

RADIAL SPACERS
ELECTROSTATIC SHIELD
EXTRA INSULATION ON END SECTIONS
END TURNS
AXIAL SPACERS
HIGH/LOW INSULATING BARRIER
HIGH-VOLTAGE WINDING
LAMINATED STEEL CORE
LOW-VOLTAGE WINDING
CORE BARRIER
COLLAR INSULATION
AXIAL SPACERS
RADIAL SPACERS
COOLING DUCTS
COOLING DUCTS

Fig. 1-57. Cutaway view of one phase of a core-form transformer. (McGraw-Edison, Power Systems Division)

for winding to be disturbed in the axial direction, Fig. 1-57.

Both high- and low-voltage windings are wound on accurately sized, rigid insulating tubes with pre-mounted axial spacers. A special machine, with a full size selection of machined mandrels, produces the tubes by rolling and continuously cementing under high pressure the high-quality insulating paper. The cementing pattern permits complete penetration of insulating oil. Exactly positioned axial insulating spacers are cemented into place, also under high pressure.

Strength is added to windings of the pancake type by a fully patented bonding process, using preapplied adhesive. Radial spacers in these windings, locked to the axial spacers, are pre-coated with the heat-reactive adhesive. When windings are cured under high pressure, the spacers bond to the uncoated conductors. Thus, the winding is bonded into a rigid structural unit, Fig. 1-58.

Windings are cured and dried in an accurately controlled oven while they are under a calibrated spring pressure calculated to resist the short-circuit forces that may occur in service. This exclusive process prevents later dimensional change that would loosen the windings after assembly and clamping in the transformer, Fig. 1-59.

Winding taps are located to main-

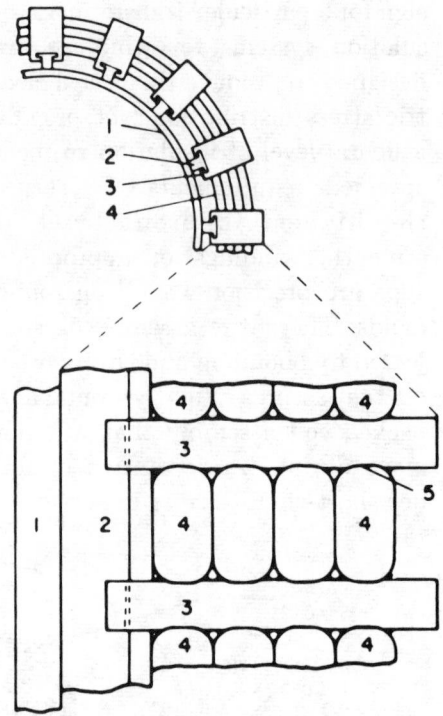

Fig. 1-58. Schematic sketch showing relation of conductors and adhesive-treated spacers in a pancake-type winding. The enlargement shows conductors compressed between spacers.

(1) **BARRIER TUBE.**

(2) **AXIAL SPACERS,** cemented to (1).

(3) **RADIAL SPACERS,** locked to (2), with top and bottom surfaces precoated with adhesive.

(4) **INSULATED CONDUCTORS.**

(5) **PRESSED DEPRESSION IN SPACER (3).**

(McGraw-Edison, Power Systems Division)

tain the electrical balance of the windings and reduce resulting axial forces to a minimum. In this way, also, impedance variation over the tap range is held to a minimum.

Fig. 1-59. High-voltage coils being placed under calibrated spring pressure preparatory to oven drying and curing. Adhesive-treated radial spacers between "pancakes" are in line with pressure blocks at ends. Coils shown are for a 30,000-kVA, 69,000GrdY/ 39,840-13,200 Δ-volt, transformer. (McGraw-Edison, Power Systems Division)

Cores. Cores are constructed of the highest quality cold-reduced, grain-oriented silicon steel, which has inherent properties of low hysteresis and eddy-current losses. The steel is chemically and thermally treated to give it a thin but effective surface insulation that is unaffected by oil or askarel. This insulation prevents the flow of eddy currents between laminations.

Laminations are accurately slit, then cut to a specified length and angle with a special oscillating-head shear, to make an efficient, mitered core joint. After being notched and punched as required, core laminations are stress-relieved by an oven-annealing process in an inert atmosphere where necessary. The temperature in the special electrically heated ovens is accurately controlled to restore the original grain orientation of the steel.

The core itself is stacked in a horizontal position, Fig. 1-60. The bottom yoke and the upright legs only are assembled together at this point. The upright legs are formed using pieces of graduated widths to form a rounded cross-section, so as to fill the cylindrical opening of the coil tube as much as possible. When the coil is lowered into place, the remaining void areas are packed out with insulating material to form a solid cylindrical support inside the coil tube.

Core-and-Coil Assembly. When the core stack is finished, top and bottom side frames are put into place, Fig. 1-61. Pressboard and masonite or lebanite blocking are used to insulate the core laminations from the framing members. Top and bot-

Fig. 1-60. First step in assembly of three-legged, three-phase core. Stacking of cores while they are in a horizontal position permits accurate assembly. (McGraw-Edison, Power Systems Division)

Fig. 1-61. Core stack with top and bottom side frames in place. (McGraw-Edison, Power Systems Division)

Fig. 1-62. Placing coils on the cores. Coil packages rest on yoke pads which are already in position. (McGraw-Edison, Power Systems Division)

tom frames are clamped and held securely by bolts which pass through the core itself and are secured outside the frame. Insulating tubes are inserted over the bolts to preserve the core insulation. Insulating washers are applied underneath the nuts on the ends of the bolts. After nuts are tightened, threads are indented by a punch to prevent nuts from jarring loose during shipment. The upright legs are retained by wrapping with fiberboard and clamping with steel strap during assembly. Heavy glass tape wrappings are then placed on the core legs. Coil pressure and packing help to retain

pressure on the upright legs of the core.

Coils are lowered into position on the core while still warm from the oven dryout, Fig. 1-62. Steel bands are cut away as the coil is lowered. The coil packages rest on yoke pads which are already in position. Circular steel clamping rings are welded to the core frame to provide support for the yoke pads and the coils.

After the coils are in position, the top yoke of the core is assembled into the upright legs, and the insulation and side frames are bolted in place, Fig. 1-63. A core grounding strap is inserted between the top

Fig. 1-63. Assembling top yoke of core into the upright legs. (McGraw-Edison, Power Systems Division)

yoke laminations and is brought to a point accessible from the top of the transformer so that it can be removed for testing core insulation in the field. Core insulation is tested for grounds before and after tanking the core-and-coil assembly.

Pressure is carefully adjusted with jack screws, clamping rods, or packing, depending on design requirement, so that the coils cannot shift during shipment and can withstand the maximum forces that may develop due to short circuits in service.

The core frame assembly is used for vertical clamping of the coils. Two methods of clamping are used. On small transformers up through 3750 kVA, steel clamping rings are welded to the top frame. These compress on the yoke pads. Tension is

developed and kept by long tie rods at the ends of the frame. Insulating tubes are used over the tie rods for extra insulation clearance.

On transformers above 3750 kVA, vertical clamping of the coils is accomplished by jacking bolts, Fig. 1-64. The top and bottom frame members are interlocked with each other by long steel strips which lay parallel to the upright core legs, on both sides of each leg. As with smaller units, clamping rings are welded to the side frames and the jacking bolts are threaded into the clamping rings. When the top frame is in position, the jacking bolts are driven down tight against clamping plates which bear on the upper yoke pads. In this way, pressure is brought to bear on the coil package

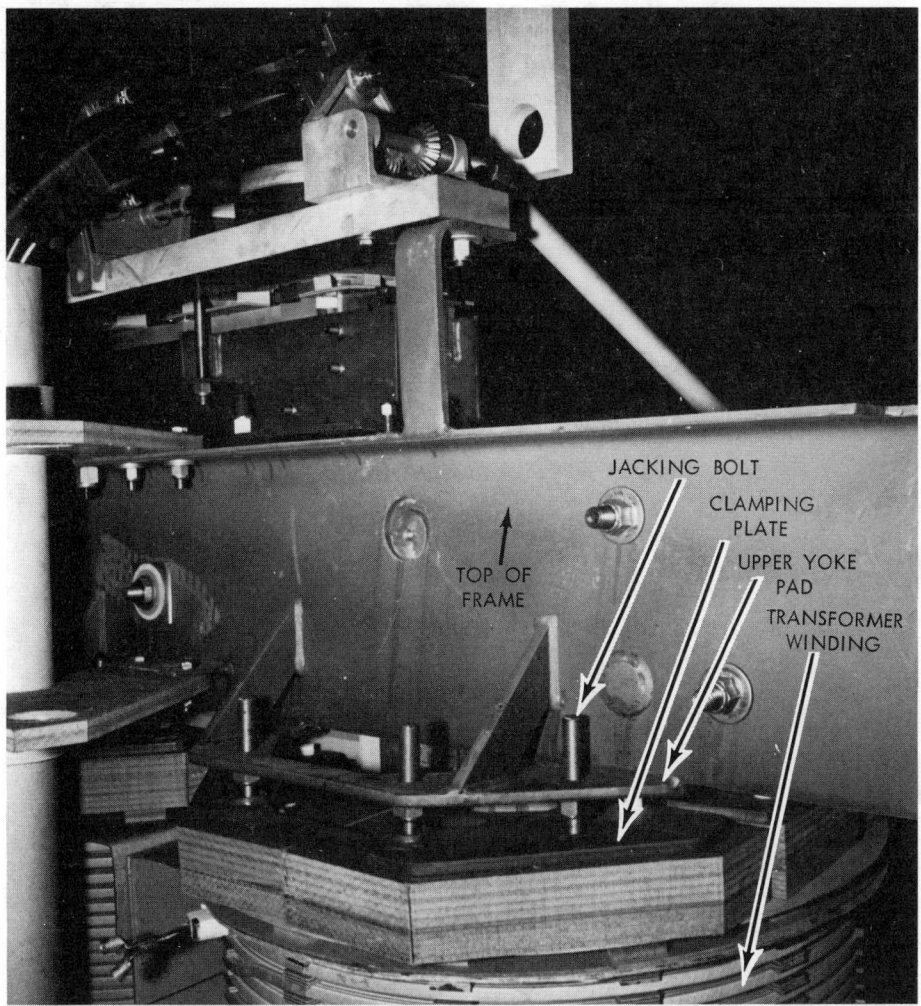

Fig. 1-64. Vertical clamping of coils is accomplished by jacking bolts which are driven down tight against clamping plates which bear on the upper yoke pads. (McGraw-Edison, Power Systems Division)

by jacking against the tension of the strip which interlocks the top and bottom side frames.

All electrical interconnections that do not have to be disconnected for operating reasons are made with either compression joints, or are brazed, to provide permanent joints of low resistance.

Leads and bus work are solidly anchored in position with insulating tubes and maple wood framework,

Fig. 1-65. Completed core-and-coil assembly with leads and bus work solidly anchored in position with insulating tubes and maple wood framework. (McGraw-Edison, Power Systems Division)

Fig. 1-65. Assembly bolts for the lead-supporting frames of higher-voltage transformers are made of special laminated-wood insulating material where there are parts that can become electrostatically charged.

Prior to tanking the completed core-and-coil assembly, each assembly is dried. Drying equipment used consists of specially designed ovens, dryer and a recirculating system. As the warm, dry air passes through the assembly in the oven, it picks up moisture from the coil assembly. After the assembly is removed from the oven, all nuts and bolts are retightened and positively locked.

Tanks. Core-form transformer tanks are constructed of specified high-quality steel plate—cut, formed and welded into rigid structures. Basic designs have been strain-gage proof-tested for strength. Covers are internally well braced, and the bottoms are braced by the external bases. Wall braces, usually of box-type design, are neat appearing, and serve the secondary purpose of providing extra gas expansion space when it is required. Use of this innovation holds down the heights of tanks and makes one-piece shipments of high-kVA and -kV units.

Optimum brace dimensions and space between tank braces have been experimentally determined so that tank resonance is eliminated. The interior sides of tank seams are welded for added strength and to prevent the pocketing of foreign materials during construction, Fig. 1-66. All exterior joints, such as at jack pads, are welded completely closed to prevent the entrance of corrosion-causing moisture into the joint.

Covers are welded on unless a bolted-on cover has been specified. The seams of welded-on covers are

Fig. 1-66. Welding main seams and parts. Tank is turned so that work can be done in horizontal position. (McGraw-Edison, Power Systems Division)

gasketed with asbestos strips to facilitate welding-on or burning-off operations.

Bolted-on covers are sealed with gaskets of nitrile rubber. The gaskets are confined in pressure-limiting grooves in a bar welded around the top of the tank wall. Gaskets of all other joints, such as for access openings and auxiliary equipment flanges, are confined, and pressure is controlled in the same way.

Runners of the new base are con-structed of wide structural channels that are fully enclosed and sealed into a box form. An inside cross channel under the center core leg is used for strengthening the bases of larger 3-phase core-form transform-ers. Bars are substituted for the in-side channel for small transformers, Fig. 1-67.

The large bearing surfaces facili-tate skidding and rolling in any di-rection, with no restriction on the location of rollers. Runner length permits an angle of transformer tilt usually well beyond ANSI require-ments. Foundations may be of any desired design of construction, since no restriction is imposed on the lo-cation or area of bearing points.

The new jack lugs, located at the 4 corners of the tank, have been de-signed to provide a maximum of free working space and jack clear-ance. Pulling cables, attached at holes in the jack lugs for pulling in any direction, stay clear of radiators and other projections from the tank. No spreaders or yokes are needed, Fig. 1-68.

Core-form transformer tanks are tested for leaks by filling them with oil and keeping them under hydro-static pressure of 6.5 psi or above for several hours. All joints are painted with a solution of, or dusted with, light blue chalk which turns dark in the presence of oil, disclos-ing even the minutest leaks, Fig. 1-69.

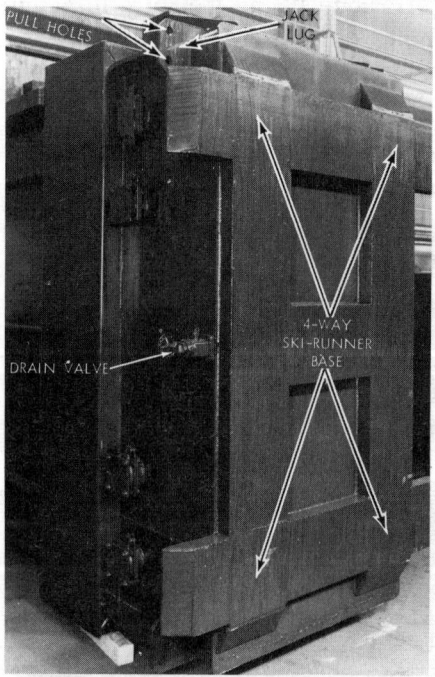

Fig. 1-67. Four-way ski-runner base for larger core-form transformers. For smaller trans-formers, bars are substituted for channel across the center of the bottom. (McGraw-Edison, Power Systems Division)

Fig. 1-68. Jacking and pulling facilities for core-form transformers are designed for easy and efficient use. (McGraw-Edison, Power Systems Division)

Fig. 1-69. Tank being leak-tested under hydraulic (oil) pressure. Light blue chalk, dusted or painted on joint, turns dark in presence of oil. (McGraw-Edison, Power Systems Division)

Tanks are steel-grit blasted inside and out to thoroughly clean them and develop a good surface for the paint. Then, the welds are chipped and ground to remove rough spots. A thorough compressed-air cleaning finishes the preparation for painting.

A red primer coat of paint—specially formulated of zinc-chromate, zinc-oxide and iron-oxide pigments, with a synthetic alkyd resin vehicle—is promptly flowed on after the tank is cleaned and before any surface contamination can occur, Fig. 1-70. This is followed, after a proper drying period, with a flowed-on ex-

Fig. 1-70. Base coat and first finish coat of paint flowed on to secure complete and uniform coverage. (McGraw-Edison, Power Systems Division)

terior coat of dark gray paint, which consists primarily of titanium-oxide pigment with an alkyd resin vehicle, and which matches ANSI No. 24, unless otherwise specified. The flow-on process is most effective in filling every crevice. The interior is painted light gray for protection, to a point below the oil line.

The final exterior coat is a low-gloss dark-gray paint, sprayed on just before shipment. Compounding of the final coat, which has proved long-lived weathering qualities, is basically like the second coat. The low-gloss finish is accomplished with a special flatting agent. The appearance of a low-gloss or satin finish is superior to a high-gloss finish, because it obscures minute irregularities inherent in transformer tank surfaces.

Assembly of Components. Tanking of the core-and-coil assembly is done immediately after the final dryout process. After the assembly is taken out of the dry- out oven, all nuts and bolts are rechecked for tightness, and the assembly is vacuumed and blown to remove any dust or foreign particles. The core-and-coil assembly is then lowered into the tank, the cover is put in place and the tank is filled

Fig. 1-71. Lowering core-and-coil assembly into tank. (McGraw-Edison, Power Systems Division)

with oil according to the filling process called for by the particular voltage class, Fig. 1-71.

McGraw-Edison developed a new principle of securing the core-and-coil assembly to the tank. In this method, the bottom of the assembly is secured by gusseted plates welded to the tank floor which engage a notch in the bottom of the core frame, Fig. 1-72. The top of the as-

Fig. 1-73. An upper fixture bolts the core frame to the tank wall. (McGraw-Edison, Power Systems Division)

Fig. 1-72. Bottom of core-and-oil assembly is secured by gusseted plates welded to tank floor which engage a notch in the bottom of the core frame (McGraw-Edison, Power Systems Division)

sembly is then securely fastened to the tank wall at each end. An upper fixture bolts the core frame to the tank wall either from inside the tank or outside, depending on the transformer size and also the external features of the tank, Fig. 1-73. For external bolting, a gusseted plate is used, and the bolt fixture bears on two large horizontal pins. One, two, and even three bolts are

used depending on the weight of the core-and-coil assembly.

The outside of the tank wall is reinforced to take the strain of horizontal shocks, Fig. 1-74. One-half-inch material is used to form a box around the bolting area. A 2½-inch-diameter bolt is used with a 6-inch washer welded to it. The washer has a gasket-retaining groove and an

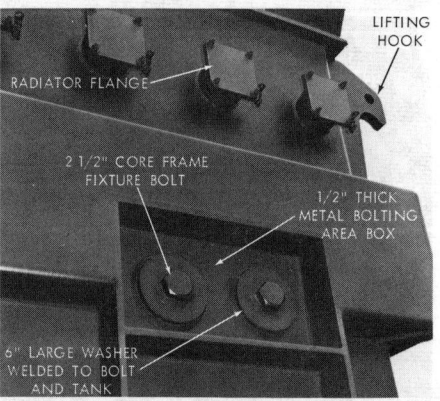

Fig. 1-74. Outside tank wall reinforced to take strain of horizontal shocks. (McGraw-Edison, Power Systems Division)

asbestos gasket is used so that the washer can be welded to the tank wall after the bolt is driven home. Should the unit later require un-tanking, the weld can be chipped away and the bolt removed.

Tests. Three complete test areas are available for making all stand-

Fig. 1-75. Portion of W. E. Kerr High-Voltage Test Center. (McGraw-Edison, Power Systems Division)

Fig. 1-76. Transformer ready for shipment with auxiliary parts loaded on other cars. (McGraw-Edison, Power Systems Division)

ard tests on core-form transformers. Two of these areas, one illustrated in Fig. 1-75, are equipped for special tests, such as measuring corona level, and extend the dielectric test range to units rated over 765 kV. These complete and modern facilities assure proved quality in transformers of all ratings.

Shipment. Core-form transformers are designed for upright, 1-piece shipment with permanent cover in place. Examples of such shipments are illustrated in Fig. 1-76.

Construction of Large Power Transformers

The shell-type core design and construction is commonly used for large-size power transformers, Figs. 1-77 and 1-78.

Fig. 1-77. Three single-phase autotransformers rated 200 mVA, 500 to 345 kV. (McGraw-Edison, Power Systems Division)

Fig. 1-78. 150/500/250 mVA (OA/FOA/FOA), 345 GrdY-138/GrdY-13.8 △ kV, LTC transformer. (McGraw-Edison, Power Systems Division)

Fig. 1-79 shows conductor wrapping machines. All insulating materials meet rigid specifications for the type of application. Each coil is shaped to exact dimensions as illustrated in Fig. 1-80.

Illustrated in Fig. 1-81 each winding package is clamped under specific spring pressure preparatory to treatment in the oven at the right.

Use of special high-grade, grain-oriented, cold rolled steel provides basic assurance of low-loss, low-exciting-current and low-sound-level cores in shell-form transformers. Core steel must be carefully handled and cut to assure a good fit so the core is efficient magnetically and strong structurally, Fig. 1-82.

Fig. 1-83 shows complete winding

67

Fig. 1-79. Two of a bank of wire insulating machines. Left: Wrapping a single conductor. Right: Wrapping a group of three insulated conductors. Heat-reactive adhesive is applied and only dried in tunnel. (McGraw-Edison, Power Systems Division)

Fig. 1-80. Finished coil, shown clamped to dimensions, is heated with circulating current to bond turns together. Integrity of conductor insulation is then voltage-tested. (Mc-Graw-Edison, Power Systems Division)

Fig. 1-81. Finished winding package shown being clamped under a pre-determined spring pressure preparatory to treatment in oven at right. The process shrinks the insulation to final dimensions and cures adhesives that bond the entire package into a structural unit. (McGraw-Edison, Power Systems Division)

Fig. 1-82. Core steel being accurately sheared and notched according to design specification. (McGraw-Edison, Power Systems Division)

Fig. 1-83. Winding package being lowered into bottom section of tank. (Clamps have been temporarily removed from near winding package.) Shunt packs and blocking members are in place in near winding pocket. (McGraw-Edison, Power Systems Division)

Fig. 1-84. Core steel being accurately stacked. Steel rests on insulated tank flange and non-magnetic "T" beam inserted through winding windows. Clamps and bands are removed as stacking progresses. Windings are kept clean, warm, and dry by means of filtered warm air being continuously blown through the winding from special heaters (not visible). (McGraw-Edison, Power Systems Division)

Fig. 1-85. Center tank section being lowered into position. Note core-clamping jack screws. When in position, coil-and-core assembly will be completely and tightly blocked, forming a totally solid structure, end to end, from tank wall to tank wall. (McGraw-Edison, Power Systems Division)

Fig. 1-86. Assembly of 345-kV, 400-mVA transformer, complete with LTC auxiliary windings and connections, being inspected prior to placement of cover section with LTC switching equipment. (McGraw-Edison, Power Systems Division)

package for one phase being lowered into bottom section of tank. Accurate stacking of the shell core is necessary to obtain the full advantages from the excellent characteristics of the steel, Fig. 1-84.

Fig. 1-85 shows the center tank section being lowered into position. When in position special screw-type clamps are adjusted to secure the core and the top portion of the windings. Shell-form transformers of this design have been remarkably free from coil or core shifting.

Fig. 1-86 shows transformer being inspected prior to placement of cover section.

Transformers– Principles and Classification

In the previous chapter a transformer was described as a stationary piece of electrical equipment which interlinks one or more independent electrical circuits with a magnetic circuit whereby the voltage factor of electrical energy may be changed from one value to another.

When a transformer receives electrical energy in the form of alternating current at a high voltage and delivers it at a lower voltage, it is termed a step-down transformer; when a transformer receives electrical energy in the form of an alternating current at a low voltage and delivers it at a higher voltage, it is termed a step-up transformer.

By using transformers, it is feasible to generate electrical power at a relatively high voltage, transmit this power at a high voltage or at a voltage even higher than that generated, and then step it down in one or more steps to a value suitable for distribution and use by the ultimate consumer. Fig. 2-1 shows how, through the use of transformers, electrical power is transmitted from the generating plant to the ultimate consumer.

Transformers serve many functions other than the transmission and distribution of electrical energy. When transformers are used for special purposes they are normally referred to by the purpose they serve such as for sign lighting, control and signaling, gaseous-discharge lamp transformers (mercury vapor or fluorescent lamps), bell ringing, instrument, constant current, series transformers for street lighting, isolating transformers which are used in oper-

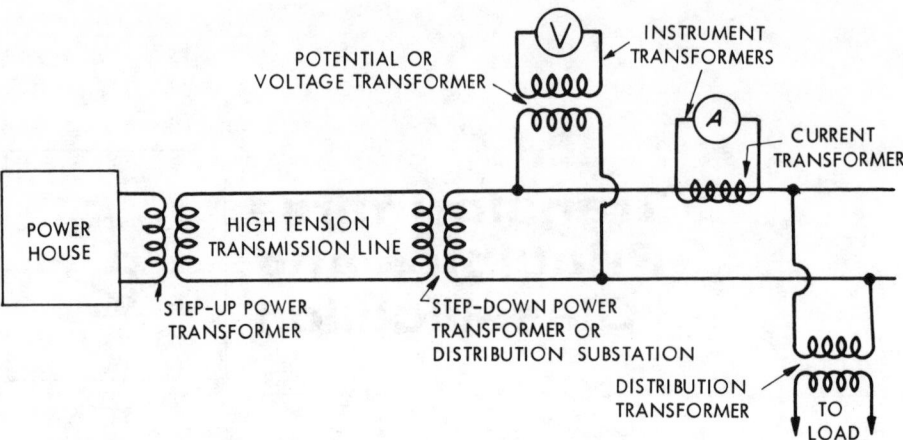

Fig. 2-1. Diagram showing relative positions of different types of transformers used in a typical power system.

ating rooms of hospitals or transformers which are used to match the characteristics of one circuit to another. Special transformers can be used to convert pulsating direct current to alternating current. An example of this type of transformer use is in the common auto radio where a vibrator converts direct current into a pulsating direct current which in turn is converted into an alternating current of the desired voltage. Another common use of the transformer is to control the starting current of motors and the dimming of residential and commercial lighting. Transformers are used in electronic and communication equipment to increase or decrease the voltage level of AC and to link AC circuits together. Although transformers serve many functions and have many uses they all operate on the same basic theory of electromagnetic induction.

How the Transformer Works

Principle of Operation and Definitions

The transformer works on the principle that energy can be efficiently transferred by magnetic induction from one set of coils to another set of coils by means of a varying magnetic flux produced by an alternating current, provided both sets of coils are linked together with this magnetic circuit which must be common to both coils.

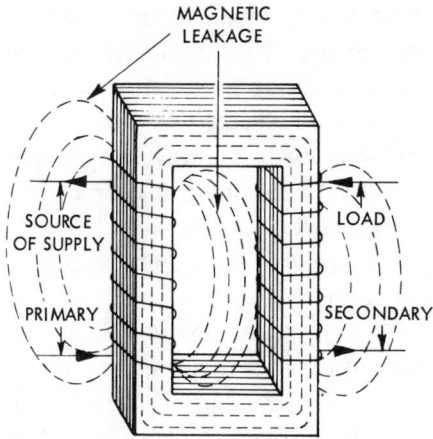

Fig. 2-2. Diagram of core-type transformer.

The coil of a transformer which is energized from a source of power in the form of alternating voltage and current is called the primary winding; and the transformer coil which delivers this alternating current and voltage to the load is called the secondary winding. (See Fig. 2-2.)

In some special applications the transformer may have a third set of coils which comprise what is called a tertiary winding. This type of transformer with three windings instead of the customary two windings known as the primary and secondary windings has several uses which will be described in the appropriate place later in this chapter.

Fig. 2-2 shows the primary and secondary windings on separate legs of the magnetic circuit. This is done so that we can more readily understand how the transformer works.

In actual practice, half of the primary and secondary coils are wound on each of the two legs shown in Fig. 2-2, with sufficient insulation between the two windings and between the windings and the core to properly insulate the windings from each other and from the core. A transformer wound with the primary and secondary coils on separate legs will function at greatly reduced efficiency because of greatly increased magnetic leakage which is shown by the stray flux lines in Fig. 2-2. Magnetic leakage is that portion of the magnetic flux which passes through either one of the coils but not through both. The greater the distance between the two coils, the longer the magnetic circuit and the greater the leakage; therefore in actual practice, insulated primary and secondary coils are wound upon each other or interleaved with one another in such a way as to keep the magnetic leakage to an economic minimum.

The terms "high-voltage" and "low-voltage" windings refer to the relative values of the normal voltages of these coils. The operation of the transformer is reversible in the sense that either the high voltage or the low voltage can be used as the primary or secondary.

Operation Without Load

When an alternating voltage is impressed on the primary winding, an alternating current flows which magnetizes the magnetic (iron) circuit first in one direction and then in the other. The reversals in the direction of this magnetism are of course the same as the alternations of the alternating current supply. This alternating magnetic flux flowing around the entire length of the magnetic circuit induces voltages in both the primary and secondary coils, and, since both coils are linked by this same flux, the induced voltage per turn of the primary and secondary coils must be of the same value and in the same direction at any given instant of time. This induced voltage opposes the voltage impressed across the primary coil. Since it is opposite in direction, it is therefore called the counter electromotive force. This counter electromotive force or induced voltage is nearly equal to the impressed voltage, the difference in the two being that voltage necessary to send the current through the primary winding. This current which flows in the primary winding (with no current flowing in the secondary winding) is just enough to establish a magnetic field of sufficient strength to counteract the applied voltage. The current flowing under these conditions is therefore called the exciting current. Since current will flow in the primary even when the secondary winding is not connected to any load, it is frequently referred

to as the no-load current of the transformer. In the design of the transformer this exciting current must not be permitted to become too large, for it is a wattless current (a wattless current is a current out of phase with the voltage by 90°) and therefore does not represent real energy.

For a given magnetic circuit the exciting current is decreased by increasing the number of turns in the exciting or primary winding. The measure of the magnetizing force of any coil in which a current is flowing is the number of ampere-turns produced by the coil. Therefore, for a required number of ampere-turns the current can be reduced by increasing the number of turns. This is the fundamental principle upon which the designer selects the proper ratio of turns and cross sectional area of the magnetic circuit.

As previously stated, due to the reversals of magnetic flux in the core, a counter electromotive force is induced in each of the turns of the primary and secondary coils on that core, and this induced voltage on the primary side is nearly equal to that of the impressed voltage. If the primary winding had no resistance and no energy were required in the reversal of the magnetic flux, and furthermore if it were assumed that the magnetic circuit had an infinite resistance with respect to the flow of eddy currents so that there would be no eddy current flow, the induced voltage would be equal to the impressed voltage. Each turn of the primary winding would then have an induced voltage equal to the impressed voltage divided by the number of turns in the primary winding, or the volts per turn would equal the total voltage divided by the total number of turns.

If we assume that the secondary coil is so closely interleaved with the primary coil that there is no magnetic leakage of the lines of flux, it is evident that in each turn of the secondary coil the flux must set up a counter electromotive force of the same value as that of each turn in the primary coil. The total induced voltage of the secondary coil must then be equal to the volts-per-turn of the primary coil multiplied by the number of turns in the secondary coil. The voltages of the primary and secondary windings are therefore directly proportional to the number of turns in the primary and secondary coils.

Therefore $\quad \dfrac{E_p}{E_s} = \dfrac{N_p}{N_s} \quad$ (1)

where E_p and E_s are respectively the induced voltages of the primary and secondary coils, and N_p and N_s are respectively the total number of turns in the primary and secondary coils.

Example. A certain transformer having a primary winding of 3000 turns and a secondary winding of 300 turns is connected to a 2400-volt source of alternating current. The voltage induced in the primary coil will be nearly 2400 volts, and the volts induced per turn will be

$$2400 \div 3000 = .8 \text{ volt.}$$

The voltage induced in the secondary coil will be

$$E_s = E_p \times \frac{N_s}{N_p} = 2400 \times \frac{300}{3000} = 240 \text{ volts}$$

and the volts induced per turn will be

$$240 \div 300 = .8 \text{ volt.}$$

Since the volts induced per turn are the same for both primary and secondary coils, the secondary voltage is equal to primary volts per turn multiplied by the number of turns in secondary coil, or

$$.8 \div 300 = 240 \text{ volts.}$$

Operation with Load

When an alternating current is caused to flow through the secondary winding, the transformer is said to be loaded. For example, if a 110-volt motor is connected across the 110-volt terminals of a 2200- to 110-volt transformer excited from a 2200-volt source of supply, the motor will draw current from the secondary winding of the transformer. The amount of current will increase as the power required by the motor to perform its work is increased. The 110-volt induced voltage of the secondary coil causes this current to flow through the circuit of the motor. Within the transformer this secondary current tends to magnetize the iron circuit in a direction opposite to that of the magnetizing action of the primary exciting current. This magnetizing action tends to lower the induced electromotive force in both primary and secondary. Since this is lowered but a very slight amount, the difference between the impressed voltage and the counter electromotive force is increased, and this permits a greater current to flow in the primary winding. Actually, the increase in flow of current in the primary winding is just that amount sufficient to establish a magnetizing force equal to the magnetizing action of the current flow in the secondary due to the load. The resultant flux in the core is therefore maintained at a constant value by the primary current, regardless of the value of the load placed upon the secondary of the transformer. If then we neglect the small amount of exciting current in the primary coil required to maintain this constant flux in the magnetic circuit, the magnetizing force of the primary winding due to the load placed upon the secondary is equal to the magnetizing force of the load current, or the ampere-turns of the primary are equal to the ampere-turns of the secondary. This can be stated as follows:

$$I_p N_p = I_s N_s \text{ or } \frac{I_p}{I_s} = \frac{N_s}{N_p} \quad (2)$$

where I_p is the increase in primary current over the no-load exciting current due to the secondary load, and I_s is the secondary load current.

Since the exciting or no-load current of a transformer is very small in comparison to its output capacity, it may be entirely neglected for practical purposes and equation (2) represents the relationship between the currents in the two windings.

From equations (1) and (2) it is evident that the ratio of the number of turns in the primary to the number of turns in the secondary is equal to the ratio of the primary to the secondary voltages, whereas the ratio of the current in the primary to that in the secondary coil is equal to the ratio of the number of turns in the secondary to the number of turns in the primary.

In equation (2) we can then substitute the ratio $\frac{E_s}{E_p}$ for $\frac{N_s}{N_p}$ and we have $\frac{I_p}{I_s} = \frac{E_s}{E_p}$ or $I_p E_p = I_s E_s$ **(3)**

$I_p E_p$ is the energy input, neglecting no-load losses, in terms of the volt-amperes of the primary coil; $I_s E_s$ is the energy output of the secondary winding due to the connected load. Thus it is seen that energy is transferred from a primary source of supply having a voltage of E_p volts to the secondary or load side at a voltage of E_s by the magnetic action of the alternating currents in these windings.

Example. A certain transformer rated at 3 kilovolt-amperes has 1760 turns of wire in its primary winding and 88 turns of wire in its secondary winding. The primary winding is connected to a source of supply having a potential of 2400 volts. Therefore the secondary electromotive force or secondary terminal voltage, from equation (**1**), is

$$E_s = \frac{N_s}{N_p} \times E_p = \frac{88}{1760} \times 2400 = 120 \text{ volts.}$$

If ten lamps, each taking half an ampere, are connected to the secondary terminals of the transformer, the total secondary current will be $10 \times .5 = 5$ amperes, and the primary current from equation (**2**) will be

$$I_p = \frac{N_s}{N_p} \times I_s = \frac{88}{1760} \times 5 = .25 \text{ ampere.}$$

The energy delivered to and supplied by the primary winding is

$$E_p I_p = 2400 \times .25 = 600 \text{ watts.}$$

The energy delivered to the lamps by the secondary coil is, from equation (**3**),

$$E_s I_s = 120 \times 5 = 600 \text{ watts.}$$

Check: Each lamp takes .5 ampere at 120 volts. Therefore each lamp is rated at $120 \times .5 = 60$ watts. Ten of these lamps would then require 60×10 or 600 watts.

Since the transformer in this example is rated at 3 kilovolt-amperes (which equals 3000 watts with a noninductive load obtained by a pure resistance load such as lamps) and is delivering a load of 600 watts only, it is loaded at $(600 \div 3000) \times 100$ (percent) = 20 percent of its rated capacity. Without exceeding the rating of this transformer we would be able to connect $100 \div 20$ or 5 times as many lamps of the same rating as given in the example. In such an instance, the current in the second-

ary would be $50 \times .5 = 25$ amperes, and the power delivered by the secondary would be $25 \times 120 = 3000$ watts or 3 kilowatts. The current in the primary would be 5 times the value obtained in the previous example or $.25 \times 5 = 1.25$ amperes and the power delivered to the primary would be $2400 \times 1.25 = 3000$ watts or 3 kilowatts.

This transfer of energy from primary to secondary without any loss represents the ideal condition, and the foregoing discussion is therefore that of an ideal transformer. Practically, such a condition cannot exist, since we know that (1) the windings, both primary and secondary, have resistance; therefore when any current is flowing through these windings there must be a drop in potential or voltage. (2) The magnetic circuit does require energy to magnetize the circuit alternately first in one direction and then in the other at a rate equal to the alternations of the supply voltage. (3) Energy is required to supply the eddy currents which are created in the magnetic circuit. (4) Magnetic leakage does exist to some extent in all transformers.

The extent to which a well-designed transformer deviates from the ideal is shown by the following results which are representative of a distribution transformer of the same rating as given in the previous example.

At no load the value E_s is 120 volts; the decrease in secondary voltage, at no load, due to exciting current flowing through the primary and the decrease due to magnetic leakage are so small that for all practical purposes they cannot be measured; the no-load primary current is .0625 ampere; the energy taken from the source of supply by this no-load current (iron loss or core loss) is 30 watts. This core loss, or iron loss, as it is more commonly known, is constant at all loads if the primary voltage is maintained constant.

When 50 lamps taking a total of 25 amperes are connected to the secondary, then E_s is 117.36 volts. The I^2R loss in the primary coil is 33 watts; and the I^2R loss in the secondary is 33 watts. Therefore the energy delivered to the lamp is $117.36 \times 25 = 2934$ watts. The power taken from the source of supply is $2934 + 30 + 33 + 33 = 3030$ watts. The efficiency of this transformer with this secondary load of 25 amperes is equal to $\frac{2934}{3030}$, or 96.83 percent.

Vector Diagram of the Transformer

Ideal Transformer

No Load. To fully understand the current, voltage and flux relationship in a transformer, it is necessary to represent each of these values in a vector diagram. A vector is a quantity involving direction as well as magnitude.

Consider first an ideal transformer in which there is neither resistance nor reactance. Reactance X, in ohms, is equal to magnetic leakage (or inductance in henrys) multiplied by $2\pi f$ where π (Greek letter pi) has a numerical value of 3.1416, and where f is the frequency of the circuit. The term $2\pi f$ is frequently expressed as ω (Greek letter omega) and inductance in henrys is indicated by the letter L. Then $2\pi fL = \omega L$. If we assume a 1-to-1 ratio of transformation for the sake of simplicity, OE_p (Fig. 2-3) represents the impressed primary voltage. O_c represents the current flowing in the primary winding to produce the flux $O\Phi$, which lags the impressed voltage by 90°, and OE_s represents the secondary induced voltage lagging the flux by 90°, therefore 180° behind E_p or in direct opposition to the impressed voltage. If the magnetic circuit were also an ideal circuit there would be no iron loss (eddy current and hysteresis loss) as indicated by the vector Oa and

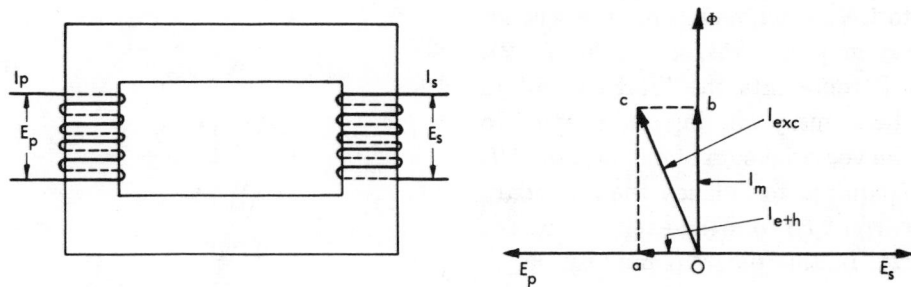

Fig. 2-3. Diagram of transformer (left); No-load vector diagram (right).

the no-load current would be equal to Ob which is the magnetizing current required to establish the total flux Φ. Since the magnetic circuit does have hysteresis and eddy current loss, the total no-load or exciting current indicated by the vector Oc is the vector sum of Oa and Ob. This no-load current lags the primary impressed voltage E_p by the angle E_pOc. The power factor of the transformer at no load is represented graphically by the cosine of the angle E_pOc or is equal to I_{e+h}, which is the iron loss current, divided by IOc, which is the exciting current. If we multiply each of these current values by the primary voltage we have the iron loss in watts and the exciting power.

No-load power factor $= \dfrac{I_{e+h}}{I_{exc}}$

$$= \frac{\text{iron loss in watts}}{\text{exciting watts}} \quad (4)$$

Since the iron loss is small compared to the exciting power, the power factor is very low at no load, as is apparent from an inspection of equation (4).

With Secondary Having Noninductive Load. In Fig. 2-4 the lines OE_p, OE_s, $O\Phi$, Oa, Ob, and Oc represent the same quantities as in Fig. 2-3. In addition to these we have a secondary load I_s amperes in phase with the secondary voltage OE_s since we are considering a load having no inductance. I_p represents the additional primary current which must flow in the primary winding due to the secondary load I_s. Since we are assuming a 1-to-1 ratio of transformation, I_p is equal to I_s and directly opposite in direction. The product I_pN_p, which represents the primary ampere turns, is equal

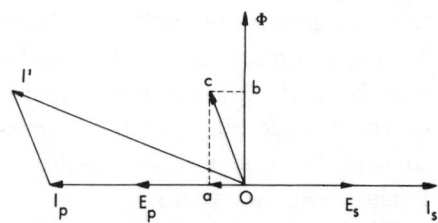

Fig. 2-4. Vector diagram of ideal transformer with noninductive load.

to I_sN_s which represents the secondary ampere turns, see equation (2). OI' represents the total current in the primary winding and is equal to the vectorial sum of the current OI_p (required to balance the secondary current I_s) and the exciting current Oc. In this diagram it must be remembered that Oc is not drawn to the same scale as OI_p but is shown much larger than the actual ratio that exists between these two currents in order to make the diagram a little clearer. *In actual practice the exciting current O_c seldom is more than 10 percent of the load current OI_p, and in very large transformers may be less than one percent.* This small value of current when added vectorially to OI_p increases OI_p a very small amount, making OI' (the total current in the primary winding) practically equal to the load current. The effect of the exciting current with respect to additional loss and additional heat because of this I^2R loss is disregarded in practice.

It should be noted, however, that this noninductive load greatly reduces the angle of lag between the primary impressed voltage E_p and the total primary current I', thereby greatly improving the power factor of the transformer which is measured by the cosine of the angle between these two vectors.

With Secondary Having Highly Inductive Load. In Fig. 2-5 all

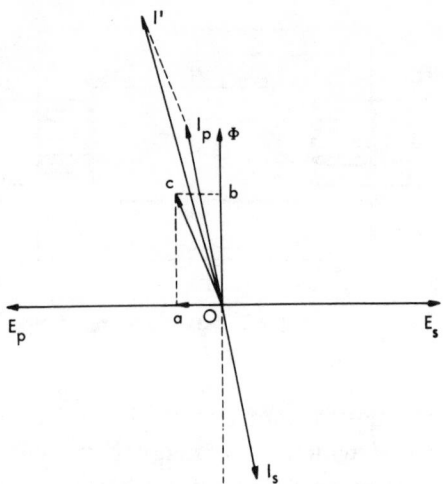

Fig. 2-5. Vector diagram of ideal transformer with inductive load.

symbols and vectors represent the same things as in Fig. 2-4. The secondary load current I_s is shown lagging the secondary voltage E_s nearly 90°, since the load on the secondary is almost purely an inductance load. With such a load, the exciting current becomes more noticeable in its effect in making the total primary current I' greater than the load current I_p. The angle E_sOI_s represents the angle of lag of the load current behind the secondary voltage and the cosine of this angle represents the power factor of the secondary connected load.

Vector Diagram of Actual Transformer

In an actual transformer carrying load, the foregoing diagrams are

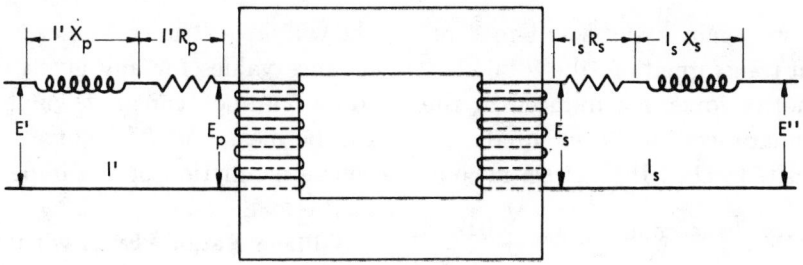

Fig. 2-6. Diagram of actual transformer showing resistance and reactance in windings.

complicated by resistance and reactance drops in both the primary and secondary windings. To get a clear picture of the effects of these two quantities it is desirable to consider them as being external to the transformer proper, as in indicated in Fig. 2-6 where X_p, X_s, and R_p, R_s represent respectively the reactance and resistance in ohms of the primary and secondary windings. $I'X_p$ and $I'R_p$ represent voltage drops in the primary winding due to reactance and resistance respectively when a load current, I_p, is flowing in this winding. The total impressed voltage, E', applied to the primary winding must (1) overcome the resistance R_p, of the primary winding, (2) overcome the reactance, X_p, due to leakage flux, and (3) balance the electromotive force induced in the primary winding by the magnetic flux Φ.

In the secondary winding, $I_s X_s$ and $I_s R_s$ represent the voltage drops due to reactance and resistance respectively when a current I_s is flow-

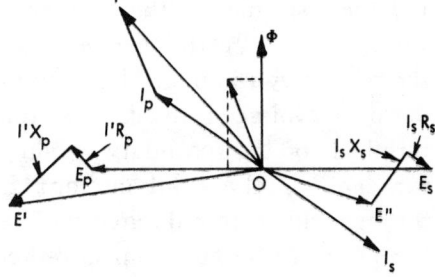

Fig. 2-7. Vector diagram of transformer having resistance and reactance.

ing in this winding. The secondary terminal voltage is E'''.

The vector OE', Fig. 2-7, represents the total impressed voltage on the primary and is the result of combining three vectors which are (1) $I'R_p$, the resistance drop, (2) $I'X_p$, the reactance drop and (3) OE_p, the voltage impressed across the coil. The resistance drop $I'R_p$ is in phase with I', the total current in the primary winding. Thus the vector $I'R_p$ is parallel to OI'. The reactance drop $I'X_p$ is 90° out of phase with I' and the vector $I'X_p$ is at right angles to OI'.

If we again assume a 1-to-1 ratio of transformation, the total electromotive force, E_s, induced in the secondary coil will be equal and opposite to OE_p. If the turn-ratio of primary to secondary is $\dfrac{N_p}{N_s}$ then the induced voltage OE_s will be $\dfrac{N_s}{N_p} \times OE_p$. A portion of this secondary induced voltage is used to overcome (1) the resistance of the secondary winding and (2) the reactance of the secondary winding. The remainder of the voltage is available at the terminals of the secondary winding for sending the load current I_s through the external circuit. The secondary resistance drop is represented by $I_s R_s$; the reactance drop is represented by $I_s X_s$. The remainder of the voltage which represents the terminal voltage when the transformer is supplying a load of I_s amperes is represented by the line OE''.

From an examination of Fig. 2-7, it is evident that the ratio of primary impressed voltage, E', to secondary terminal voltage, E'', is no longer 1 to 1, and these voltages are no longer in direct opposition to each other. This ratio and the phase angle between E' *and* E'' are dependent upon the relative values of IR and IX in both primary and secondary windings. These in turn are dependent (1) upon the resistance and reactance of the transformer and (2) upon the current in

the windings. The resistance and reactance values for any given transformer are of constant value and can be measured. The current is of course a function of the impedance of the load.

Voltage Ratio. Even with all of these foregoing variables affecting the voltage ratio, it is common practice in distribution and power transformers to consider the voltage ratio equal to the turn ratio of the primary and secondary windings. The error is considered negligible, but it is an error which must be taken into account for measuring devices such as voltmeters, ammeters, wattmeters and varmeters. This error is compensated for in such transformers by slightly changing the turn ratio.

Summary. From the foregoing discussion it can be seen that the no-load or exciting current is the only factor that affects the ideal relationship between primary and secondary currents as expressed in equation (**2**). However, the relationship between primary and secondary electromotive forces as expressed in equation (**1**) are affected by the resistance of both primary and secondary windings and by the total magnetic leakage or reactance of these windings. With a noninductive load, only the resistance affects this ratio, and the secondary terminal voltage decreases with increase of load. This is due almost entirely to the *IR* drop of the primary and secondary coils. With an inductive load, the resistance drop is less perceptible and the reactance becomes a major factor in lowering the secondary voltage. This can be seen by a short study of Fig. 2-7.

If the secondary load is highly inductive, the load current I_s and the primary current I' will be as shown in Fig. 2-5. The $I_s R_s$ and $I' R_p$ drops will be in phase with these respective currents and therefore nearly at right angles to the voltages E_s and E_p. Therefore when they are added and subtracted respectively from E_s and E_p they have very little effect. However the $I' X_p$ and $I_s X_s$ drops, which are at right angles to I' and I_s respectively, will be nearly parallel to E_p and E_s and they therefore add and subtract almost directly from E_p and E_s respectively by an amount equal to their arithmetical value. Actually, it is impossible to measure the magnetic leakage of the primary and secondary separately, and it is therefore common practice to refer to the total magnetic leakage in terms of either primary or secondary winding alone. If this reactance is expressed in ohms with respect to the primary winding, it can be expressed in ohms with respect to the secondary winding by multiplying this value in ohms by the square of the ratio of secondary to primary

turns. Conversely, if this reactance is expressed in terms of ohms with respect to the secondary winding, it can be expressed in terms of ohms with respect to the primary winding by multiplying this reactance by the square of the ratio of primary to secondary turns.

Example. The primary of a certain 10-kilovolt-ampere transformer having a primary voltage of 1,000 volts and a secondary voltage of 100 volts (no-load turn-ratio of 1,000 to 100 or 10 to 1) has a primary resistance of 1.5 ohms and the secondary coil has a resistance of .015 ohm. The leakage reactance or ωL, as referred to the primary winding, is 5 ohms.

The secondary coil delivers 100 amperes to a noninductive load and, neglecting the no-load losses, the primary then takes 10 amperes from the 1,000-volt source of supply. The *IR* drop of electromotive force in the primary coil is therefore 10 (amperes) $\times$ 1.5 (ohms) = 15 volts.

Since the load is noninductive, the current I_s and I_p, neglecting no-load losses, will be parallel to E_s and E_p respectively. Therefore, this 15-volt drop, represented by $I' R_p$ in Fig. 2-7, will be parallel to E_p and must be added directly to E_p, Fig. 2-8, to obtain the im-

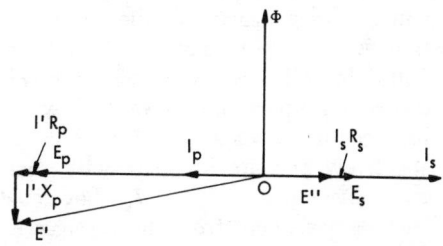

Fig. 2-8. Vector diagram for illustrative example.

pressed voltage E'; or E_p, which represents the primary induced voltage, will be equal to $E' - I'R_p$ and in this example equal to $1{,}000 - 15 = 985$ volts. The IX drop due to the leakage reactance of 5 ohms represented by $I'X_p$ in Fig. 2-8 is equal to 10 (amperes) $\times$ 5 (ohms) $=$ 50 volts. However this 50 volts is nearly at right angles to E', Fig. 2-8, therefore when subtracted vectorially it does not appreciably lower its value. The total electromotive force induced in the secondary coil will be

$$\frac{N_s}{N_p} \times 985 = \frac{100}{1000} \times 985$$
$$= 98.5 \text{ volts.}$$

The IR drop of electromotive force in the secondary coil is equal to 100 (amperes) $\times$ .015 (ohm) $=$ 1.5 volts. This is represented by the vector $I_s R_s$ in Fig. 2-7 and it will be parallel to E_s since we have a noninductive load. Therefore, it will subtract directly from the secondary induced voltage E_s, Fig. 2-8, or the secondary terminal voltage will be equal to $98.5 - 1.5 = 97$ volts.

If we assume that the secondary is delivering 100 amperes of current to a highly inductive circuit, then this 100 amperes represented by I_s in Fig. 2-7 is nearly 90° behind E_s, and the primary current of 10 amperes represented by I_p is nearly 90° behind E_p in phase relationship. Therefore, the drop due to leakage reactance, which is equal to 50 volts, is very nearly parallel to E_p so that E_p is very nearly equal to 1,000 volts -50 volts $= 950$ volts. Since the IR drops of primary and secondary windings equal 15 volts and 1.5 volts respectively and are in phase with their respective currents I_p and I_s, they must then be nearly 90° from the voltages E_p and E_s. Therefore they will have no appreciable effect in lessening these volt-

ages. Since we have expressed the total leakage reactance as 5 ohms when referred to the primary winding, the drop $I_s X_s$ shown in Fig. 2-7 will not exist in this example. Therefore, the secondary terminal voltage E'' will be nearly equal to

$$\frac{N_s}{N_p} \times 950 = \frac{100}{1000} \times 950 = 95 \text{ volts.}$$

If we were to express this leakage reactance in terms of secondary winding its value would be

$$\left(\frac{N_s}{N_p}\right)^2 \times 5 = \left(\frac{100}{1000}\right)^2 \times 5$$
$$= \frac{1}{100} \times 5$$
$$= .05 \text{ ohm}$$

The drop due to this leakage reactance with 100 amperes flowing would equal .05 (ohm) $\times$ 100 (amperes) $=$ 5 volts.

Since the IR drops are negligible in both windings, the secondary induced voltage will be

$$\frac{N_s}{N_p} \times 1000 = \frac{100}{1000} \times 1000$$
$$= 100 \text{ volts.}$$

The secondary terminal voltage will be equal to the secondary induced voltage minus the drop due to leakage reactance, or 100 volts $-$ 5 volts $=$ 95 volts.

The method used in measuring this leakage reactance, in practice, determines whether it is expressed in terms of primary or secondary winding. It is most commonly expressed in terms of the high-voltage winding which is the primary winding in the case of a step-down transformer and the secondary winding in case of a step-up transformer.

Classification of Transformers

Classification With Respect to Service

Transformers may be classified into five general categories:

1. Voltage Transformation
2. Voltage Regulation
3. Current Regulation
4. Metering and Protection
5. Accessory.

Each of these five general categories may be subclassified into a number of types, such as:

(a) Instrument
(b) Constant-Current
(c) Series transformers for street lighting
(d) Small power
(e) Control and signal
(f) Electric sign
(g) Chime and bell ringing
(h) Generator step-up
(i) Neon sign
(j) Distribution
(k) Grounding

While this type of classification will be helpful in a general way, it must be understood that each of these groups is not sharply defined. From the point of view of design and construction, there is a general similarity in the transformers of all of these groups. Further, the dividing line, for example, between distribution, and substation transformers, is more arbitrary than fundamental.

Instrument Transformers.. The purpose of instrument transformers is to step down the voltage or current of a circuit to a low value that can be effectively and safely used for the operation of instruments, such as ammeters, voltmeters, wattmeters and varmeters, relays used for various protective purposes and telemetering used for indications at remote areas and dispatching energy. They also perform the necessary function of insulating the instrument, relay scheme, or telemetering equipment from the higher voltage power circuit.

Instrument transformers, including those for either current or voltage transformation, have a small volt-ampere capacity, which is necessary to provide the energy required by the measuring instrument with which they are used. The requirement here is for the utmost accuracy of voltage and current transformation, particularly when the transformers are used in connection with the metering of power. The errors that must be compensated for are of two kinds: those in the ratio of transformation and those pertaining to the relative phase position of impressed and delivered electromotive forces and currents.

Although instrument transformers function on the same basic principle as power transformers, their construction may be quite different. The units in the lower ranges of voltage, particularly those for switchboard application, are operated without oil, while higher voltage units are immersed in oil for insulation purposes. While most instrument transformers are small in size, the higher voltage units become very large compared with the devices to which they are connected. Current and voltage

mary, usually one or more turns, is connected in series with the line. When the primary has a large current rating, the primary winding may consist of a straight conductor passing through the current of the magnetic circuit as shown in Fig. 2-9. This single conductor may be a portion of the current carrying bus or any conductor that we are interested in measuring or protecting.

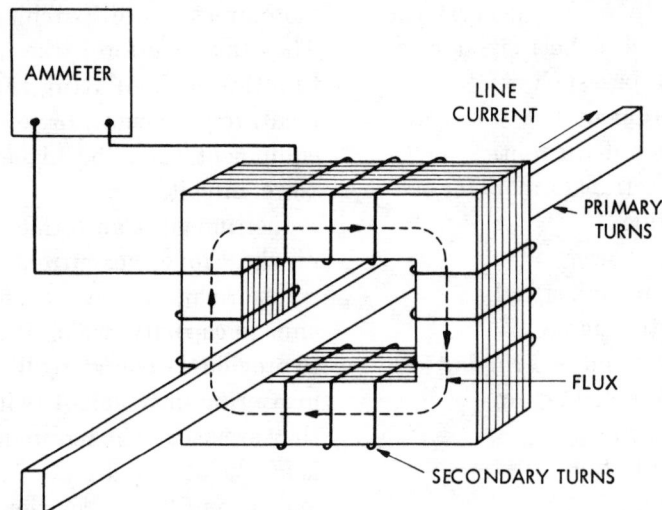

Fig. 2-9. Current transformer.

transformers of very high voltage are used for measuring the interchange of power from one system to another when primary transmission systems are tied together.

Current Transformers. The current transformer is also called a series transformer because the pri-

The secondary winding, consisting of many turns of insulated wire, is wound around the iron core. It is standard practice that the secondary of a current transformer be designed to produce 5 amperes when the rated current is flowing in the primary. The secondary of a current

Fig. 2-10. Typical current instrument transformers. (Westinghouse Electric Corp.)

transformer is always rated at 5 amperes, no matter what the ampere rating of the primary may be, Fig. 2-10. This facilitates the production of standardized current devies which are rated at 5 amperes. The ratio of current transformation is approximately the inverse ratio of the number of turns.

The current ratings of the primary winding of a current transformer are determined by the maximum value of the load current to be measured. Assuming that the current rating is 400 amperes. The secondary winding will have a rating of 5 amperes. Thus, the ratio between the primary and secondary is 400 to 5 or 80 to 1. This means that the secondary winding will have 80 times as many turns as

the primary. If the primary winding has five (5) turns, the secondary winding will have 400 turns. We can now see that the ratio of the primary to the secondary current is, as previously mentioned, inversely proportional to the ratio of primary to secondary turns. For example, using a 400 to 5 ampere current transformer, the following load currents can be tabulated to indicate the current in the secondary:

(1) For a load current of 400 amperes, the secondary of the current transformer would have 5 amperes. (2) For a load current of 300 amperes, the secondary current would be 3.75 amperes. (3) For a load current of 100 amperes, the secondary current would be 1.25 amperes.

The ratio of a current transformer required will depend on the maximum line current it is expected to carry. Ratings may vary from 5 to 5 and 8000 to 5 amperes.

The foregoing assumes an ideal transformer in which (1) there is neither resistance nor reactance in the primary and secondary winding; (2) no energy is required to energize the core; and (3) the load on the secondary, usually the current coil of an ammeter, has no impedance. Since these conditions all actually exist in the current transformer circuit, they must be compensated for, otherwise, the errors introduced, especially when metering large quantities of power with an integrating meter whose current coil is connected to the secondary of the instrument transformer, become of great importance and may result in a considerable amount of power being unaccounted for.

These factors which introduce errors are partially compensated for in the design of instrument transformers by using a magnetic circuit having steel of very high permeability at the flux density at which the transformer is operated. Furthermore the flux density is kept at a low value so that the disturbing element of magnetizing current will be low. Additional compensation is obtained by decreasing the number of turns of the secondary below that determined by the inverse ratio of currents in primary and secondary so that a greater current will flow in the secondary and establish the desired current ratio at the particular secondary load for which greatest accuracy is desired.

Note that the secondary current is not determined by the impedance load connected across the secondary as is the case of a voltage transformer, but is entirely dependent upon the amount of current flowing through the primary, which is in a series circuit. If, therefore, the secondary circuit is open-circuited, all the current flowing through the primary winding must be a magnetizing current. Since the magnetic circuit is designed for low magnetizing current when the transformer is loaded, this large increase in magnetizing current will build up an enormous flux in the magnetic circuit and cause the transformer to act as a step-up transformer, inducing an excessively high voltage across the terminals of the secondary. *Therefore, a current transformer should always have its secondary short-circuited when not connected to an external load.*

A manually or automatically operated secondary short-circuit device is supplied as a standard accessory by most manufacturers with present day current transformers, Fig. 2-11. However, one must take extreme care to make sure that before working on a current circuit

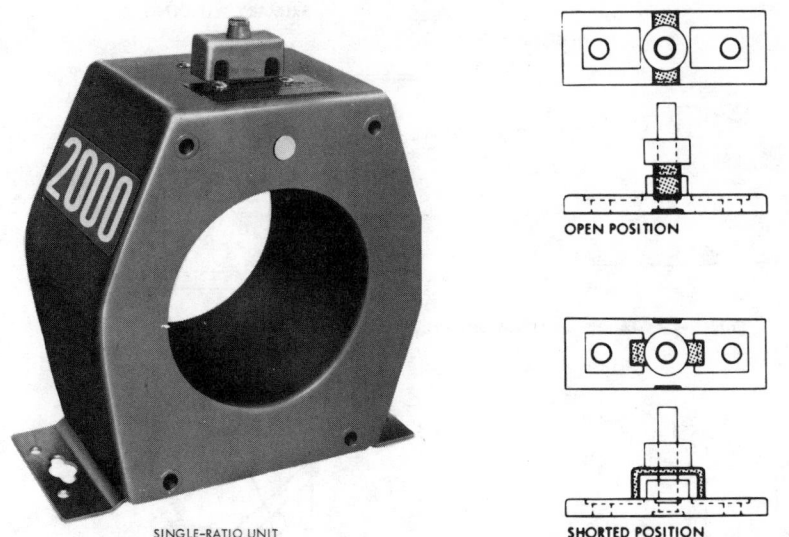

OPEN POSITION

SHORTED POSITION

SINGLE-RATIO UNIT

Fig. 2-11. Typical current transformer and manually operated short-circuiting device. Available for 600 volts, indoor or outdoor, 10kV BIL (Basic insulation level), Primary Amperes 800 through 4000, 25 to 60 hertz. Designed for high current metering and relaying on low voltage systems. Suitable for use with uninsulated bus bar or cable up to 600 volts or with insulated primary conductors at higher voltages. (Bus bar passes through hole in center of transformer.) (Westinghouse Electric Corporation)

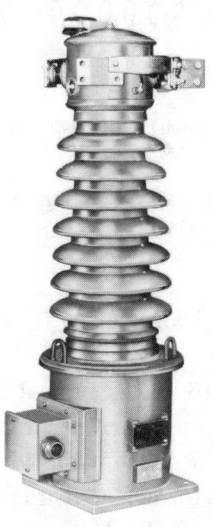

Fig. 2-12. Wound-type current transformer. (Westinghouse Electric Corporation)

that the circuit has been jumpered out and the series path re-established ahead of the point being worked upon.

There are three types of current transformers in general use: the wound, the window and the bar.

A wound-type current transformer has separate primary and secondary windings mounted on a laminated core as shown in Fig. 2-12. This current transformer is designed so that its primary winding consists of one or more turns of large cross-section wire connected in series with the circuit to be measured, Fig. 2-13.

Wound-type current transformers

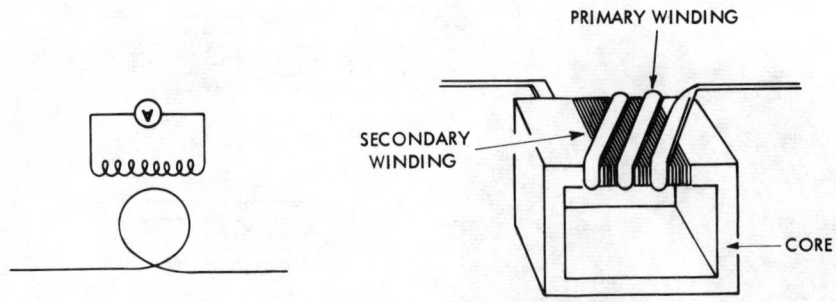

Fig. 2-13. Basic wound-type current transformer usage.

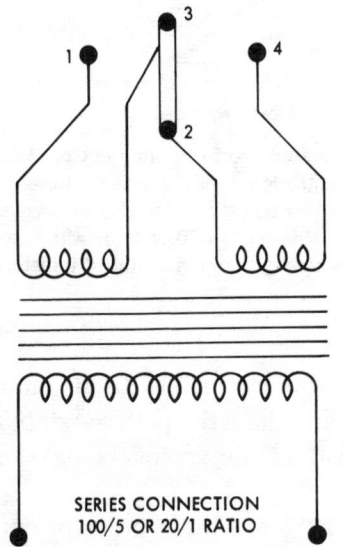

Fig. 2-14. Dual ratio current transformer—series connected.

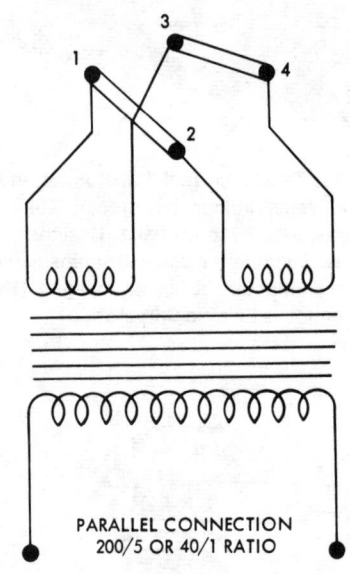

Fig. 2-15. Dual ratio current transformer—parallel connected.

may be constructed so they have a dual ratio. This is done by having two primary windings as shown in Fig. 2-14, which may be connected in series or parallel by links. The external primary leads are connected to terminals 2 and 3. For the highest ratio, the windings are connected in parallel by connecting terminals 1 to 2 and 3 to 4, Fig. 2-15.

Fig. 2-16 shows typical window or through-type current transformers. Fig. 2-17 represents a window current transformer as it might be seen mounted in the bushing of an oil-circuited breaker. This type of cur-

Fig. 2-16. Typical window-type current transformers. (Westinghouse Electric Corporation)

rent transformer consists of a built-in cylindrical-ring core of thin iron lamination like a stack of washers as shown in Fig. 2-18. Around the core is wound copper wire which forms the secondary winding. Fig. 2-19 reveals how the taps are brought out from the winding.

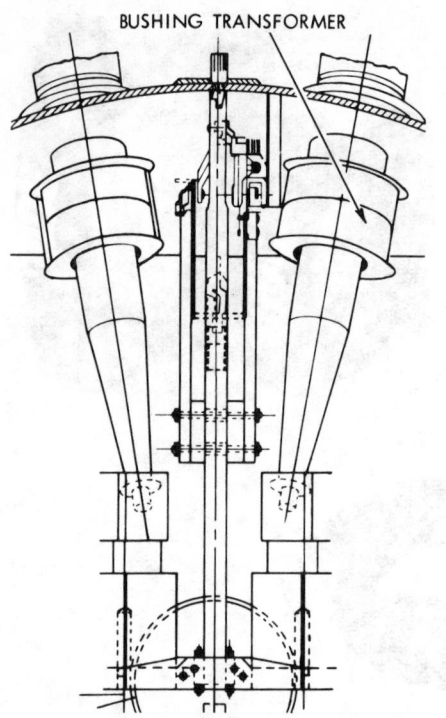

BUSHING TRANSFORMER

Fig. 2-17. Application of bushing current transformer. (Saskatchewan Power Corporation)

The high-tension (voltages) conductor, which is the lead to the terminal on the breaker bushing, forms the one primary turn of the transformer. A through or window-type current transformer will be somewhat less accurate at the lower ratings than the wound-type.

Other forms of this type of current transformer have their cores and secondary coils enclosed in molded insulating cases. This special current transformer consists of a circular iron core with many secondary turns wrapped around it, Fig. 2-10. The

primary conductor that is to be measured is fed through the window and serves as the current-transformer's primary winding. The ratio of this type of current transformer may be changed by varying the number of wires passing through the window of the transformer. The resulting *ratio* is not affected by the direction of wires in the window, but the direction of the current in the conductors does affect the metering and, therefore, it is important.

The operating ratio of a window-type current transformer may be determined as follows:

$$\text{Operating ratio} = \text{nameplate ratio} \times \frac{1}{\text{no. of wires in window}}$$

Assume a 300/5 ampere-current transformer has one conductor through its window, therefore its ratio is 60:1 shown on its nameplate. However, if two conductors are passed through the same 300/5 ampere current transformer, the new operating ratio is:

$$\frac{300}{5} = \frac{1}{2} \times \frac{300}{10} \text{ or } 30{:}1$$

This means that with 150 amperes in the conductor, there will be 5 amperes in the secondary of this connection.

The window-type current transformer has subtractive polarity as does any other instrument transformer, and the primary and second-

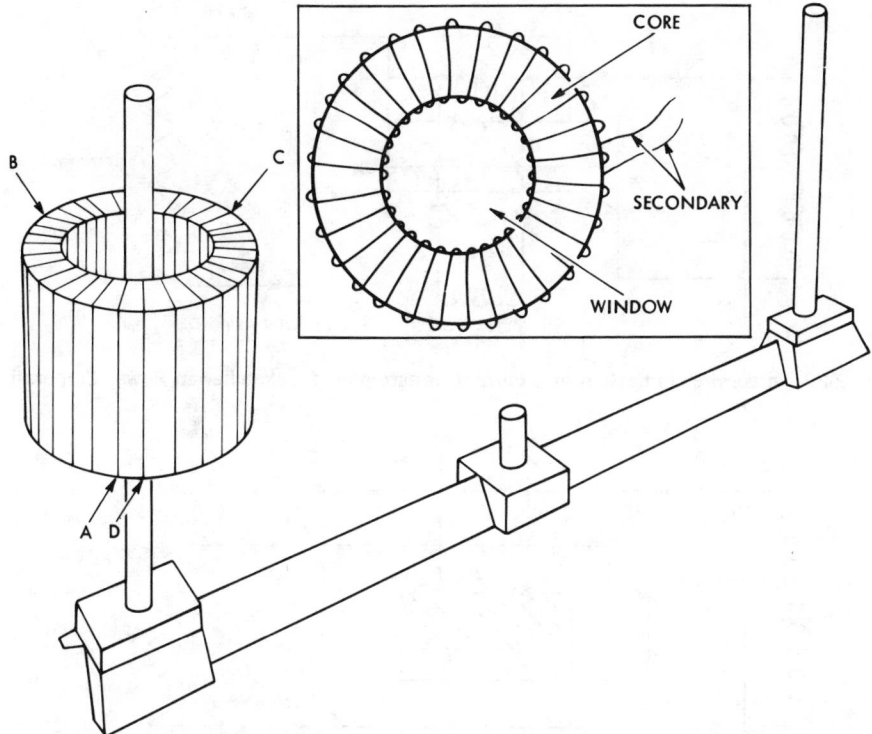

Fig. 2-18. Construction of window-type transformer. (Saskatchewan Power Corporation)

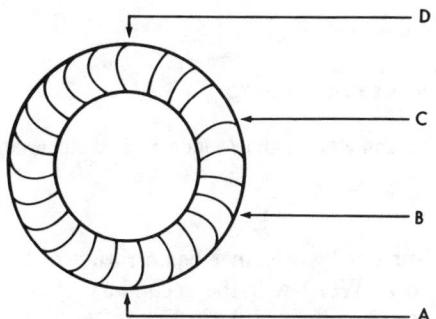

Fig. 2-19. Taps of window-type transformer. (Saskatchewan Power Corporation)

ary polarity marks are indicated on the current transformer.

If the primary conductor enters the window from the end of the

current transformer marked with the polarity mark and the current in this conductor goes in this same direction, then the secondary current will leave the terminal of the window-type current transformer marked with the polarity mark. This is subtractive polarity.

A typical type window type current transformer is the General Electric JKP-O which may be used in both single-phase (two- or three-wire) and three-phase (three- or four-wire) circuits.

Figs. 2-20 and 2-21 illustrate a special connection which uses two

95

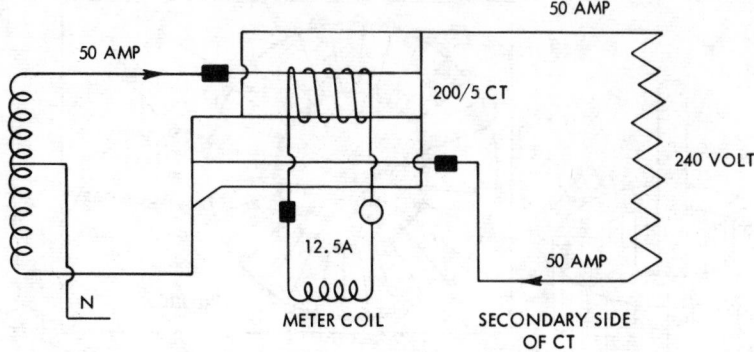

Fig. 2-20. Three-wire connection of a current transformer. (Saskatchewan Power Corporation)

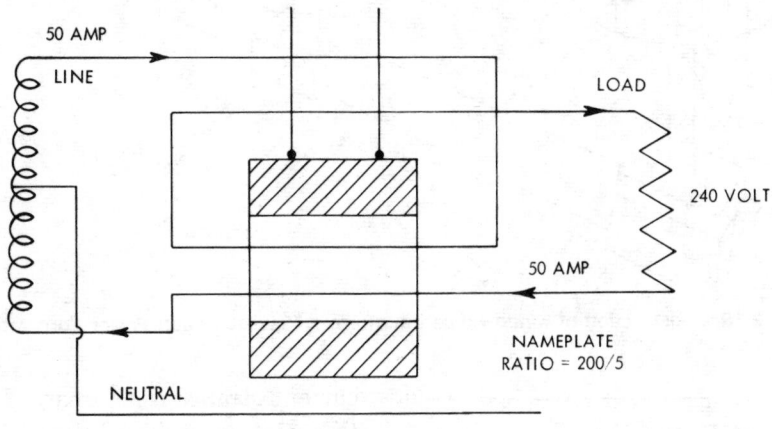

JKP-0 CT
CONNECTED AS 3-WIRE CT CROSS SECTION VIEW TO
SHOW WINDOW

Fig. 2-21. JKP-O CT connected as 3-wire CT cross-section view to show window. (Saskatchewan Power Corporation)

conductors, one from each phase, to make a three-wire current transformer. Each conductor must pass through in an opposite direction to ensure proper registration for this type of metering.

Example. Assume 50-ampere, 240-volt load. Assume connection as shown in Fig. 2-21 and assume that the JKP-O current transformer has a rating of 200 to 5. What will the secondary current in the meter coil be?

The operating ratio $= 200/5 \times \frac{1}{2}$
$$= 200/10 = 20/1$$

Thus for 50 amperes in each phase, the secondary current is

$50 \times 1/20 = 2.5$ amperes (Fig. 2-20)

If the current had been 100 amperes, there would be 5 amperes in

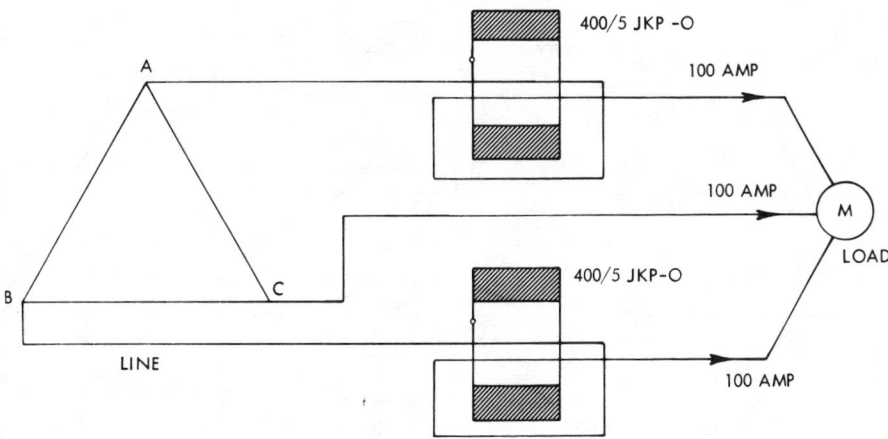

Fig. 2-22. Three-wire, three-phase current transformer connection. (Saskatchewan Power Corporation)

the secondary coil of current transformer and meter coil.

In the case of a three-wire, three-phase connection as illustrated in Fig. 2-22 the current transformer is a standard two-wire current transformer. Two current transformers are used with the same nameplate ratio and each must have the same number of turns through the window. Also the direction of the phase conductor through the window of each current transformer must be in the same direction.

Example. Assume 240-volt, three-phase, 100-ampere motor load with no lighting load; assume connections as shown in Fig. 2-22. Note that both JKP-O's have the same nameplate ratio 400:5 and both have two turns through the window (of the same phase). That is, phase A current transformer has two turns of conductor A and no other phase conductor is involved. The operating ratio for both current transformers is:

$$\frac{400}{5} \times \frac{1}{\text{no. of wires in window}} = \frac{400}{5}$$

$$= \frac{1}{2} = \frac{400}{10} = \frac{40}{1}$$

The secondary current in the meter coil will be $100 \times 1/40 = 2.5$ amperes in each current transformer if the motor load draws 100 amperes as indicated.

In the case of three-phase, four-wire delta (120/240 volt) connected as shown in Fig. 2-23, you will find another special application using one JKP-O as a three-wire current transformer and one as a two-wire current transformer.

The bottom-type, JKP-O current transformer is the three-wire current transformer and the top-type JKP-O current transformer is the two-wire current transformer as shown in the figure. For standard purposes, the nameplate ratio of *both* current transformers is the same. To avoid ratio errors, it is

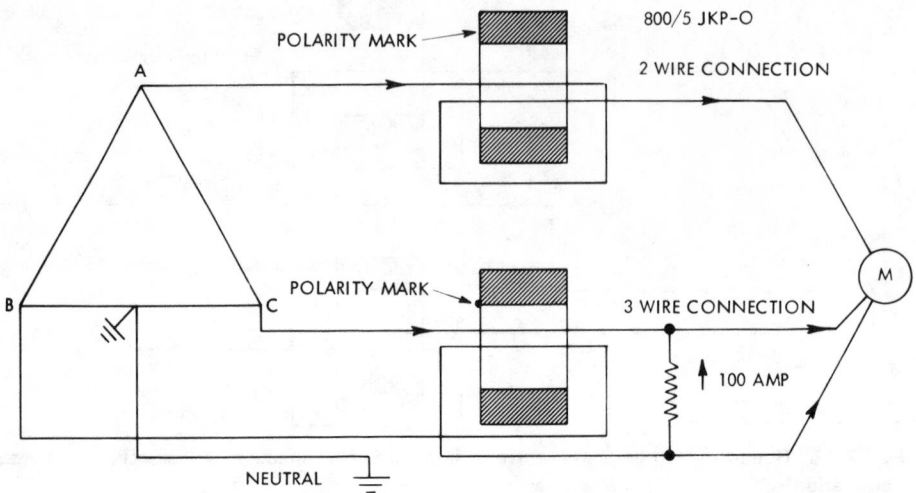

Fig. 2-23. Four-wire, three-phase delta current transformer connection. (Saskatchewan Power Corporation)

necessary to have the same number of turns in the window of each current transformer. Also, the direction of the conductor through the window must be correct.

Example. Assume 120/240, four-wire, delta load with 400-ampere, three-phase balanced load and a 100-ampere, single-phase 240-volt load; assume no lighting load (120 volt). The operating ratio of the two-wire JKP-O or three-wire JKP-O.

The operation ratio =

$$\frac{\text{name-}}{\text{plate ratio}} \times \frac{1}{\text{no. of turns in window}}$$

$$= \frac{800}{5} \times \frac{1}{2} = \frac{800}{10} = \frac{80}{1}$$

Note that in both current transformers there are two turns: in one case both turns are from the same phase (two-wire current transformer), and in the other case, there is one turn from two phases (three-wire current transformer).

For the three-phase, 240-volt motor load, the current transformer secondary current will be 400 amperes $\times \frac{1}{80} = 5$ amperes in both the three-wire and two-wire current transformer.

In the single-phase, 240-volt load, the current transformer secondary current in the three-wire transformer only will be: $100 \times 1/80 = 1.25$ amperes. In the three-wire current transformer, the secondary current will be in the vector sum of the single-phase current of 1.25 amperes and the three-phase current of 5 amperes; and the magnitude depends upon the power factor of each of these loads. This current flows through one current coil of a two-element, three-phase, three-wire meter, while the 5-ampere output of the two-wire current transformer flows through the other current coil of the meter. By placing a solid bar

Fig. 2-24. Bar-type current transformers. (Westinghouse Electric Corporation)

for indoor or outdoor use. The application will determine the type to be used.

Usually either the dry or compound-filled type is used for voltages below 22,000 and either the compound or oil-filled for voltages above 22,000. As the voltage is increased, insulation becomes of greater importance and complicated equipment results.

The amount of insulation and its type will depend upon the voltage at which the transformer is to be used. As the voltage is increased, the primary winding has to be more highly insulated from the secondary.

Polarity Markings. When current transformers are used with meters or relays which rely not only on magnitude but also on phase position, polarity is of importance. Polarity is indicated by a marking on one primary and one secondary terminal, usually by a white dot. By definition, when current is flowing toward the marked primary terminal, it is flowing away from the marked secondary terminal. This marking corresponds to the H_1 and X_1 terminals as used on power transformers. As mentioned previously all standard instrument transformers are all subtractive polarity.

Current Transformer Precautions. It is important to understand that the current transformer differs from potential and other transformers in that the primary winding is de-

permanently through the hole in the window-type current transformer, we now have a transformer known as the bar primary type, Fig. 2-24. This type of construction is particularly suited to withstand the stresses of heavy overcurrent, and the tendency to assume a circular shape is eliminated. In order to avoid magnetic stresses that could destroy the bus and damage the transformer, care must, therefore, be taken to properly mount these transformers with respect to adjacent conductors.

Classes of Current Transformers. Current transformers may be classified as dry, oil, or compound-filled signed for connection in series with the line current at all times.

The secondary circuit of a current transformer should never be open-circuited when there is current flowing in the primary winding. Under open-circuit conditions, the primary current becomes an exciting current which will cause a high voltage to be induced in the secondary winding. This voltage may be sufficient to puncture the insulation, and it presents the hazard of a dangerous high-voltage shock to anyone who may come in contact with the open-circuit secondary.

Therefore, those working with current transformers must always make certain the secondary winding circuit of a current transformer is closed.

All current transformers should have their secondary winding grounded in service because an electrostatic field will build up in the secondary winding.

Accuracy of Current Transformers. A current transformer differs from the ordinary transformer in that its primary current is determined entirely by the load on the system and not by its own secondary load. As previously stated, a current transformer must change the magnitude *only* of the current being measured. Thus, its accuracy must be known so that any errors can be included in the computation of the over-all measurement, or er-

rors must be within the limits of a specified small value so that they may be disregarded as insignificant.

The design and construction of the current transformer itself, the circuit condition such as current and frequency, and the burden (load) imposed on the secondary circuit of the current transformer all affect its accuracy. In general, the greater the burden, (load) the greater the error.

Small errors classified as ratio and phase angle are present in all current transformers. Since a portion of the primary current is required to magnetize the core and to supply the core losses, the secondary current will be proportionately less than the primary current.

Voltage (Potential) Transformers. The potential transformer, Fig. 2-25, operates on the same principle as a power or distribution transformer. The main difference is that the capacity of a potential transformer is relatively small as compared with power transformers. Potential transformers have ratings of 100 to 500 volt-amperes. The low-voltage side is usually wound for 120 volts. The burden on the low-voltage side consists of the potential coils of various instruments. In some cases, potential coils of relays and other control equipment are also connected to the secondary of the potential transformer. In most cases, the load is relatively light, and it is not necessary to have a capacity of

600 VOLTS, INDOOR OR OUTDOOR

1200 VOLTS, INDOOR OR OUTDOOR

1200 VOLTS, INDOOR FUSED

2400 THROUGH 14,400 VOLTS, INDOOR USE

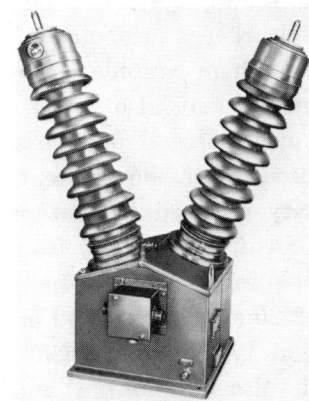

TWO HIGH-VOLTAGE BUSHINGS, 14,400 THROUGH 69,000 VOLTS, OUTDOOR USE, OIL-FILLED (SUITABLE FOR METERING OR RELAYING

TYPICAL INSTALLATION OF TYPE APT OUTDOOR POTENTIAL TRANSFORMER

Fig. 2-25. Type EMPL potential transformers, 600 and 1200 V, indoor or outdoor. Primary volts: 240 through 600, 60 hertz. (Westinghouse Electric Corporation)

potential transformers greater than 500 volt-amperes.

The primary windings of the potential transformer are designed to be connected in parallel with the circuit in question, and, as previously stated, the secondary is designed to a standard value, usually 120 volts. In this way standard instruments and relays can be used and the worker is protected against dangerous high voltage. Potential transformers may also be used for isolation purposes. They are sometimes used to provide a potential of a polarity reversed to that of the power circuit ratio range from 1:1 to 345 kV to 120 volts.

Fig. 2-26 shows a method of connecting a typical instrument load through voltage (potential) and current transformers to a high voltage single-phase line. The load on these instrument transformers includes an ammeter, a wattmeter, a watt-hour meter, and a voltmeter.

Types of Potential Transformers. Where high voltages are encountered, care must be taken in choosing a potential transformer which is suitable for the conditions to be encountered. Usually, on circuits greater than 25,000 volts, a potential transformer designed for outdoor service is used, and this may be of either the one-bushing or two-bushing type. On circuits of 25,000 volts and lower, either indoor or outdoor type transformers are employed.

More detailed information may be found in American National Standard Institute's (ANSI) bulletin C57.13-1968 or Canadian Standards Association's bulletin C13-1958.

Where phase-to-ground voltages are being measured, the single-bush-

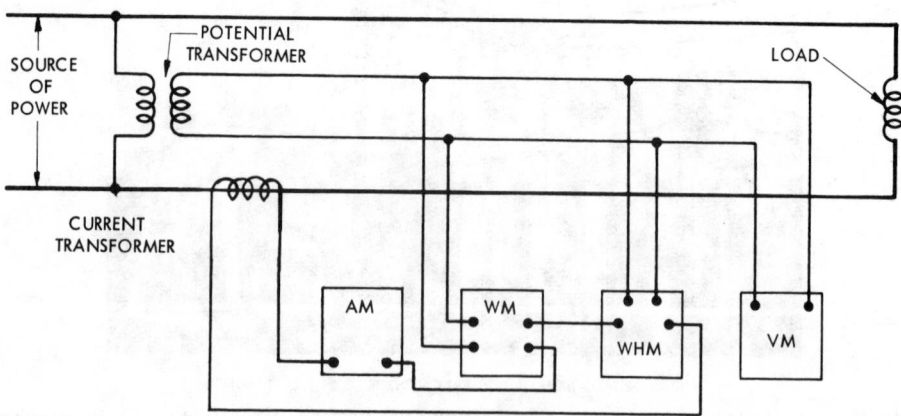

Fig. 2-26. Diagram showing uses of instrument transformers.

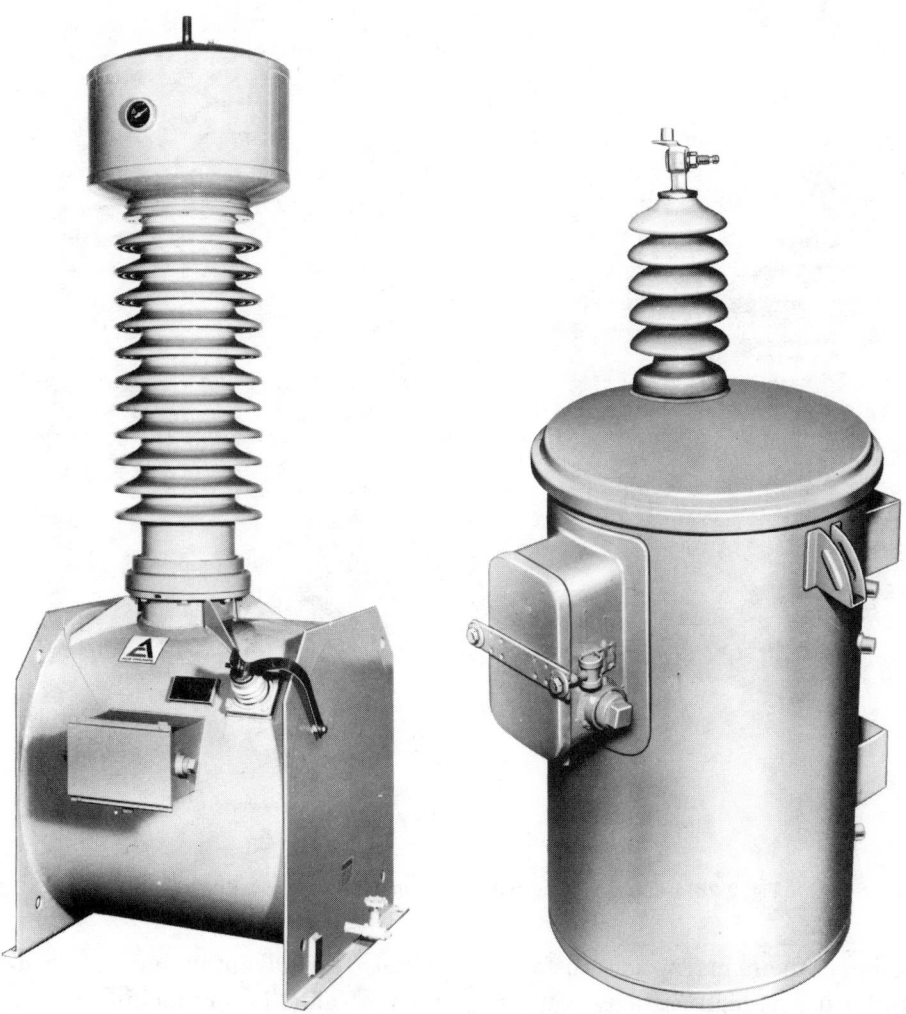

Fig. 2-27. Single-bushing potential transformers. (Allis-Chalmers)

ing potential transformer is used, Fig. 2-27.

A double-bushing potential transformer has two bushings on the high-voltage winding capable of being connected to a circuit, the voltage of which is the same as that on the nameplate of the transformer. It may be connected either phase-to-phase or phase-to-neutral. On a three-phase system, this potential transformer may be connected delta, wye, or open-delta. Fig. 2-28 illustrates a typical double-bushing potential transformer.

Because capacitors are easier to

103

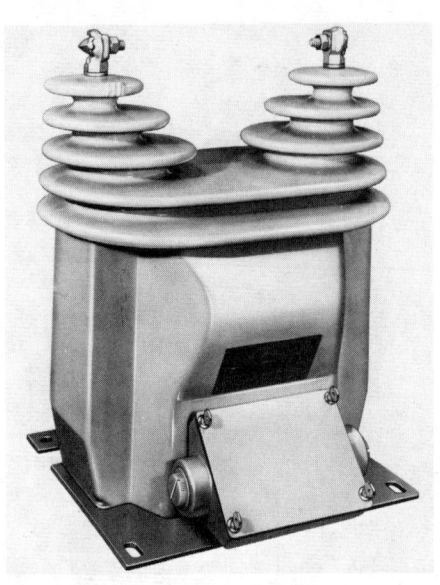
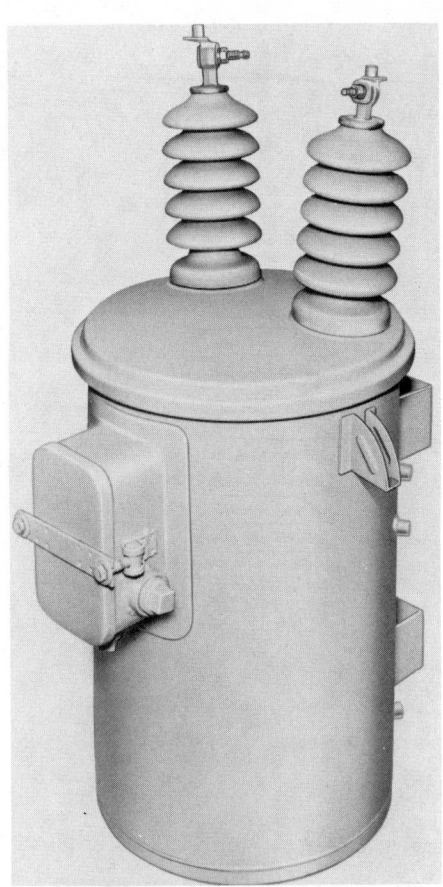

Fig. 2-28. Double-bushing potential transformers. (Allis-Chalmers)

design than are magnetic transformers for use at high voltages, capacitor voltage transformers are now generally being used on system having a voltage that exceeds 100 kV. Fig. 2-29 illustrates this type of potential device.

The capacitor voltage transformer is made up of a capacitive potential divider and a potential transformer. By stacking a number of capacitors in series according to the desired voltage, the advantages of standardization and lower manufacturing cost are obtained. Each is provided with taps in order to accurately adjust the output according to phase and magnitude. The spark gaps protect both the bushing and secondary instruments or relays.

Accuracy of Potential Transformers. There are two causes of errors in potential transformers.

(1) The exciting current necessary

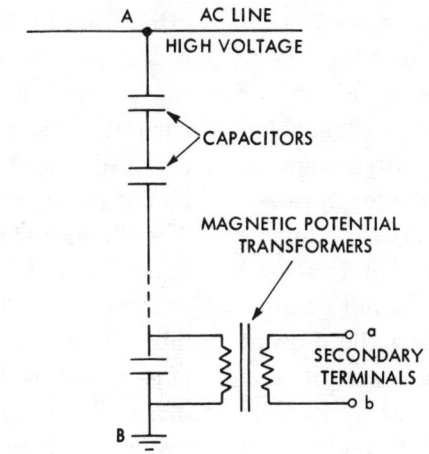

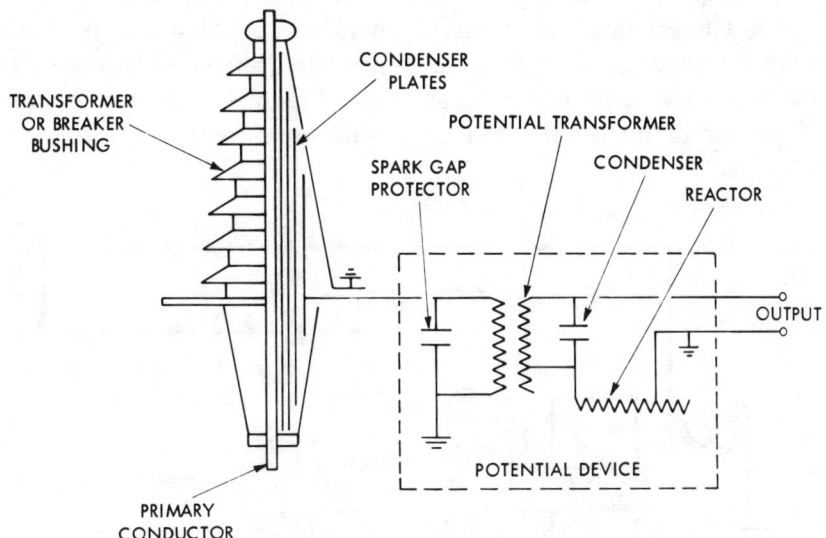

Fig. 2-29. Capacitor voltage transformer.

to magnetize the core causes an impedance drop in the primary winding.

(2) The load current drawn by the burden causes an impedance drop in the primary and secondary windings.

The secondary voltage is, therefore, slightly less than the ratio would indicate with a phase angle between the primary and secondary voltage reversed.

Constant-Current Transformers. The constant-current transformer is

105

extensively used to supply a constant current for series street lighting utilizing either incandescent or gaseous-discharge lamps connected in series. A constant-current source is one in which the voltage impressed on the secondary series circuit is automatically increased or decreased as the current tends to fall or grow larger, thereby keeping the current of the secondary series circuit at a constant value. The usual method of feeding a constant-current circuit is by the use of a constant-current moving-coil regulator as shown in Fig. 2-30. The primary coil is energized by a constant voltage source.

This type of transformer is built for either air or oil cooling. Oil is often used because it has proved to be an excellent agent for insulation, cooling and lubrication. Constant-current transformers are constructed for outdoor-pole-mounting, subway-mounting and indoor-station use. The indoor-station and subway-mounting types are built for indoor use and normally serve from 15 to more than 200 lamps. The outdoor pole-mounting type is usually built in smaller sizes and is usually installed close to the area served by the street lighting circuit. (See Figs. 2-31 and 2-32.) They are made weatherproof and are normally oil insulated, as the conditions of installation require. The primary coil of a constant-current transformer is nor-

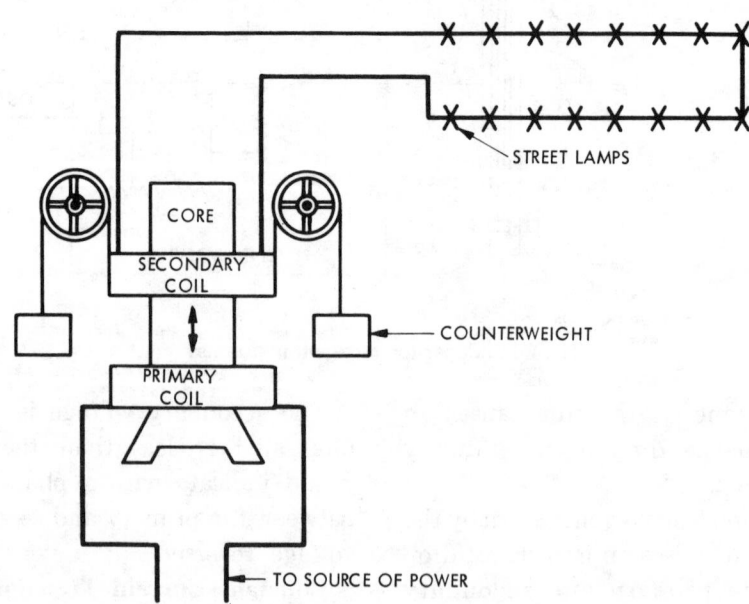

Fig. 2-30. Constant-current moving-coil regulator.

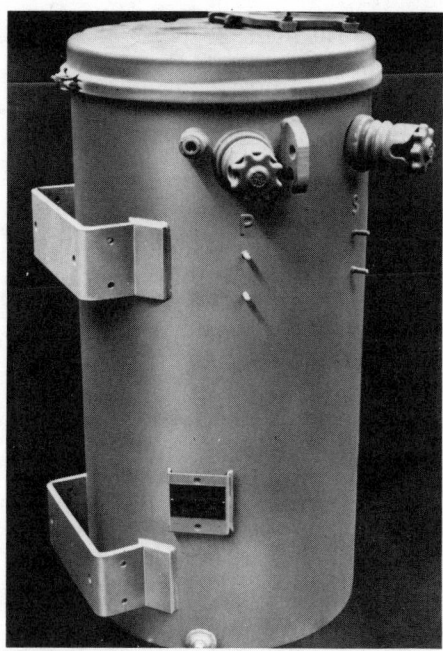

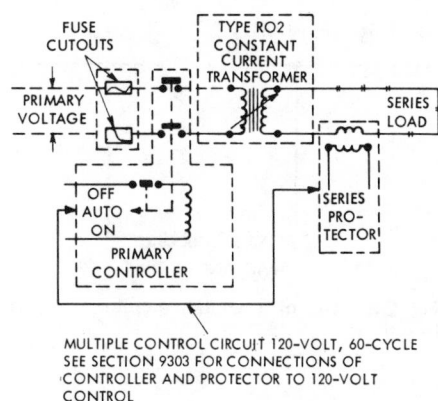

Fig. 2-32. Typical wiring for series lighting circuit. (General Electric Co.)

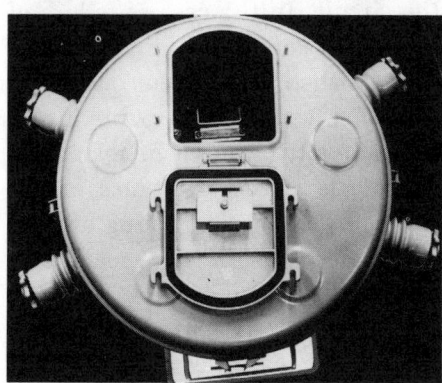

Fig. 2-31. Type RO-2 constant current transformer (pole-type); 10 to 30 kW; 2,400 to 7,200 volts primary; 6.6 to 20 amperes secondary; 60 hertz; oil-insulated, self-cooled. (General Electric Co.)

mally wound for 2,400 volts but can be wound for any reasonable higher voltage of up to about 10,000 volts,

while the secondary is wound for the voltage required for operating the number of lamps in the series circuit. The constant current in the secondary circuit is normally 6.6 amperes. The standard ratings of this type of transformer are 5, 10, 15, 20, 25, 30, 35, 40, 50, 60 and 70 kVA.

The efficiency of a constant current transformer is high, being about 96% at full load for a 100-lamp transformer but the power factor which depends upon the magnetic leakage is low at all loads, reaching 75 to 80% at full load and decreasing almost proportionally for lower loads.

When one of the lamps in a series circuit burns out, the circuit must be re-established in some manner or else all of the lamps in the circuit will not burn.

This circuit may be re-established by the use of a film cut-out, or else

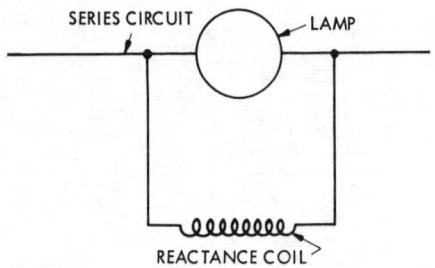

Fig. 2-33. Use of a reactance coil in a series circuit.

the continuity of the circuit is maintained by a small reactance coil shunted around the lamp as shown in Fig. 2-33. The film cut-out is a device in the base of the lamp which punctures and re-establishes the circuit when subjected to a voltage above a certain critical value, Fig. 2-34. The device consists of two metal contacts separated by a thin

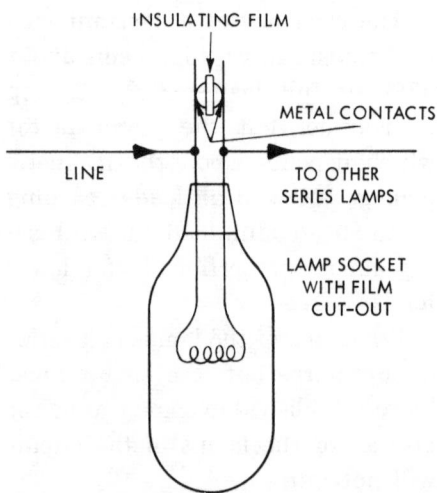

Fig. 2-34. Typical series lamp used on a series street lighting circuit.

film of insulating material. When the lamp burns out, the entire voltage of the circuit is momentarily impressed across the contacts of the cutout device, puncturing the insulation between the metal contacts and thus shorting out the lamp and re-establishing the continuity of the series circuit.

A system of regulation is sometimes employed without the use of a moving-coil regulator. This consists of a constant voltage transformer in series with a reactance coil; both reactance coil and the transformer have taps by which the voltage of the circuit, as well as the amount of reactance, can be adjusted to the number of lamps in the circuit. In the case of failure of individual lamps, the circuit will be reestablished by film cut-outs. The reactance coil is used to produce a circuit relatively high in reactance. A change in the resistance of the circuit by adding or taking away a lamp or two will not materially change the flow of current, since the impedance of the circuit is largely made up of reactance.

An objection to series circuits, in general, is the low power factor resulting from the reactance coil of the reactance-coil-voltage-transformer combination, or from the leakage reactance of the moving-coil regulator.

Operation of Constant Current Regulator. The construction of the moving-coil regulator is such that the primary coil and the secondary coil can move with respect to each other. Either the primary or the secondary coil may be made movable. Figs. 2-30, 2-32 and 2-35 show a fixed primary coil and a movable secondary coil. The secondary coil is suspended from a lever which is counterweighted. Transformers are usually equipped with a dashpot, in order to prevent rapid changes or any "hunting" action of the movable coils.

Assume that the secondary coil is "floating," that is, free to move either up or down, and is delivering a certain amount of current to the lamps on the secondary side. The magnetic flux Φ, which at a given instant passes through the primary coil, flows partly through the secondary coil as the useful flux U, and

partly leaks across between the primary and secondary coils as the leakage flux L. Note that the magnetic circuit is such that the coils are mounted on a central leg of the core and the useful flux U has two parallel paths through which it completes its circuit. These two parallel paths are the upper and lower yoke sections and the outer legs. The current flowing in the secondary is opposite in direction to that of the primary and there is therefore a repulsion between these two coils due to the leakage flux L of both coils setting up a strong magnetic field in the air space between the two coils. The counterweight is so adjusted that it nearly balances the weight of the coil, the remainder of the coil weight being balanced by the upward push of the leakage flux produced with a given amount of

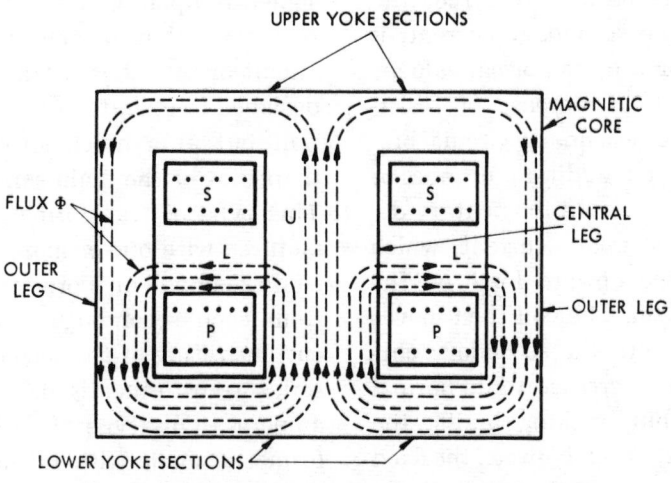

UPPER YOKE SECTIONS

MAGNETIC CORE

FLUX Φ

CENTRAL LEG

OUTER LEG

OUTER LEG

LOWER YOKE SECTIONS

Fig. 2-35. Magnetic circuits in a constant-current transformer.

current flowing through the secondary coil.

Assume now that the secondary load changes, for example, that it decreases. This change of load would be produced by short-circuiting one or more lamps, causing a decrease in the load resistance. Because of the decreased load resistance, the secondary current tends to increase, which in turn tends to increase the primary current. This increases the leakage flux and produces an increased repelling force between the two coils. The secondary coil, due to the unbalanced action of the counterweight, then moves away from P, Fig. 2-35, at the same time the useful flux U is decreased by the action of the increased secondary current. The induced electromotive force in S therefore becomes smaller in value and the upward movement of the secondary coil continues until the induced electromotive force, and therefore the secondary current, is again restored to its normal value.

Similarly, an increase of resistance in the secondary circuit produced by the addition of one or more lamps causes a momentary decrease of secondary current, which decreases the upward force on the secondary coil. The coil then moves downward until the secondary current is again increased to its normal value. Within working limits, the magnetic repulsion between the fixed and moving coils of the regulator for a given position of coils is proportional to the current flowing in the coils, which makes the transformer capable of adjustment so that any desired current may be maintained simply by changing the amount of counterweight.

The value of secondary current in a series street-lighting system is normally 6.6 amperes. The foregoing action between primary and secondary coils is such that this value of current is maintained practically constant over the entire working range of the regulator.

Series Transformers. Series transformers operate on the same principle as instrument current transformers. The primary winding is connected in series with the main series circuit and the secondary is used to feed series incandescent or mercury street lamps which are designed to operate at a different current than that of the main series circuit or to isolate a lamp which is designed to operate at the same current but at a much lower voltage than that of the main series circuit. This type of transformer is manufactured with one or more secondaries to feed one or more series lamps from each secondary. The normal current rating of the secondary of a series transformer is 6.6, 15 or 20 amperes. This type of series transformer is made with ratings of .25, .5, 1, 2, 3, 4, 5, 6, 7.5 and 9 kW.

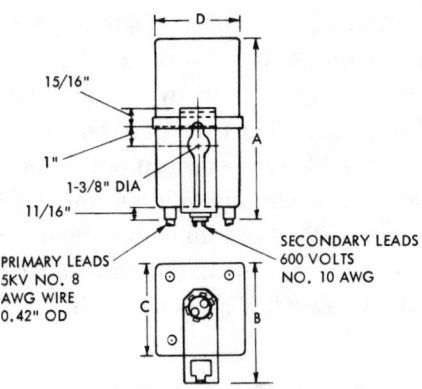

PRIMARY LEADS
5KV NO. 8
AWG WIRE
0.42" OD

SECONDARY LEADS
600 VOLTS
NO. 10 AWG

Small Power Transformers.
This type of transformer is constant-potential, self-air-cooled and is available in small standard sizes of 75, 150, 225 and 300 watts. A com-

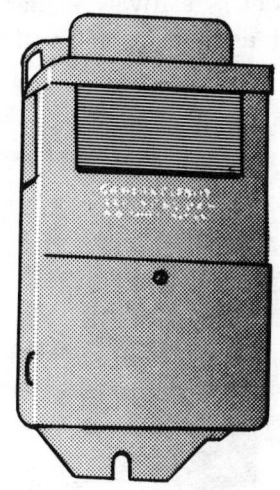

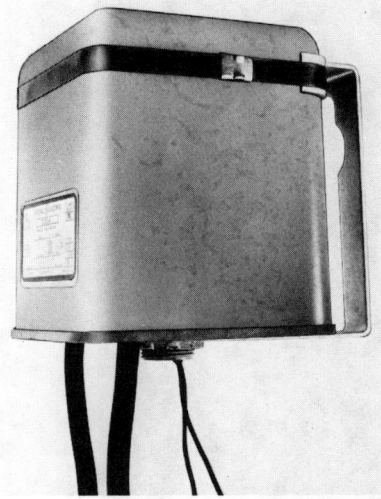

Fig. 2-36. Series ballast and transformer, aerial mount. (For pole-base operation remove mounting handle.) (General Electric Co.)

Fig. 2-36 shows a series ballast and transformer used to supply a mercury lamp from a series street-lighting circuit. This type of transformer is used to supply a single lamp from a series, 6.6 or 20 ampere, main circuit.

Fig. 2-37. Small power transformers, self-air-cooled.

mon use of the small power transformer is to step-down the supply voltage to lamps which serve localized lighting at a machine to a safe value, such as, 115, 64, 32 or 6 volts when the supply circuit to the machine is of a relatively higher voltage such as 3-phase, 460 volts, Fig. 2-37.

Another use of the small power transformer is to serve underwater lighting systems in swimming pools, Fig. 2-38. This type of transformer is available for weatherproof outdoor service in various voltage and kVA ratings. This special transformer is a two-winding insulated transformer which electrically isolates the sec-

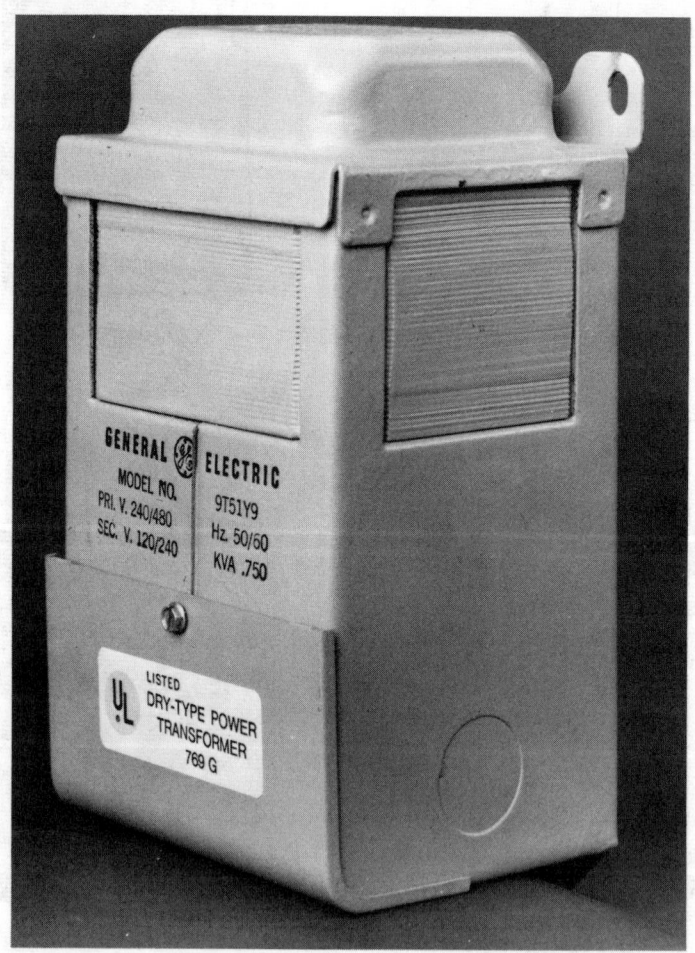

Fig. 2-38. Small power, self-air-cooled transformer. Used for low-voltage underwater lighting fixtures in swimming pool. (Specialty Transformer Dept., General Electric Co.)

ondary swimming pool lighting system from the primary supply system.

Control and Signal Transformers. This type of transformer, Fig. 2-39, is self-air-cooled, constant-potential and is used to step-down the voltage to supply signal circuits or control circuits of electrically operated switches. A common use of this type

TYPE MTA

TYPE AP

TYPE AP-200

TYPE SC

Fig. 2-39. Control transformers for machine tools. Single phase, 60 and 50/60 hertz "Black Line." (Westinghouse Electric Corporation)

of transformer is to step-down the motor-control circuit voltage to a safe value of 120 volts when the motor and its controller are supplied from a 460-volt, 3-phase power circuit.

Still smaller control transformers operating at 120/25 volts are commonplace in residential building, supplying control voltages for heating and air conditioning.

Electric Sign Transformers. This type of transformer is of the constant-potential type and is used to step-down a 120/208- or 115/230-volt circuit to a lower voltage to serve low-voltage incandescent lamps in signs. They are normally single-phase, air-cooled and are manufactured in sizes up to 5 kVA.

Bell-Ringing and Chime Transformers. This type of transformer is a specially designed small-capacity, constant-potential transformer and is used to operate the common door bell, buzzer chime, annunciators, fire or burglar alarms and industrial signaling. It functions with normal primary voltages of 120 or 240 volts and with single or multiple secondary voltages of from 8 to 24 volts.

Neon-Sign Transformer. The function of this type of transformer is to operate the gaseous-discharge lamps commonly used with electric signs. This type of transformer is of the varying-voltage type and the windings are designed so that the

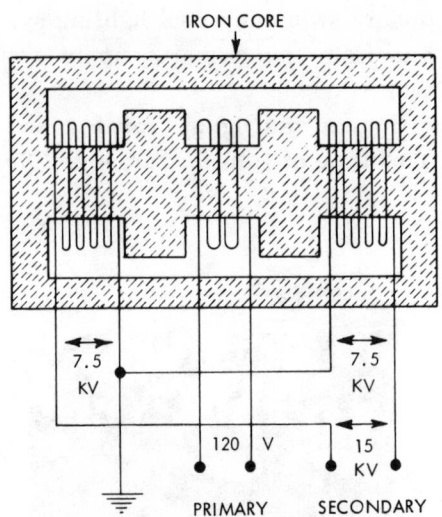

Fig. 2-40. Cross-section of a neon-sign transformer, showing the primary and secondary windings and the flux leakage paths in the magnetic circuit.

secondary voltage decreases with load. Fig. 2-40 shows the three windings of a neon-sign transformer with the magnetic leakage paths. The center coil is the primary and the two outside coils are the secondary. The secondary voltage decreases as the load increases because of the increased leakage flux and greater voltage drop of the primary. Neon-sign lighting transformers are rated according to the secondary voltage and the short-circuit current in mA (milliamperes).

The primary voltage rating is normally 120 or 240 volts while the secondary voltage varies from 2000 to 15,000 volts depending upon the diameter and length of the glass

114

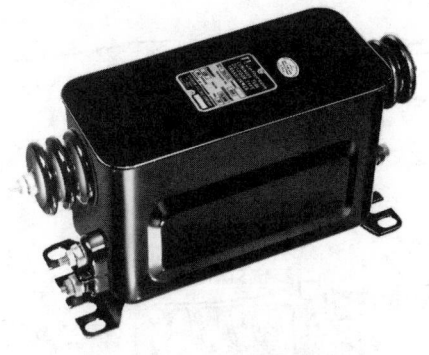

Fig. 2-42. Cold cathode transformers produced in both 120 and 300 mA secondary ratings. The 120 mA transformers have secondary ratings of 7,500 to 15,000 volts. The 300 mA units have secondary ratings that range from 1,800 to 3,500 volts. (Jefferson Electric Co.)

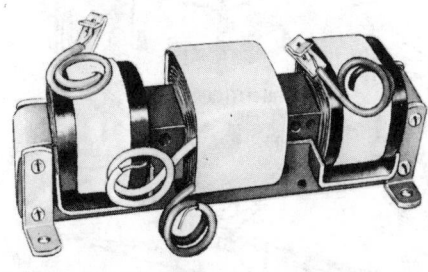

Fig. 2-41. Neon sign lighting transformers. *Top,* luminous tube transformer available with secondary voltage ranging from 3,000 to 15,000 volts at 30 and 60 mA. All units designed to operate at 120 volts at a frequency of 60 hertz. *Center,* this compact transformer is used for individual illuminated letters. Secondary voltage ratings range from 5,000 to 9,000 volts at 30 mA. Normal primary voltage is 120 volts. *Bottom,* this unenclosed transformer is adaptable for small portable signs used indoors. Secondary voltage ratings range from 2,000 to 7,000 volts at 20 mA. (Jefferson Electric Co.)

tubing and the type of gas used in the tubing.

When the secondary voltage exceeds 7,500 volts the two secondary coils usually are connected together and to ground which will give a secondary three-wire single-phase circuit of 7,500/15,000 volts. The current rating is in the range of from 30 to 120 mA.

These transformers are of the self-air-cooled type and are normally mounted inside the sheet-metal sign enclosure, Figs. 2-41 and 2-42.

Generator Step-Up, Substation and Distribution Transformers. There is basically no electrical difference between these transformers except for the purpose they serve and their size. Generator step-up

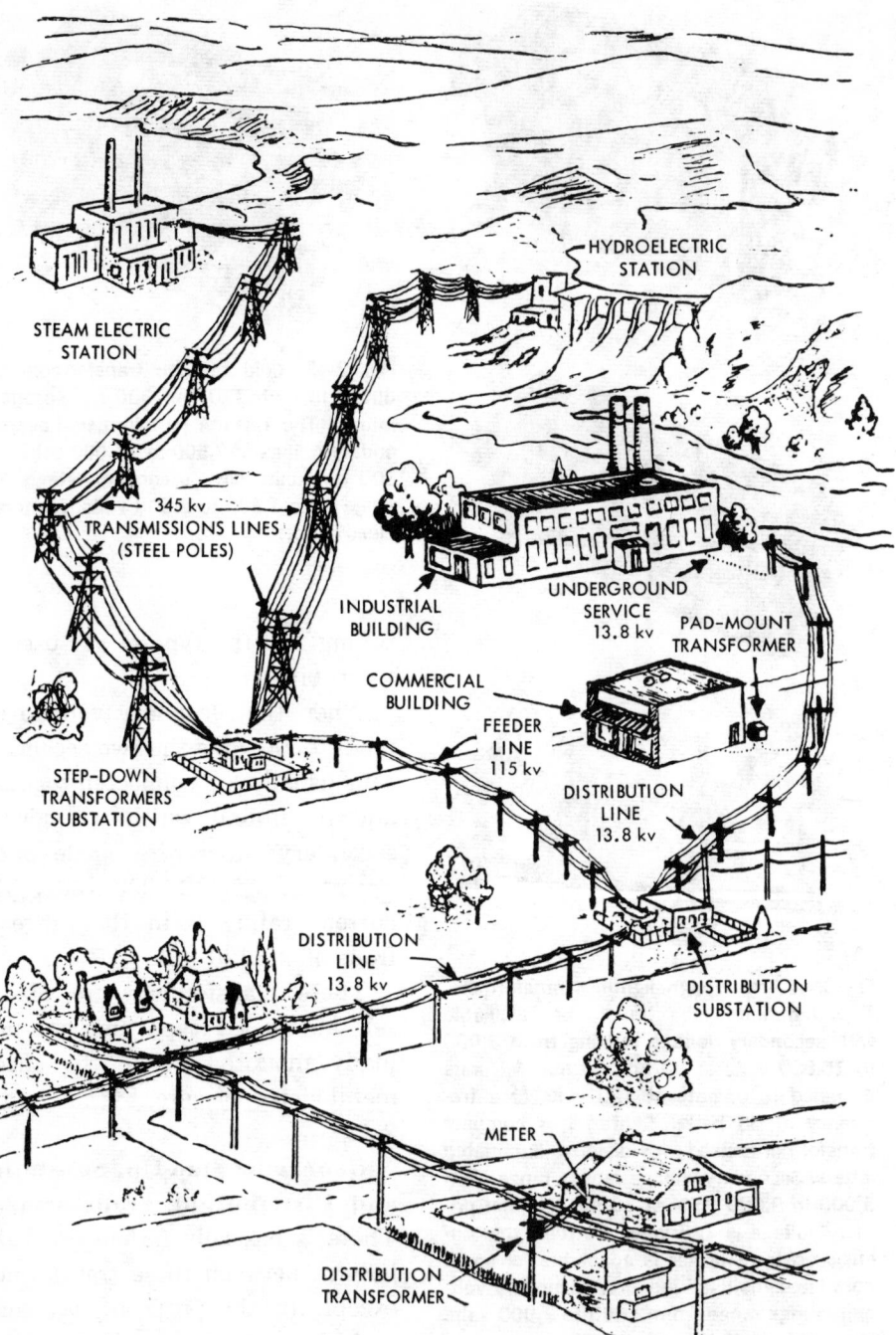

Fig. 2-43. Transmission of electrical power from the generating station to the consumer.

transformers, sometimes referred to as power transmission transformers, are used by the utility to raise the generator voltage to a higher voltage for transmission purposes. Substation transformers are normally used to transform the transmission voltage to the voltage level of the primary feeders. Distribution transformers change the primary feeder voltage to the utilization voltage required by the consumer.

Fig. 2-43 shows how electrical power is transmitted from the steam or hydroelectric station to the ultimate consumer. The alternators at the steam or hydroelectric generating plants produce power at a nominal voltage in the range of 22 kV. The transmission transformers step-up this voltage to 345 kV at which voltage the power is transmitted to a central substation where the power from both generating plants is brought together. Large step-down transformers reduce this voltage to a lower voltage in the range of 23 to 115 kV as the electrical power nears the city. The feeder lines deliver the electrical power to a distribution substation where the voltage is again stepped-down to a voltage in the range of 13.8 kV. The electrical power is delivered to the ultimate consumer by distribution lines at 13.8 kV as shown in Fig. 2-43, (Note: distribution voltage could be one of several voltages; 13.8 is only one.) where it is again

stepped-down by distribution transformers to 115/230 volts for residential and small commercial buildings. Some large industrial plants receive electrical power at a voltage of 13.8 kV where in-plant distribution transformers step down the voltage to the desired level for in-plant use.

Generator Step-Up Transformers. Generator step-up transformers for electrical power transmission are installed at the generating plant or source of power. As a class they are used for stepping up the voltage for purposes of economical power transmission. The standard nominal voltage ratings for this class of transformer are 13.8, 23, 46, 69, 115, 230, 345, 500 and 765 kV. Voltages of 345 kV and higher are classed as EHV (Extra High Voltage) and are used to transmit power over long distances. Experimental work is in progress with voltages as high as one megavolt (1,000,000 volts).

The kVA and kV ratings of generator step-up transformers are largely determined by the amount of power produced by the generator and the distance that this power must be transmitted. Although these units are manufactured in very large sizes, most utilities endeavor to limit the rating to 500 mVA (500 megavolt-amperes = 500,000,000 volt amperes). One primary reason for this limit is the physical size of one large transformer. Also if only one large transformer is installed and a malfunction should develop, the entire system would be without power; whereas if there are two smaller units and one should go down, the other unit would continue to supply power at a reduced value to the system.

The various methods used to cool these large transformers will be discussed in detail, in the appropriate place, later in the chapter.

Substation Transformers. As previously explained there is little electrical difference between a substation transformer and a distribution transformer. The term substation is basically defined as an assemblage of equipment which is installed for the purpose of switching and/or changing or regulating the voltage of electricity. Most manufacturers do not actually define a substation transformer until its intended purpose and the component equipment that is to be used with the transformer has been determined. In other words, if a transformer with its associated equipment is to be used by a utility, in its function as a utility, to step down transmission voltages to an intermediate voltage, it would be called a substation transformer. On the other hand, if a transformer and its associated equipment were installed by a large

Fig. 2-44. Primary substation transformer rated 501-25,000 kVA. (Medium Transformer Dept., General Electric Co.)

industrial plant and used to step down a high voltage from a utility transmission line to an intermediate or low-voltage, which in turn was transmitted and distributed throughout a large area owned and operated by the plant, it would normally be called an industrial distribution transformer.

The substation as shown in Fig. 2-44 is used by the public utility to step down the high-voltage of the transmission or sub-transmission line to a lower value which in turn is distributed to the consumer.

The substation and its associated equipment is manufactured for both indoor and outdoor installations. Fig. 2-44 shows an outdoor installation of substation transformers.

Mobile Substations. The purpose of a mobile substation, Fig. 2-45, is to restore service quickly in emergencies such as floods, fires, storms or lightning disasters. Using a mobile to bypass a complete substation also permits planned maintenance during regular hours, with no loss in service, and with greater safety to personnel.

Fig. 2-45. Mobile substation. (Minnkota Power Cooperative, Inc.)

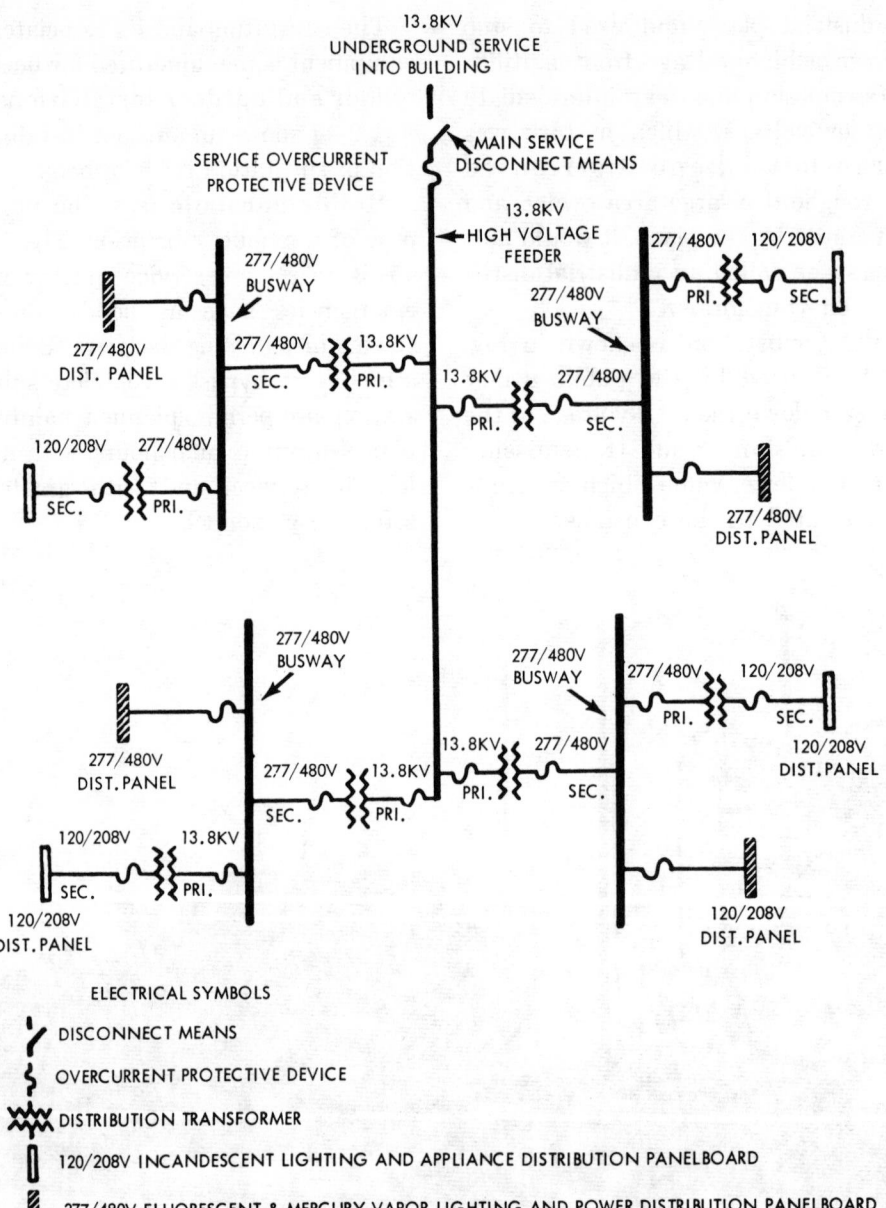

Fig. 2-46. The above one-line schematic shows how distribution transformers are used in an industrial plant where the public power utility serves the plant with an underground service at 13,800 volts, similar to the industrial plant shown in Fig. 2-43. The 120/208-volt panelboards normally serve incandescent lamps, small portable motors, and appliances while the 277/480-volt panelboards serve fluorescent and mercury vapor fixtures and larger motors.

Mobile substations rated up to 25,000 kVA and mobile transformers rated up to 57,000 kVA are now being used. They also permit a higher quality of maintenance.

Distribution Transformers. The normal function of distribution transformers is to take power from the primary lines and deliver it directly to the consumer.

Because of the increased demand for electrical power and the recent technical advances in the development of special high-voltage cable (in the 15 kV range) the use of high-voltage feeders in industrial plants is becoming more common. It is the function of distribution transformers to step down the high voltage of the feeder to such a value as is required for the loads to be served, Fig. 2-46.

The obvious advantage of this type of electrical power distribution system is that large amounts of power can be delivered to different areas in the plant at a high voltage and a low current thus resulting in a much smaller size feeder conductor.

Many public power utilities limit the size of the pad-mounted transformers that supply power to a consumer to 1,000 kVA at 120/208 volts and 2,000 kVA at 277/480 volts. Most public utilities classify transformers operating under 23 kV as distribution transformers although technically a transformer

TABLE I. STANDARD KILOVOLT-AMPERE RATINGS FOR OVERHEAD TYPE DISTRIBUTION TRANSFORMER

Single-Phase		Three-Phase	
5	75	15	150
10	100	30	225
15	167	45	300
25	250	75	500
37 1/2	333	112 1/2	
50	500		

could function at 69 kV and still be classed as a distribution transformer, provided that the rating does not exceed 500 kVA. Table I lists the standard kVA ratings for overhead type, distribution transformer.

The construction of the outside metal enclosure of distribution transformers depends upon the method of mounting and the location of installation. They may be installed indoors or exposed to the weather or they may be installed on a pole, or in a special enclosed area either inside or outside a building, in a manhole or vault, on a pad outside the building, on a switchboard or in other special locations.

Transformers which are auxiliary to motors and generators form a subdivision of the distribution class, and cover a wide range as to type of design and rating. Those built for starting duty since the starting cycle covers a relatively short period of time, are physically small compared to their rating. Some of the smaller low-voltage units are cooled by air, while oil is used for insulating and cooling purposes in those of higher voltage and larger output ratings.

Classification of Transformers With Respect to Purpose

Transformers can be classified into four classes according to the current or voltage transformation for which they are employed.

Current Transformer. This type of transformer as previously explained is designed to change the current of a system. The primary winding is connected in series with a main circuit whose current value it is desired to change. The most common application of this class of transformer is for street-lighting installations and for instrument transformers.

Constant-Current Transformer. This type of transformer is designed to supply a constant current to the secondary circuit. The primary is supplied by a constant voltage source. The voltage of the secondary will increase or decrease according to the load but the secondary current will remain constant.

Constant-Potential Transformer. This is the most commonly used type of transformer and is designed to operate at a reasonably constant primary and secondary voltage from no load to full load. Transmission and distribution transformers are examples of this type of transformer.

The constant-potential type of transformer is used for the transmission and distribution of electrical power.

Varying-Potential Transformer. This type of transformer is necessary for specific uses, such as for gaseous-discharge lamps where it is necessary to reduce the secondary with an increase of load. The primary is connected to a constant potential supply circuit normally in the range of 120 or 240 volts.

Classification of Transformers With Respect to Installation Requirements

As previously outlined, transformers may be classed in respect to the service or purpose for which they are used. A classification along entirely different lines may also be used based on location and method of installation as follows:

(a) On switchboards
(b) Indoors
(c) Out-of-doors
(d) In underground manholes or vaults
(e) There are also a number of special applications which cannot be included in the above general divisions.

On Switchboards. Transformers used with switchboards or panelboards fall into two groups, namely, current transformers and voltage transformers. They are used for the operation of instruments, such as ammeters, voltmeters, wattmeters

Fig. 2-47. Outdoor metal-clad switchgear shown with the door open. (Medium Transformer Dept., General Electric Co.)

and relays for various protective purposes. Such transformers are usually of the dry type, that is, they operate without oil and are normally mounted on the inside of the switchboard enclosure, Fig. 2-47. When higher voltages are involved, these transformers are immersed in oil for insulating purposes.

Indoors. Transformers for indoor service are very similar to outdoor transformers, the basic difference being that because they are not subject to the elements of na-

ture, such as rain and snow, they do not have to be constructed so that water cannot enter into them, Fig. 2-48. The following requirements are contained in the National Electrical Code (NFPA No. 70) and are mandatory when transformers are installed indoors.

Dry-type Transformers Installed Indoors. Dry-type transformers may be installed indoors when they are rated at not more than 112½ kVA and not more than 600 volts and the transformer is completely

Fig. 2-48. Ventilated dry-type indoor transformer rated through 2,500 kVA with maximum of 15 kV primary voltage. When located in clean, dry indoor locations, no special fireproof vaults or outdoor venting pipes are required. (Westinghouse Electric Corp.)

enclosed except for ventilating openings, Fig. 2-49. Dry type transformers rated at more than 112½ kVA may be installed indoors if they are installed in a transformer room of fire-resistant construction unless they are constructed with Class B (80°C rise) or Class H (150°C rise) insulation and are properly separated (6 feet horizontally and 12 feet vertically) from combustible material or are separated from combustible material by a fire resistant heat-insulating barrier.

All dry-type transformers that are rated at more than 35,000 volts are required to be installed in an approved vault.

Askarel-Insulated Transformers Installed Indoors. Askarel as used in the National Electrical Code is defined as a synthetic nonflammable insulating liquid which when decomposed by the electric arc, involves only nonflammable gaseous mixtures. Askarel is the proper name for non-combustible (fire-resistant) liquid insulation. New askarel is clear and water white to light straw in color. It has a minimum dielectric strength of 35 kV, a minimum resistivity of 100 ohm-cm X 10^9 and a maximum acid (neutralization) number of 0.014.

There are other insulating oils used in transformers that also meet the same specifications as askarel and are permitted to be used. Transformers installed indoors containing this type of insulating oil which are rated in excess of 25 kVA shall be provided with a pressure-relief vent. When transformers are placed in a

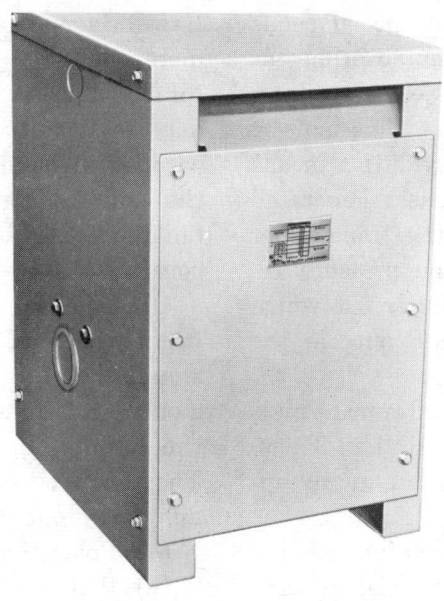

Fig. 2-49. Dry-type transformer with Class H insulation, 150° rise. 37½ through 250 kVA, single-phase. (Westinghouse Electric Corp.)

poorly ventilated place special means must be provided for absorbing or removing any gases generated by arcing inside the case. This may be accomplished by providing a means for absorbing any gases generated inside the case of the transformer or by connecting the pressure relief vent to a chimney or flue which will carry such gases outside the building.

All askarel-insulated transformers that are rated for more than 35,000 volts shall be installed in an approved vault.

Oil-insulated Transformers Installed Indoors. Special precautions must be taken when oil-insulated transformers are installed indoors because this type of oil is inflammable. The National Electrical Code basically requires that this type of transformer shall be installed in a specially constructed transformer vault that serves no other purpose.

If the total rating of the transformers does not exceed 112½ kVA, the transformer vault may be constructed of not less than 4 inches of reinforced concrete which is less than the requirements for vaults that contain transformers of over 112½ kVA ratings. A vault is not required for oil-insulated transformers installed indoors if they function at a voltage of not more than 600 volts and the total rating does not exceed 10 kVA. Oil-insulated

transformers that operate at not more than 600 volts and have a total rating not exceeding 75 kVA may be installed inside a building without further fire protection where the surrounding structure of the building is classified as fire-resistant construction.

Out-of-Doors. Transformers for installation out-of-doors are constructed with different types of metal enclosures depending upon the requirements of installation. They may be divided into several different groups:

(a) Pole mounting
(b) Pad mounting
(c) Out of doors protected by a fence or other effective enclosures.

The pole-mounted transformer is used primarily for direct distribution of electrical power to the consumer. When a single transformer is mounted on a pole the transformer is either secured directly to the pole or it is supported by a separate mounting bracket which is secured to the pole. The separate mounting bracket is widely used because the transformer can be easily removed for replacement or servicing. Single-transformers in the range of 500 kVA can be satisfactorily mounted on a single pole by this method, Fig. 1-38. When several transformers are mounted on a pole they are supported by a cluster-mounting hanger which is secured

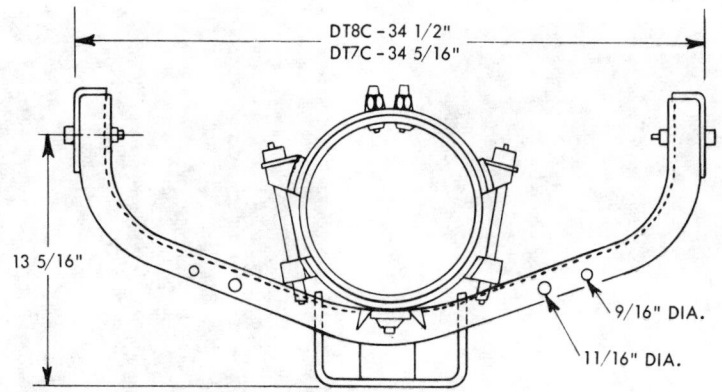

Fig. 2-50. Cluster-mounting hanger. (McGraw-Edison, Power Systems Division)

to the pole, Fig. 2-50. Three single-phase transformers of 250 kVA each can be satisfactorily mounted in this manner, Fig. 2-51. The old method of placing several larger transformers on a platform supported by two or more poles is being largely replaced by cluster-mounted transformers on a single pole or by a

Fig. 2-51. Single-phase transformers, 250 kVA each, mounted on a pole by cluster-mounting hanger.

pad-mounted transformer for larger transformers.

Pad-mounted Transformers. Because of the need for larger transformers and the common use of underground services supplied by the utilities this type of distribution transformer is becoming extremely popular. The pad-mounted transformer is designed for mounting directly on a concrete pad outside of a building and is of the enclosed type with no live parts exposed. It is primarily used for the direct distribution of electrical power from the utility to the consumer. These transformers are available in small sizes in the 10 kVA range for residential use, Fig. 2-52, and in larger sizes in the 5,000 kVA range for commercial and industrial use, Fig. 2-53. (See Fig. 2-43 for the typical use of a pad-mounted distribution transformer supplying power to a commercial store.) This type of

127

Fig. 2-52. New Mini-Pad Transformer. These transformers *(top)* are used for underground residential distribution and are designed for installation on front or back lot lines. They are available in ratings of 15 through 167 kVA, single-phase at a normal high voltage of 15 kV. *Bottom,* larger residential transformer. (Distribution Transformer Dept., General Electric Co.)

Fig. 2-53. New Compad II Transformer. This transformer is used for commercial and industrial installations. It is available for three-phase power in ratings of 75 through 2,500 kVA at a normal high voltage of 35 kV. (Distribution Transformer Dept., General Electric Co.)

transformer is available over a wide range of kVA ratings and is either air- or oil-cooled depending upon the size and voltage.

Out-of-doors installation of transformers for both distribution and transmission of power, together with the necessary switching equipment and protective devices comprising an outdoor substation, are commonly used. When transformers and their associated equipment are installed out-of-doors, the installation must be weatherproof and, if not of the totally enclosed type, must be made inaccessible to unauthorized personnel. For small capacity installations the transformer and its associated equipment may be contained in a common weatherproof enclosure with no live parts exposed, but for larger installations, where some hazards would exist if the installation were accessible to the public, a barrier or a fence is installed to eliminate the entrance of all but authorized personnel.

Normally transformers that are installed out-of-doors are oil-insulated, and because the oil is inflammable special precautions must be taken. The National Electrical Code requires that oil-insulated transformers shall be installed and located so that a fire originating in the transformer will not spread to adjacent buildings or combustible material or will not in any way block or conflict with doors, windows or fire escapes. Special safeguards that are recognized by this code are: space separation, fire-resistant barriers, automatic water spray systems and special enclosures which confine the oil of a ruptured transformer tank. Oil enclosures may consist of curbed areas or basins, fire resistant dikes, or trenches filled with coarse crushed stone.

In Underground Manholes or Vaults. In many areas where space is limited, whether in older urban areas beset by problems of dense, congested overhead wiring or in new developments where appearance is of utmost importance, distribution systems are being installed underground.

Subsurface transformers, normally referred to as URD (underground residential distribution) transformers are specially designed for underground installation to serve underground distribution systems. Such transformers are subject to complete submergence in water and are therefore constructed so that all joints (such as between tank and cover bushing and tank wall or bushing and cover) are water-tight under varying heads of water pressure. For underground residential distribution systems a submersible distribution transformer similar to that shown in Figs. 2-54 to 2-57 would be used. For underground commercial distribution systems, a submersible distribution transformer, normally referred to as a subway transformer, would be used.

In larger cities where large amounts of power are required in small areas, the transformer may be mounted in underground vaults in the same area where the power is to be used, Figs. 2-58 and 2-59. To maintain constant service, a number of network transformers in various parts of the city may be interconnected in a parallel network, each unit being provided with a short-circuiting switch which short-circuits and grounds the primary winding while the transformer is disconnected from the circuit, Fig. 2-60. If any fault should occur in any one of the circuits, this circuit may be isolated from other circuits in the network by this switch.

Fig. 2-54. This submersible distribution transformer is designed to provide economical, reliable distribution service primarily for residential areas. Installed in a below-grade vault, it is completely safe and out of sight. This type of transformer is available for a single-phase power in ratings of 25 through 167 kVA at a normal high voltage of 15 kV. (Distribution Transformer Dept., General Electric Co.)

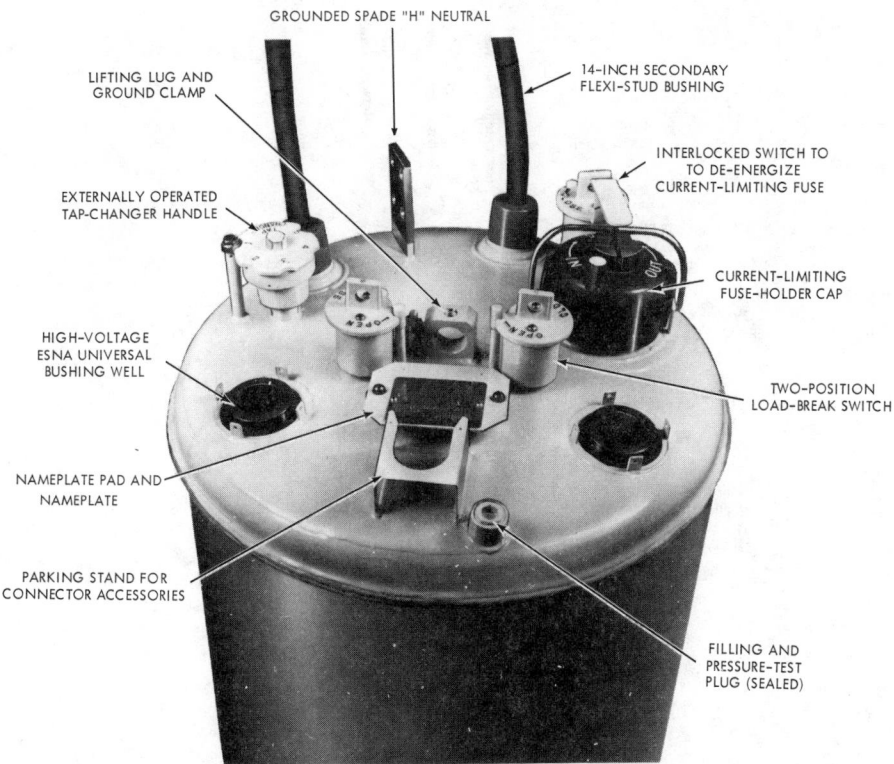

GROUNDED SPADE "H" NEUTRAL

14-INCH SECONDARY
FLEXI-STUD BUSHING

LIFTING LUG AND
GROUND CLAMP

INTERLOCKED SWITCH TO
TO DE-ENERGIZE
CURRENT-LIMITING FUSE

EXTERNALLY OPERATED
TAP-CHANGER HANDLE

CURRENT-LIMITING
FUSE-HOLDER CAP

HIGH-VOLTAGE
ESNA UNIVERSAL
BUSHING WELL

TWO-POSITION
LOAD-BREAK SWITCH

NAMEPLATE PAD AND
NAMEPLATE

PARKING STAND FOR
CONNECTOR ACCESSORIES

FILLING AND
PRESSURE-TEST
PLUG (SEALED)

Fig. 2-55. Top view of *Leapfrog II*, residential subsurface transformer (RST). (General Electric Co.)

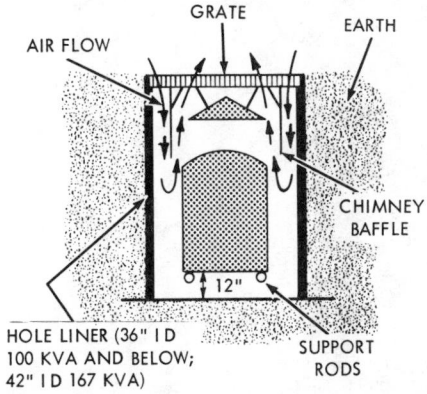

GRATE

EARTH

AIR FLOW

CHIMNEY
BAFFLE

12"

HOLE LINER (36" I D
100 KVA AND BELOW;
42" I D 167 KVA)

SUPPORT
RODS

Fig. 2-56. Diagram of *Leapfrog II* RST system. (Distribution Transformer Dept., General Electric Co.)

The system shown in Fig. 2-61 consists of a grid of interconnecting cables operating at utilization voltage. The grid is energized at many points so that the loss of any one point of supply will not cause loss of service. When a source is lost, the load formerly supplied at that point will be absorbed by the other sources of power.

Residential Subsurface Transformer (RST). This subsurface transformer (Figs. 2-55 to 2-56) is designed to provide economical, re-

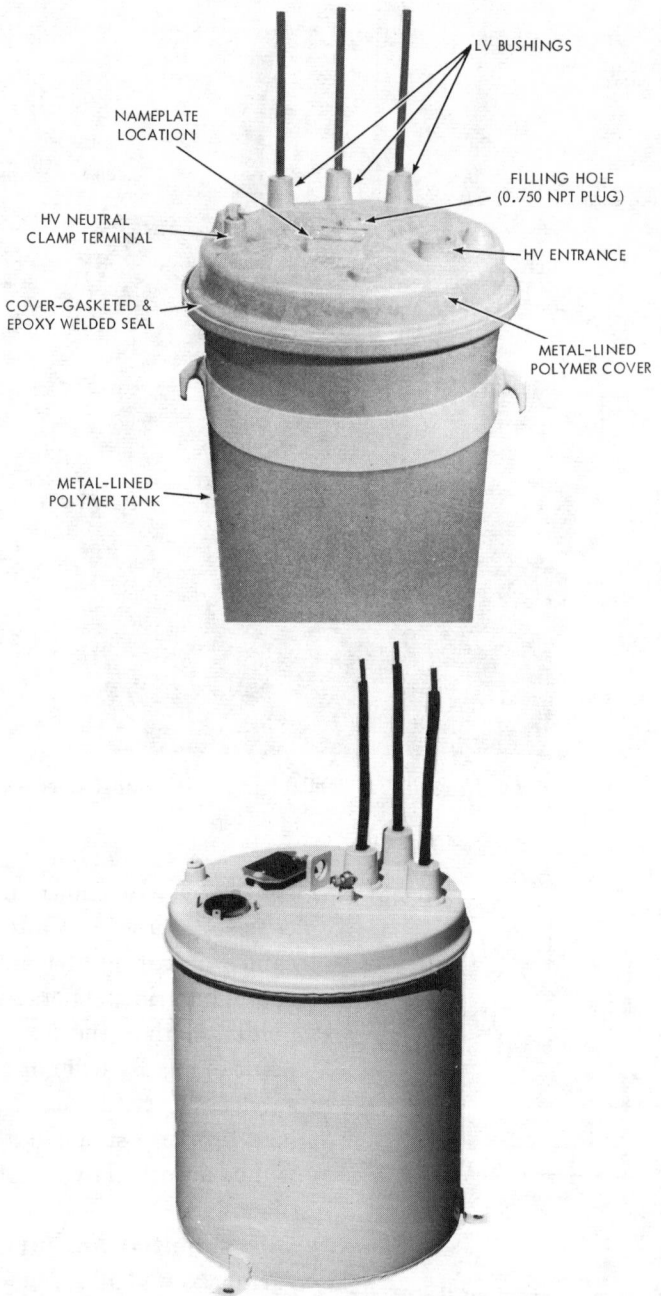

Fig. 2-57. Direct-buried transformer for URD (underground residential distribution) systems. (Distribution Transformer Dept., General Electric Co.)

Fig. 2-58. Typical three-phase, 500 kVA commercial subsurface transformer (CST). This subsurface transformer is a companion to the three-phase, pad-mounted transformer and is designed for underground vault installations. This type of installation is used where space is expensive, appearance is of utmost importance and the risk of damage is great. This type of transformer is available for three-phase power in ratings of 150 through 2,500 kVA with standard high voltage ratings up to and including 16.3 kV. (Distribution Transformer Dept., General Electric Co.)

Fig. 2-59. Installing a subsurface transformer. (Distribution Transformer Dept., General Electric Co.)

liable distribution service primarily for residential areas. Installed in a below-grade vault, it is completely out of sight. The Leapfrog® II RST is designed with the necessary cooling surface to operate at 65°C in an enclosure when used with an efficient air-directing baffle and operating at rated load. This type of transformer is available for single-phase power in ratings of 15 through 167 kVA at a normal voltage of 15 kV and below.

The specially designed chimney baffle utilizes the natural chimney effect that causes cool outside air to

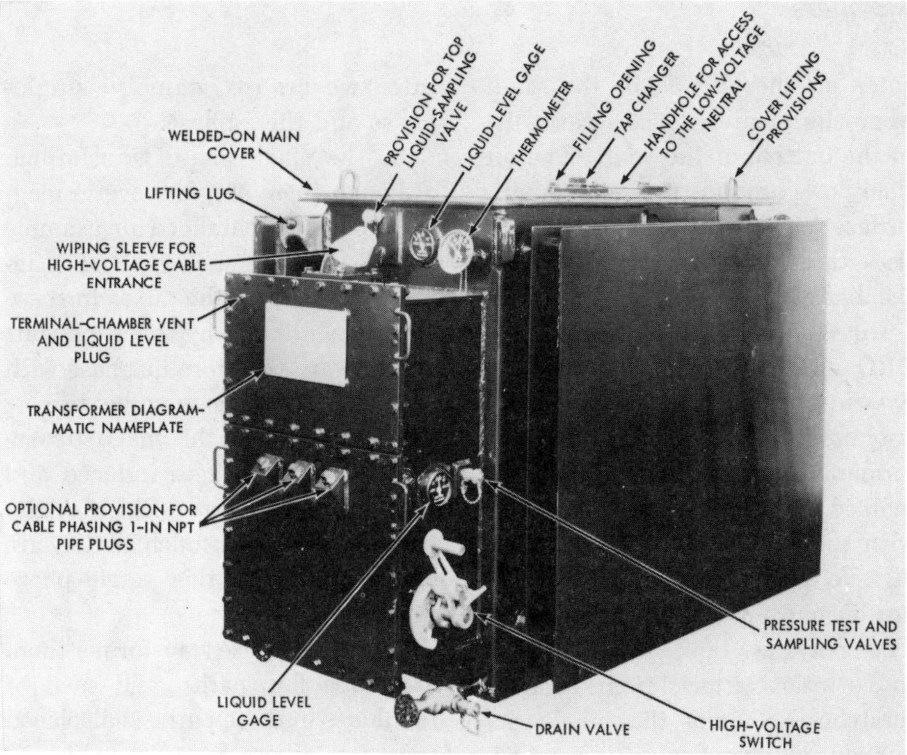

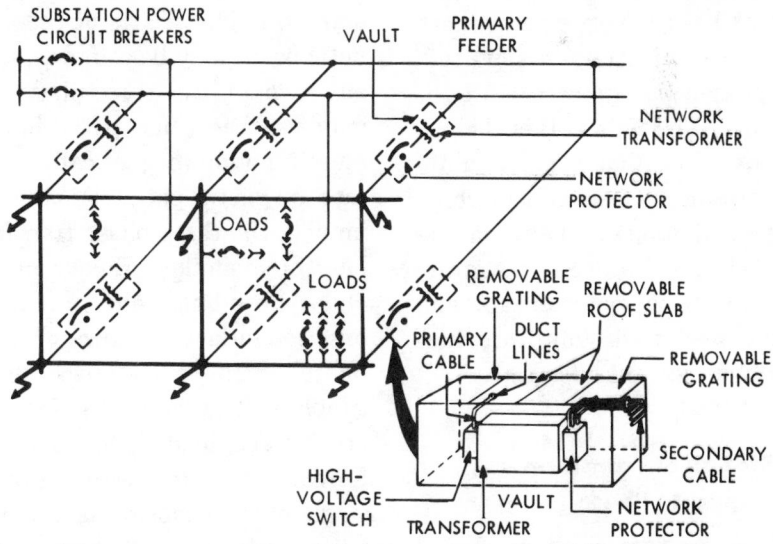

WELDED-ON MAIN COVER

LIFTING LUG

WIPING SLEEVE FOR HIGH-VOLTAGE CABLE ENTRANCE

TERMINAL-CHAMBER VENT AND LIQUID LEVEL PLUG

TRANSFORMER DIAGRAM-MATIC NAMEPLATE

OPTIONAL PROVISION FOR CABLE PHASING 1-IN NPT PIPE PLUGS

PROVISION FOR TOP LIQUID-SAMPLING VALVE

LIQUID-LEVEL GAGE

THERMOMETER

FILLING OPENING

TAP CHANGER

HANDHOLE FOR ACCESS TO THE LOW-VOLTAGE NEUTRAL

COVER LIFTING PROVISIONS

PRESSURE TEST AND SAMPLING VALVES

LIQUID LEVEL GAGE

DRAIN VALVE

HIGH-VOLTAGE SWITCH

Fig. 2-60. Network transformer showing high-voltage end. (Distribution Transformer Dept., General Electric Co.)

SUBSTATION POWER CIRCUIT BREAKERS

VAULT

PRIMARY FEEDER

NETWORK TRANSFORMER

NETWORK PROTECTOR

LOADS

LOADS

REMOVABLE GRATING

REMOVABLE ROOF SLAB

PRIMARY CABLE

DUCT LINES

REMOVABLE GRATING

SECONDARY CABLE

HIGH-VOLTAGE SWITCH

VAULT

TRANSFORMER

NETWORK PROTECTOR

Fig. 2-61. One-line diagram AC secondary network system. (Distribution Transformer Dept., General Electric Co.)

enter on the outside of the vault where its greater density carries it to the bottom of the vault. The air then flows around the transformer's surface picking up heat and then rises through the inner opening of the baffle.

Direct-Buried Transformer for URD (Underground Residential) Systems. Fig. 2-57, *top*, shows a polymer-tank direct-buried transformer. The tank and cover are molded from reinforced polyester with the high-voltage bushing well, high-voltage neutral, and low-voltage bushings molded into the cover. The polymer tank is impervious to corrosion attack from scrapes, scratches or bumps that might occur during installation. Fig. 2-57, *bottom*, shows a mild-steel tank direct-buried transformer. The mild-steel tank has a superior paint finish that is grit-blasted, passivated, epoxy-primed and painted with the finish coat. The unit is then baked. The epoxy finish has superior abrasive resistance (resists scratching during installation) and also has excellent chemical and temperature stability. Both the polymer tank and the mild-steel tank units are hermetically sealed, oil-filled units.

Classification of Transformers With Respect to Phase

All transformers, without regard to service applications, are divided into two groups, namely, single-phase and three-phase.

In the single-phase transformer, although there may be several magnetic paths in parallel (for example the distributed shell-type of magnetic circuit), all the fluxes in these paths are induced by the same voltage and are therefore in phase with each other. There may be two or more windings on this magnetic circuit, but the voltages induced and the currents flowing will be single-phase currents inasmuch as they are all produced by this single-phase flux.

In a three-phase transformer there are three flux paths, and each of the fluxes in these paths is displaced from the others by 120°, which is the three-phase relationship of the voltages impressed upon the primary windings. Parts of these magnetic paths may be common, that is, any two of the three-phase fluxes may use the same path, but they must function magnetically as separate magnetic paths. As the name implies, the three-phase transformer has three similar primary and secondary windings, one primary and one secondary winding for each phase of the three-phase primary supply voltage and secondary delivery voltage. The three-phase transformer is therefore essentially three single-phase transformers built on one magnetic circuit. In some cases the three-phase transformer may actually be an assembly of three separate single-phase transformers in one tank, with the primary and secondary coil leads of each unit arranged on a terminal board so that a three-phase connection is obtained with only three leads from the primary and secondary sides of this terminal board.

The majority of transformers made in this country are of the single-phase type. One of the obvious reasons for the large number of single-phase transformers is their use for residential consumers. Another contributing factor is that when three-phase power is supplied to a consumer from an overhead distribution line, three single-phase distribution transformers would normally be mounted on a pole instead of one large three-phase transformer because of the ease of handling and mounting of the three single-phase units, Fig. 2-51. Another advantage of using three single-phase transformers to feed a three-phase lighting and motor load is where the single-phase lighting load is much larger than the three-phase motor load (such as an office building where the only three-phase load is air-conditioning). In this type of installation, one large transformer would be installed to serve the lighting and air-conditioning load and the other two transformers need only be large enough to serve the three-phase load for the air-conditioning.

When transformers are used to transform three-phase power (with the exception of pole-mounted transformers) normally a three-phase transformer would be used. A three-phase transformer requires three times as much copper as a single-phase unit but, in comparison with three single-phase units, a three-phase transformer requires less than three times as much iron for the core. In comparing three single-phase units to a three-phase transformer, the three-phase transformer is lower in initial cost, costs less to install, has simpler connections, has a higher efficiency and requires a much smaller space.

Standards for Transformers

If costs were not a factor it would be desirable to design each transformer for the specific power system and load that the transformer would serve. It is most uneconomical however, to design each transformer specifically for each power system and load served and it has become universal practice to standardize many transformer designs with the understanding that they will be used throughout the country.

Many construction details are standardized, primarily to make it easier for the users of transformers. Many ratings have been standardized in great detail, especially for transformers in the small power and distribution class. The guiding principle of standardization is to make the standard specification such that the total cost of manufacture and use of transformers will be lower.

Standard Kilovolt-Ampere Ratings

In Table II are given the preferred ratings for distribution and power transformers. In most cases the ratings are such that three single-phase transformers may be grouped in a three-phase bank to form the same rating as one three-phase transformer.

ing surges or lightning surges. The calculation of probable over-voltages or impulse voltages actually appearing at the transformer is basically determined from the characteristics of the system with the elements of nature, such as lightning, taken into consideration. Normally over-voltages caused by lightning are of sufficient magnitude to flash over or break down transformer insulation and therefore create the greatest potential hazard to transformer insulation. At or near a lightning stroke, the voltage increases very rapidly until it reaches its peak, perhaps to 1,000,000 volts in 1/1,-000,000 sec (1,000 kV per second). Special precautions are normally taken which will limit the lightning voltage which could appear at the transformer terminals. These precautions would normally consist of:

TABLE II. PREFERRED KVA RATING FOR SINGLE-PHASE AND THREE-PHASE DISTRIBUTION AND POWER TRANSFORMERS							
SINGLE-PHASE				THREE-PHASE			
3				9			
5	100	1,667	12,500	15	300	3,750	25,000
10	167	2,500	16,667	30	500	5,000	33,333
15	250	3,333	20,000	45	750	7,500	37,500
25	333	5,000	25,000	75	1,000	10,000	50,000
37 1/2	500	6,667	33,333	112 1/2	1,500	12,000	60,000
50	833	8,333		150	2,000	15,000	75,000
75	1,250	10,000		225	2,500	20,000	100,000

Basic Impulse Insulation Level (BIL)

The insulation of a transformer is subject from time to time to momentary over-voltages that may be caused by system faults, switch-

(1) overhead ground wires which shield the overhead line conductors, (2) reduced line insulation or gaps near the transformer so that the lightning will flash over to limit the voltage, (3) lightning arresters in-

stalled as near as possible to the transformer which will discharge the lightning impulses to ground.

From the foregoing it is evident that the transformer insulation must be able to withstand the normal op-

erating voltage, the maximum fault voltage and the probable impulse voltage.

It can be basically stated that the total voltage that could appear at the terminals of the transformer is

INSULATION LEVEL APPLICATION			Chopped Wave Impulse Test		Line Bushings Included in Developed List Price: Voltage Class (BIL)	Phase-to-Phase Test: KV
Nominal System Voltage: KV ②	Basic Impulse Level: KV	Low Frequency Test to Ground: KV	KV	Minimum Time to Flashover: Ms		
1.2	45	10	54	1.5	15	①
2.5	60	15	69	1.5	15	①
5.0	75	19	88	1.6	15	①
8.7	95	26	110	1.8	15	①
15	110	34	130	2.0	15	①
25	150	50	175	3.0	23	①
34.5	200	70	230	3.0	34.5	①
46	250	95	290	3.0	46	①
69	350	140	400	3.0	69	①
115	550	230	630	3.0	115(550)	230
115	450	185	520	3.0	115(550)	185
115	350	140		3.0	115(550)	185
138	650	275	750	3.0	138(650)	275
138	550	230	630	3.0	115(550)	230
138	450	185	520	3.0	115(550)	230
161	750	325	865	3.0	161(750)	325
161	650	275	750	3.0	138(650)	275
161	550	230	630	3.0	115(550)	275
230	1050	460	1210	3.0	230(1050)	460
230	900	395	1035	3.0	196(900)	395
230	825	360	950	3.0	196(900)	395
230	750	325	865	3.0	161(750)	395
230	650	275	750	3.0	161(750)	395
345	1300	575	1500	3.0	345(1300)	575
345	1175	520	1350	3.0	345(1175)	575
345	1050	460	1210	3.0	345(1050)	575
345	900	395	1035	3.0	345(1050)	575
500	1675	750	1930	3.0	500(1675)	850
500	1550	690	1780	3.0	500(1550)	825
500	1425	630	1640	3.0	500(1425)	825
500	1300	575	1500	3.0	500(1425)	825
725	2050	920	2360	3.0	700(2050)	1155
725	1925	860	2220	3.0	700(2050)	1155
725	1800	800	2070	3.0	700(2050)	1155

① Induced phase-to-phase test will be 2 times normal operating voltage.
② Maximum operating voltage not to exceed 105% of nominal system voltage.
Note: The application of insulation level depends on use of appropriate arresters. Arrester application depends on overvoltage conditions on systems as determined by degree of grounding and overvoltage which may result with loss of load, switching, etc.

determined from the characteristics of the system. It would be impractical to design each transformer specially to withstand the voltages which may reach it, especially as the impulse voltages which can reach the transformer terminals will be different for each location on the system. Instead of designing each transformer to withstand the specific voltages for a given installation, all transformers are designed to withstand one of a number of standard impulse tests, and from consideration of the voltages which may actually appear at the transformer, one of the standard test strengths (generally known as Basic Insulation Levels) is chosen. These Basic Insulation Levels with the associated standard impulse tests are shown in Table III. The Basic Insulation Level for the transformer is then selected so that its strength is 15 to 20 percent above the impulse voltages which can reach it.

Standard Voltage Ratings

In Table IV are given the standard rated circuit voltages upon which transformers are operated. These values are open circuit values and represent the voltage on the high-voltage side of the transformer.

Nominal voltages for secondary distribution systems are given in Table V.

TABLE IV. STANDARD HIGH-VOLTAGE RATINGS

HIGH-VOLTAGE TAPS (SINGLE-PHASE)

Rated High-Voltage, Volts	Rated kVA High-Voltage Taps, Volts
2400/4160Y	2520/2460/2340/2280 4360Y/4260Y/4055Y/3950Y
4800/8320Y	5040/4920/4680/4560 8720Y/8520Y/8110Y/7900Y
6900/11950Y	7245/7070/6730/6555 12550Y/12250Y/11650Y/11350Y
7200/12470Y	7560/7380/7020/6840 13090Y/12780Y/12160Y/11850Y
7620/13200Y	8000/7810/7430/7240 13860Y/13530Y/12870Y/12540Y
12000	12600/12300/11700/11400
13200	13860/13530/12870/12540
13800	14400/14100/13500/13200
22900	24100/23500/22300/21700
26400	27800/27100/25700/25000
34400	36200/35300/33500/32600
43800	46200/45000/42600/41400
67000	70600/68800/65200/63400
115000	120750/117875/112125/109250
138000	144900/141450/134550/131100

NOTE: All voltages are Δ unless otherwise indicated.

HIGH-VOLTAGE TAPS (THREE-PHASE)

Rated High-Voltage, Volts	Rated kVA High-Voltage Taps, Volts
2400	2520/2460/2340/2280
4160	4360/4260/4055/3950
4800	5040/4920/4680/4560
6900	7245/7070/6730/6555
7200	7560/7380/7020/6840
12000	12600/12300/11700/11400
13200	13860/13530/12870/12540
13800	14400/14100/13500/13200
22900	24100/23500/22300/21700
26400	27800/27100/25700/25000
34400	36200/35300/33500/32600
43800	46200/45000/42600/41400
67000	70600/68800/65200/63400
115000	120750/117875/112125/109250
138000	144900/141450/134550/131100

NOTE: All voltages are Δ

TABLE V. NOMINAL RATED CIRCUIT VOLTAGES

120/208Y	6,900
120/240	11,500
240/480	8,000/13,800Y
277/480Y	23,000
600	34,500
2400	46,000
2,400/4,160Y	69,000
4,800	115,000

Forms of Construction

The form of the transformer construction is determined by the relative position of the magnetic and current circuits as they are linked together to form a transformer.

In Chapter 1, we found that two general forms of construction are used, the shell form and the core form. There are several modifications of these two forms, but they may all be classified in (1) the one group known as the shell form, in which the magnetic circuit surrounds the current circuit, Fig. 2-62; or in (2) the second group known as the core form, in which the cur-

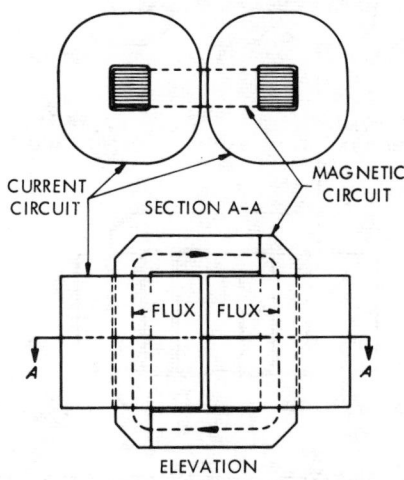

Fig. 2-63. Simple core construction of magnetic circuit.

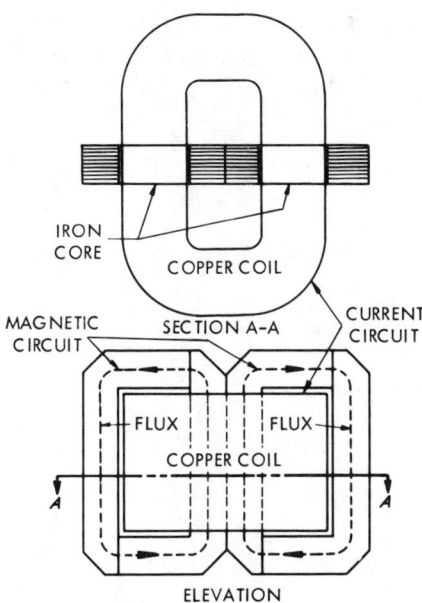

Fig. 2-62. Simple shell construction of magnetic circuit.

rent circuit surrounds the magnetic circuit, Fig. 2-63. Fig. 2-64 shows a three-phase rectangular core type of magnetic circuit and Fig. 2-65 shows a three-phase cruciform core type of magnetic circuit.

The shell and core forms of construction each have their own inherent characteristics which make them especially adaptable to certain voltage and capacity ratings. In considering which form is to be used, the following factors must be considered:

(1) Economy of material for a given performance
(2) Adaptability to heavy currents; to high voltages
(3) Ventilation or cooling

141

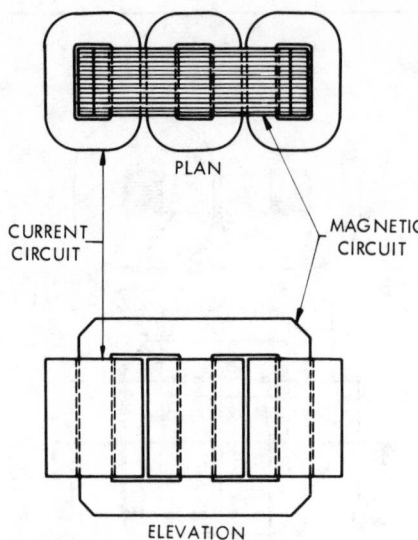

PLAN

CURRENT CIRCUIT

MAGNETIC CIRCUIT

ELEVATION

Fig. 2-64. Three-phase rectangular core-type of magnetic circuit.

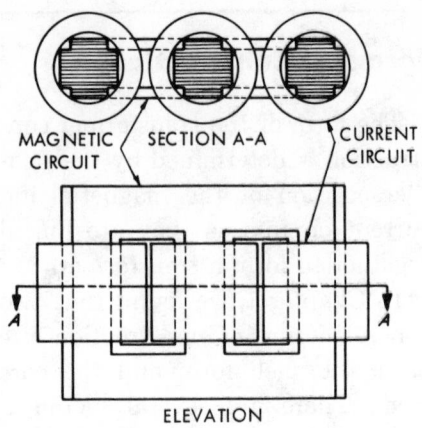

MAGNETIC CIRCUIT / SECTION A-A / CURRENT CIRCUIT

ELEVATION

Fig. 2-65. Three-phase cruciform core-type of magnetic circuit.

(4) Mechanical stresses

(5) Repairs.

For each voltage and frequency class there is a transformer of a certain output rating, above which the space factor makes the shell form cheaper, and below which the core form uses less material. In general, small and medium power transformers are of core construction (Figs. 1-49 and 1-50) and large power transformers are of shell form. (Figs. 1-77 and 1-78)

Circuitry and Cooling

The Magnetic Circuit

Function

The purpose of the magnetic circuit in the transformer is to provide a path of low resistance to the flow of flux lines produced by the current in the primary or exciting winding. Fig. 3-1 shows three simple magnetic circuits, one a coil with an air core, one a magnetic circuit composed of iron, and one a magnetic circuit composed of iron with an air gap.

A voltage E is impressed across each coil (see Fig. 3-1). A current I will be set up in each coil in the direction shown by the arrows on the leads. This current will cause a flux to flow in the magnetic circuit (dotted lines) in the direction

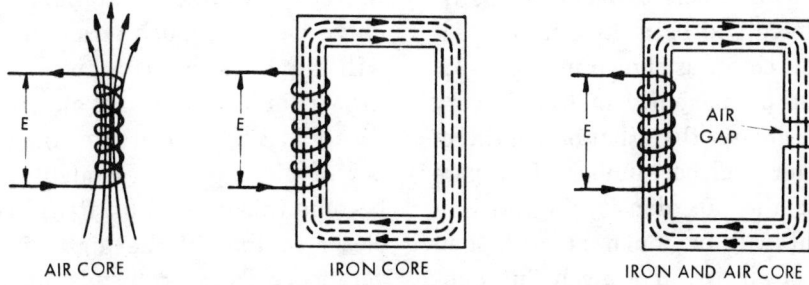

AIR CORE   IRON CORE   IRON AND AIR CORE

Fig. 3-1. Various types of magnetic circuits.

shown, and of a magnitude given by the formula

$$\Phi = \frac{.4\pi NI}{\dfrac{l}{\mu A}} \qquad (1)$$

where

$\Phi =$ magnetic flux.

$N =$ number of turns in the coil.

$I =$ current through the coil.

$l =$ mean length of magnetic path

$\mu =$ permeability of the material in the magnetic circuit.

$A =$ cross-sectional area of the magnetic path.

As the ampere turns (NI) of the exciting winding are varied, either by varying the turns, or the current, or both, the flux varies in each of the three magnetic circuits. The permeability (μ) of air is constant and the flux will be directly proportional to the ampere turns of magnetizing force. The number of lines of flux per unit of cross-sectional area of the magnetic circuit, or flux density, will therefore be proportional to this magnetizing force. A curve showing the value of flux density for varying values of ampere turns of magnetizing force will be a straight line for the air cores, as shown in Fig. 3-2.

The permeability of iron is very much greater than that of air; therefore for a given number of ampere turns, the flux density in the iron circuit will be much greater than that of air, or for a given flux density the number of ampere turns re-

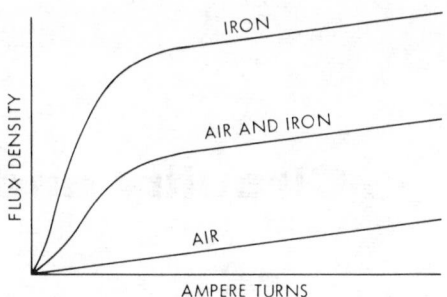

Fig. 3-2. **Magnetization curves for air, an air gap and iron core.**

quired to produce the total flux will be much smaller in the iron circuit than in the air circuit. The permeability of iron however is not constant and, as the number of lines per unit area or flux density in the iron circuit is increased, the iron tends to become saturated and the curve for the iron circuit flattens out and becomes parallel to that of the air core. At these very high flux densities the iron core is but little better than the air core magnetic circuit.

If an air gap, Fig. 3-1, is introduced in the iron circuit, the reluctance of the magnetic path is greatly increased and the magnetization curve becomes much straighter and will lie between the two curves drawn for the iron and air circuits (see Fig. 3-2). The greater the number of air gaps, the straighter will become this curve, and it will come closer to that of the air core. Reactance coils for transmission lines are made with an air core because this straight line characteristic is especially desirable at high flux densities.

Construction of Magnetic Circuit

In the transformer the magnetic circuit is made entirely of iron, in the form of laminated sheet steel, in order to carry the required flux with a minimum amount of magnetizing force or ampere turns. This laminated steel is cut from silicon-iron sheets most commonly having a thickness of 14 mils, and both sides of these laminated sheets are coated with an enamel forming an insulating coating approximately one-half mil in thickness. The length of the lamination is in the direction of the magnetic flow.

Aging and Annealing

The sheet steel as used in the transformer of today contains a sufficient amount of silicon to prevent the steel from aging. Aging is that characteristic of the magnetic circuit tending to increase the iron loss as the transformer continues to operate over a period of time. Before the introduction of silicon in the steel, no steel had been found which did not have this undesirable feature of aging. This sheet steel, as it first comes from the rolls at the steel mill, has a very high loss and low permeability. These two factors are greatly improved by an annealing process.

During the punching or shearing of the steel into laminations of the desired length and width, strains are set up which again increase the loss and decrease the permeability. If low losses are desired it is necessary to anneal the steel after the punching or shearing operation to relieve these strains. Hammering or bending the lamination so as to give it a permanent set will likewise set up stresses which increase the losses. Therefore when a unit must be torn apart for repairs, it is often necessary again to anneal the laminations before rebuilding the unit if the iron loss after rebuilding is not to exceed the original loss of the transformer.

Exciting Current

The exciting current of a transformer is the current which it draws from the line when it is excited at normal voltage and frequency with no load on the secondary. The current may be measured by an ammeter placed in the circuit.

The exciting current, therefore, represents largely a flow of current into the magnetic circuit and out again. This current, however, does produce a loss in the generator supplying the current, and a loss in the lines leading to the transformer, and it is therefore objectionable.

The total ampere turns of magnetizing force, Fig. 3-2, are equal to the exciting current in amperes multiplied by the number of turns in the excited winding. Therefore for a given transformer having a fixed number of turns, the total flux can be varied only by a change in exciting current. Because the iron circuit becomes saturated at high flux densities, the exciting current at such densities increases very rapidly. For this reason the magnetic circuit must be worked at a density low enough to limit the exciting current to a reasonable value and at the same time permit the transformer to be operated at an increased voltage without greatly increasing this current.

Iron Loss

If, while measuring the exciting current of a transformer, a wattmeter is connected in the circuit, the power in watts supplied to the transformer measured by the wattmeter is the iron loss. If compared with the volt-ampere input of the transformer when it is excited at normal voltage and with no secondary load, this iron loss will be found to be less than the volt-ampere input, thus indicating that the exciting current is not in phase with the impressed voltage. The ratio of the iron loss in watts to the volt-ampere input is the power factor of the transformer at no load; it will be found to be of a very low value, and a transformer with high exciting current contributes toward a poor power factor in the circuit.

The iron loss is made up of two components as follows:

(a) The hysteresis loss
(b) The eddy-current loss

The hysteresis loss represents the energy required to reverse the direction of the flux in the magnetic circuit. It is proportional to the weight of iron in the circuit, to the frequency of the flux, and, approximately, to the 1.6 power of the magnetic density.

The flux changing in the iron circuit generates voltages in the laminated paths which set up a flow of currents at right angles to the direction of the flux path, known as eddy currents. The eddy-current loss is proportional to the total weight of the iron, to the square of the frequency of the flux, and to the square of the magnetic density.

Variation of Iron Loss and Exciting Current

In a given transformer with other factors constant, the iron loss of a transformer varies with the frequency, not directly as the frequency, but at a mathematical power somewhere between the first and second (square of a number) power.

This mathematical power depends upon the ratio of hysteresis loss to the eddy-current loss. The hysteresis loss changes directly as the frequency, while the eddy-current loss will change directly as the square of the frequency. For example, let us assume that the ratio of the hysteresis loss to the eddy-current loss is such that their sum, which equals the total iron loss, varies as the 1.2 power of the frequency. With the frequency constant, the total iron loss varies at a mathematical power of the magnetic density between 1.6 and 2.0. For the magnetic flux densities at which transformers ordinarily work it will be found that this iron loss varies almost directly as the square of the magnetic density.

Assuming the impressed voltage constant, let us consider the variation of iron loss with change of frequency. For example, let the frequency be reduced from 60 to 25 hertz. In the first place, due to this change in frequency, the total iron loss would be less than half its former value if the magnetic density remained constant. Due to the reduced frequency, however, the magnetic density increases in the ratio of 60 to 25. The total iron loss, in turn, increases approximately as the square of the magnetic density. The net change therefore is that the iron loss increases and is approximately

doubled in going from 60 to 25 hertz.

With the frequency constant and considering the change in iron loss which is produced by an increase or decrease in the impressed voltage, it is evident that the iron loss is roughly proportional to the square of the voltage. For example, an increase in the impressed voltage of 10 percent will increase the iron loss something like 20 percent. A corresponding reduction occurs when the impressed voltage is lowered.

It is very difficult to get a definite idea of the change in the exciting current with variations of frequency and voltage. This is due to the shape of the magnetization curve, see Fig. 3-2. In the neighborhood of the bend or "knee" of this curve the change in exciting current is very rapid with variation of either the frequency or the voltage.

A rough idea of the change in exciting current with the impressed voltage may be secured from the fact that an increase of approximately 10 percent of the impressed voltage will mean practically doubling the exciting current.

The Current Circuit

Coil Conductors

The current circuit of the transformer comprises the primary and secondary windings which may assume various forms as explained a little later. The copper or aluminum conductors used in these windings may be round, square, or rectangular in cross-section. Small conductors up to and including size No. 11, approximately .0907 inch in diameter, are usually round. Square wire is not made much smaller than .0907 inch on a side; furthermore these small sizes are difficult to wind because of their tendency to turn on edge and thereby cut through the insulation between turns and between layers. Rectangular conductors of very small area are also undesirable because their thin edges are liable to cut through the insulation. Such thin conductors are also uneconomical because the insulation on the conductors occupies a greater amount of space than would be required if the conductor were made thicker and narrower. The maximum size of rectangular conductor is limited by the ease with which it can be bent or formed into the desired coil shape.

Eddy currents flowing in the conductor create eddy-current losses which cause additional heating and decreased efficiency; these losses are proportional to the square of the thickness of the conductor. The thickness of the conductor is therefore dependent upon the maximum allowable eddy currents in the conductor.

The various kinds of insulating material used on the winding conductors is basically determined by the average temperature rise of the aluminum or copper conductors inside of the coil windings. The most commonly used coil winding insulations are a high-temperature, heat-resistant enamel or a specially treated, stabilized paper tape which has been chemically modified to resist aging. Depending upon the requirements of a particular design, some winding conductors may require several layers of the special paper tape. See Fig. 1-51.

Types of Winding

Considering assembled transformer windings as a whole, they may be divided into two general classes; namely, concentric and interleaved. By concentric, as shown in Fig. 1-42, is meant that the various coils forming the completed winding are assembled concentrically. "Interleaved" windings are formed by assembling the individual coils side by side, rather than concentrically, as shown in Fig. 3-3. The individual coils forming either the concentric or the interleaved types of winding may be either rectangular or circular in shape. In general the concentric form of winding is used with the core type of construction, and the interleaved with the shell-type. At times the interleaved winding is used for large core designs and concentric winding with small shell-type units.

Concentric Windings. The rectangular form of concentric coil is shown in Fig. 1-42. In the circular

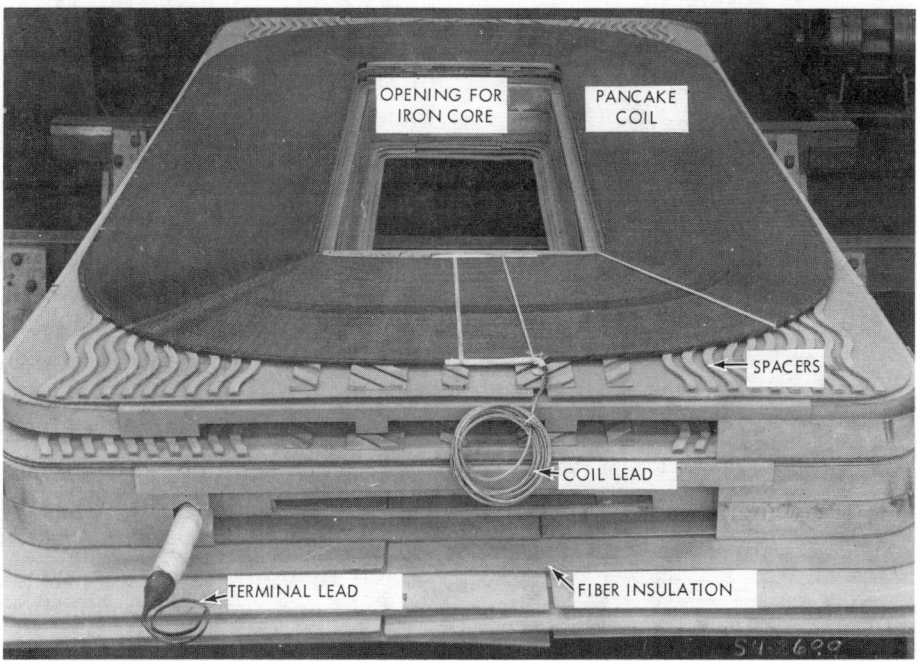

OPENING FOR
IRON CORE

PANCAKE
COIL

SPACERS

COIL LEAD

TERMINAL LEAD

FIBER INSULATION

Fig. 3-3. Pancake coil and assembly insulation partially insulated, showing spacers between coils for support of turns.

form of winding, the coil conductors are wound on a tubular form. Two types of circular coil construction are shown in Figs. 3-4 and 3-5. The cylindrical wound coil (Fig. 3-4) consists of one or more layers of insulated conductors wound on a tubular form. The windings are wrapped with epoxy glass tape which, after baking, forms a strong band around the coil. This type of coil normally functions at voltages up to *8.7 kV (95 kV BIL)* and currents up to 2500 amperes. The wind-conductors of a continuous-wound pancake coil (Fig. 3-5) consist of from one to five rectangular straps

Fig. 3-4. Circular coil construction: cylindrical-wound coil. (Westinghouse Electric Corp.)

149

Fig. 3-5. Circular coil construction: continuous-wound pancake coil. (Westinghouse Electric Corp.)

winding, but relatively few of the core-type use it. The individual coils of the interleaved windings are sometimes called pancake coils. Such a coil with insulation between coils is shown in Fig. 3-3. Pancake coils are usually wound with square or rectangular conductors, one or more conductors in parallel, depending upon the current to be carried. Winding the coils in this form with the type of insulation between coils, as shown in Fig. 3-3, exposes the maximum amount of conductor surface to the cooling medium. This enables the heat to be conducted rapidly from the coils to the cooling medium and prevents overheating.

Individual concentric coils are wound on micarta or heavy pressboard cylinders. The cylindrical wound coil is normally wound directly on a heavy pressboard cylinder with oil impregnated pressboard insulation between adjacent layers. The continuous wound pancake coil is normally wound over vertical insulation strips on a heavy pressboard cylinder. The individual coils are wound so that when assembled together there is sufficient space between the coils to form adequate ducts for the circulation of oil for cooling of the coil. The core and coil assembly must be rigidly braced to withstand the mechanical forces under line fault conditions and to resist vibration and shock forces during shipment.

in parallel, wound continuously on a tubular form. The coils are pressed hydraulically during construction to the calculated short circuit forces that they must withstand when in service. This type of coil is suitable for substation use and normally functions at voltages above *8.7 kV (95 BIL)*.

Interleaved Windings. Interleaved windings may be used with either the core- or shell-type of construction. Practically all the larger shell-type transformers use this

Individual pancake coils are assembled with flat insulating barriers between coils, usually with spacers of pressboard to permit the circulation of oil. These individual coils are assembled into high-voltage and low-voltage groups and these groups in turn assembled together to form the completed high-voltage and low-voltage winding.

Copper Loss

The copper loss of a transformer is the energy loss in the windings when the transformer is loaded. It is composed of two parts. The first is the product of the square of the current flowing in the windings and their resistances, and the second is due to the small eddy currents flowing in the copper. This eddy-current loss is due to the magnetic leakage of the flux through the air.

This eddy-current loss may be as high as 15 percent of the total copper loss. The copper loss due to the resistance of the windings is dependent upon the temperature of the winding, since the resistance of the copper varies with temperature. As the temperature rises the resistance increases, and as the temperature falls the resistance decreases. On the other hand the eddy-current loss decreases with increase of temperature, since the actual value of the eddy currents decreases as the resistance increases. It is therefore necessary to state the temperature

at which the copper loss is specified. The total of these two losses may be found by short-circuiting the secondary winding and measuring the input when enough voltage is applied to either winding to cause normal current to flow. If this input is divided by the square of the primary current flowing, a resistance value is obtained which may be called the effective resistance value of the transformer.

The voltage impressed on the transformer to circulate normal full-load current is called the impedance voltage. This impedance voltage is the sum of the resistance drops through the primary and the secondary windings and the reactive drop. The regulation of the transformer is dependent to a considerable degree upon this impedance voltage which is usually expressed as a percentage of the rated primary voltage.

Reactive Voltage

Leakage lines of force which pass only through the primary coil are not effective in inducing voltage in the secondary coil. Likewise, lines of force which pass through the secondary coil and not through the primary are not effective in producing a counter electromotive force opposing the impressed electromotive force in the primary winding. These two elements together form the total reactive drop through the

151

Transformers

transformer winding. The reactive drop may be calculated after the impedance and the copper loss are known.

Example. A certain 100 kVA, 2400 volts to 240 volts, single-phase, 60-hertz transformer has its secondary short-circuited. With an ammeter in the primary side it is found that a voltage of 120 volts across the primary winding is necessary to send full-load current through this winding. The wattmeter in the circuit reads 3000 watts. The copper loss is therefore 3000 watts which, expressed in percentage of kVA rating of the transformer equals 3 percent. The impedance equals 120 volts and, expressed in percentage of normal voltage of winding across which the 120 volts are measured, is $120 \times \dfrac{100}{2400} = 5$ percent impedance. The reactive drop expressed in percent is therefore

$$\sqrt{5^2 - 3^2} = \sqrt{16} = 4 \text{ percent}$$

This reactive drop in percent may be expressed in volts by multiplying by the rated voltage. In terms of the primary winding the reactive volts drop would be 4 percent of 2400 volts = 96 volts. In terms of the secondary winding the reactive volts drop would be 4 percent of 240 volts = 9.6 volts.

Likewise the impedance volts in terms of the secondary winding would be 5 percent of 240 volts = 12.0 volts.

Regulation

The regulation of a transformer is defined as the difference in voltage between no load and full load expressed as a percentage of the full-load voltage. This regulation is dependent upon the reactive voltage and the total copper loss of the transformer. When supplying a non-inductive load, the copper loss forms the major portion of the voltage drop. As the power factor of the load decreases, the reactive voltage increasingly adds to the total voltage drop. When good regulation is required, it is necessary that the copper loss and reactive voltage be kept to a low value.

Efficiency

The efficiency of a transformer is the ratio of the output to the input. It is also equal to the ratio of the output to the output plus losses. That is:

$$\text{Efficiency} = \frac{\text{output}}{\text{input}} =$$

$$\frac{\text{output}}{\text{output} + \text{copper loss} + \text{core loss}}$$

For the ordinary power transformer, the efficiency is generally from 96 to 99 percent. The losses are due to the copper losses in both windings and the hysteresis and eddy-current losses in the iron core; the copper losses vary as the square of the current in the winding and the winding resistance. The core losses consisting of the hysteresis and eddy-current losses caused by the alternating magnetic flux in the core are approximately constant from no load to full load with rated voltage applied to the primary.

Methods of Cooling

The dissipation of the heat losses generated in a transformer is of great importance, since the rating of the transformer is determined not only by these losses but also by the temperature rise of the transformer when delivering a certain load. Any means that can economically be used to dissipate more readily the heat generated in the coils and magnetic circuit will permit us to increase the rating of a transformer having a definite physical size, or will permit us to decrease the dimensions of a unit having a definite kilovolt-ampere rating.

There are many methods of cooling transformers. Some of the most common methods in the United States are listed as follows:

(a) Self-air-cooled (dry type) (AA)

(b) Oil-immersed self-cooled with air (OA)

(c) Oil-immersed self-cooled with water (OW)

(d) Oil-immersed forced-oil-cooled with forced-water cooler (FOW)

(e) Oil-immersed self-cooling with air-blast for additional cooling, permitting an increased rating (OA/FA)

(f) Self-cooled/forced-air cooled (dry type) (AA/FA) (often referred to as air-blast-cooled).

Standard Abbreviations in the United States for Methods of Cooling Transformers

When reading descriptive bulletins or material relating to the cooling methods of transformers, the following symbols or abbreviations are used to indicate the various methods of cooling employed.

(a) AA Dry-type self-cooled

(b) AA/FA Dry-type self-cooled with forced-air cooling as transformer is loaded

Liquid-immersed Transformers

(c) OA Oil-immersed self-cooled

(d) OA/FA Oil-immersed self-cooled with forced-air cooling as transformer is loaded

(e) OA/FA/FA Oil-immersed, self-cooled with two stages of forced-air cooling as transformer is loaded

(f) FOA Oil-immersed, forced-oil, forced-air cooling. Oil pumps and fans are employed as cooling method (the FOA type of cooler normally has no kVA rating without fans and oil pumps in operation)

(g) OA/FA/FOA Oil-immersed, self-cooled with forced-air-forced-oil cooling as transformer is loaded

(h) OA/FOA/FOA Oil-immersed self-cooled with two stages of forced-air-forced-oil-cooling as transformer is loaded. This type of cooling would be employed where the daily load cycle varies widely. The first stage

153

of FOA gives 33⅓ percent load increase over the OA rating, the second stage of FOA gives 66⅔ percent load increase over the OA rating.

(i) FOW Oil-immersed forced-oil, forced-water cooling.

Standard Abbreviations in Canada for Methods of Cooling Transformers

Standard abbreviations used in Canada for the methods of cooling a transformer are established by the Canadian Standard Association. Some of the more popular abbreviations and examples of their use are listed below:

transformers. The coil and core are enclosed in a sheet metal enclosure that is provided with louvers or grates, Figs. 2-48 and 2-49. The cooling of the coil windings is accomplished by natural convection of the surrounding air and by the radiation of heat from the different parts of the transformer structure, Fig. 3-6. Self-air-cooled transformers have been in use for many years, but because of the heat problem they have been restricted to the smaller sizes. Modern technical developments of special high temperature coil insulating materials that resist deterioration under heat have ex-

STANDARD ABBREVIATIONS IN CANADA		
MEDIUM IN WHICH THE CORE IS LOCATED	METHOD OF CIRCULATING MEDIUM	METHOD OF TRANSFERRING HEAT TO SURROUNDING AIR
O (OIL)	N (NATURAL CONVECTION)	S (SELF–COOLED)
A (AIR)	F (FORCED CONVECTION)	P (FAN COOLING)
L (SYNTHETIC LIQUID)		W (WATER COOLING)
G (GASES)		

EXAMPLES:

OIL-IMMERSED, NATURAL CIRCULATION, SELF–COOLED	ONS
OIL-IMMERSED, NATURAL CIRCULATION, WATER–COOLED	ONW
OIL-IMMERSED, NATURAL CIRCULATION, FORCED-AIR-COOLED	ONP
OIL-IMMERSED, FORCED OIL, WATER–COOLED	OFW
OIL-IMMERSED, FORCED OIL, FORCED-AIR-COOLED	OFP
OIL-IMMERSED, NATURAL CIRCULATION, FORCED-AIR-COOLED (SECOND STAGE OF FORCED-AIR-COOLING)	ONPP

Self-Air-Cooled, Dry-Type

Self-air-cooled transformers are normally referred to as dry-type

tended the size of these units to the 3,750 kVA capacity range at 15 kV. Dry-type distribution transformers offer many advantages over

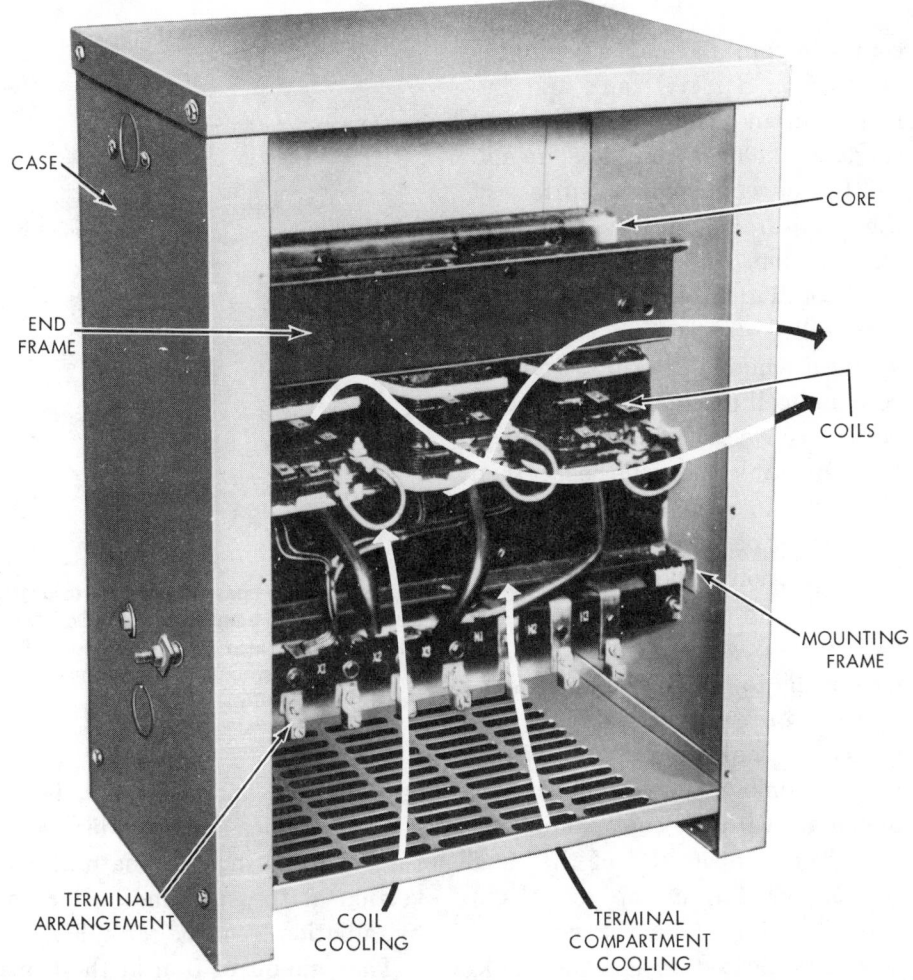

CASE

CORE

END
FRAME

COILS

MOUNTING
FRAME

TERMINAL
ARRANGEMENT

COIL
COOLING

TERMINAL
COMPARTMENT
COOLING

Fig. 3-6. Self-air-cooled dry-type transformer (with louvered enclosure cover removed). The location of the terminals in the bottom portion of the transformer case places them in an area where cool ambient air enters the case; therefore, high-temperature cable insulation or over-size cables are not required for safe, long-life operation. (Westinghouse Electric Corp.)

liquid-filled transformers. They are ideally suited for indoor installation because they are safe and cannot explode. No toxic gasses can be released and fire hazards are negligible. Some of the small transformers such as bell-ringing, control, signaling, neon-sign-lighting or current

and voltage instrument transformers are of the totally enclosed type and the small amount of heat from the coil windings is dissipated by radiation into the surrounding air.

Oil-Immersed Self-Cooling

In this type of transformer the

complete core and coil assembly is enclosed in a metal tank and immersed in an insulating oil. The oil conducts the heat from the core and coil by convection to the surface of the enclosure where it is dissipated by radiation.

Plain Tank Wall. In small liquid-filled distribution transformers the oil conducts the heat from the core and coil assembly by means of natural convection to the surface of the metal tank where the heat is then dissipated into the surrounding air.

All types of liquid cooling depend on the motion or flow of the liquid from the core and coil assembly upward and outward to the cooler surface of the tank. The liquid in self-cooled transformers rises when it is heated by the core and coil assembly and sinks again when it comes in contact with the cooler tank surface. The temperature of the tank surface varies from top to bottom depending on where the core and coils are located inside the tank. When the core and coils are located near the bottom of the tank, the tank as a whole will be hotter, and more heat will be dissipated at a given maximum temperature rise than if the core and coils are located near the top of the tank.

In the case of small distribution transformers such as the pole-mounted distribution transformers commonly used in residential areas, the smooth tank wall used to en-

Fig. 3-7. Single-phase pole-type residential distribution transformer with oil-filled plain-wall tank. (See also Figs. 1-39 and 1-40 in Chapter 1.) (Distribution Transformer Dept., General Electric Co.)

case the core and coil assembly has enough surface area to dispose of the generated heat by natural convection of the surrounding air and by radiation from the tank, Fig. 3-7.

The amount of iron in the transformer increases as the cube of its dimensions, whereas the tank surface increases as the square. Therefore, as transformers are built larger and have increased kVA ratings, the amount of iron used in the construction of the core and, consequently, the loss in the transformer will increase at a faster rate than the physical dimensions of the tank surface. A point is soon reached where the tank surface cannot dissipate the heat, and means to increase its ability to do so must be applied.

The heat-dissipating surface area of the tank can be increased by making the tank taller than is actually necessary to encase the transformer.

Tank Wall with Fins or Plates. When more heat must be dissipated than is permitted by the plain wall tank, additional radiating surface area can be obtained by welding radiating fins or plates to the walls of the tank, thus increasing the total surface area of the tank that will be exposed to the surrounding air.

Tank Wall, with Hollow Fins or Tubes. As the kilovolt-ampere rat-ing increases, the problem of cooling increases also. The heat generated in a transformer is proportional to the volume of material or the product of the three dimensions. The cooling ability of the unit is proportional to the external surface or to the square of the dimensions. It is therefore apparent that as the kilovolt-ampere rating (or the volume) increases, the heat-dissipating ability of the unit increases also, but at a much slower rate. Therefore it becomes necessary to provide additional means of cooling.

In medium-size transformers, additional radiating surface is obtained

Fig. 3-8. Pad-mounted, distribution-type CTP underground transformer. (Westinghouse Electric Corporation)

Fig. 3-9. Pole-mounted distribution transformer with hollow tubes welded to the top and bottom of the tank. (Westinghouse Electric Corp.)

by welding hollow radiating fins or tubes to the tank. The hollow fins or tubes are welded to the top and bottom of the tank, Figs. 3-8 and 3-9. They are welded at the top and bottom of the tank by means of a nipple at each end which permits the oil in the tank to circulate by natural convection through the hollow fins or tubes.

Tank Wall with External Radiators. Liquid-immersed, self-cooled (OA) power transformers use external radiators for additional ra-

diating surface, Fig. 3-10. The type and number of cooling units with a self-cooled (OA) rating will depend on the shipping clearance, rating of the unit, and other physical factors and will be optimized at the time the transformer is designed. The radiator consists of a group of hollow cooler tubes which are connected to headers at the top and bottom, Fig. 3-11. These headers are either welded to the tank wall or are attached by means of flanged connections through which the oil flows. The radiators for large units are usually connected to the tank by flanged connections so they may be removed for shipment whereas the radiators for smaller units are welded to the tank wall. In the large units, shut-off valves are normally installed at the inlet and outlet connections so that should any leaks develop, the radiator may be removed and repaired without removing all of the oil from the tank.

Oil-Immersed Water-Cooling

The large self-cooled power transformer, due to its great size, represents a considerable investment. Where water is cheap and available in sufficient quantities, and where a somewhat lower operating efficiency of the transformer is permissible, some operating companies prefer the water-cooled type. A copper cooling coil, wound in a single or several parallel sections (depending upon

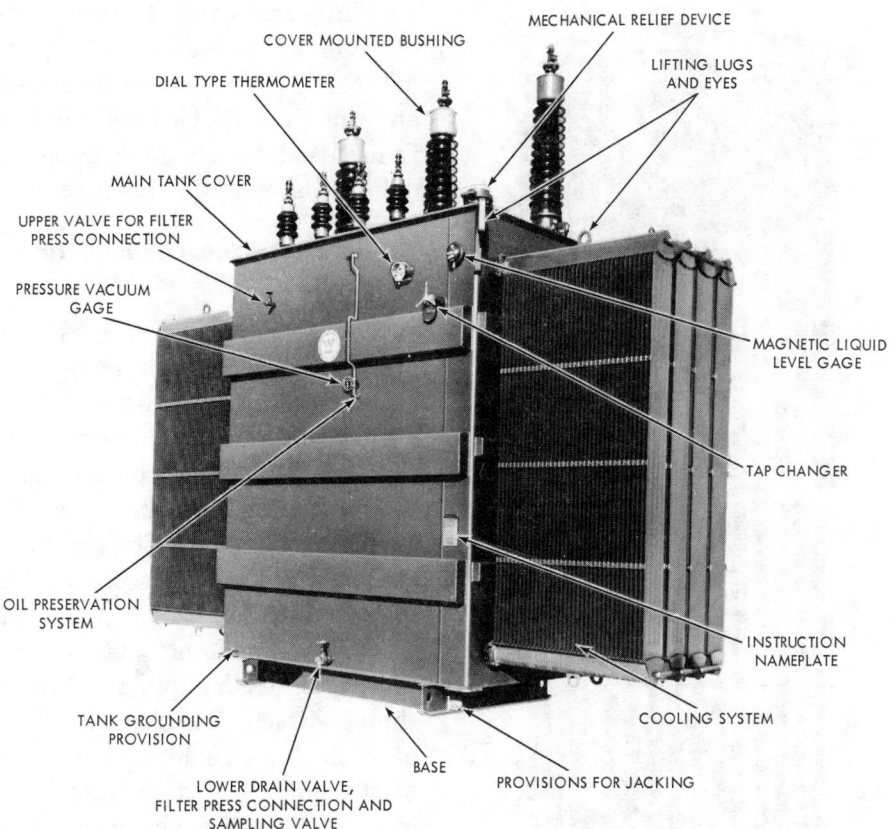

COVER MOUNTED BUSHING

MECHANICAL RELIEF DEVICE

DIAL TYPE THERMOMETER

LIFTING LUGS AND EYES

MAIN TANK COVER

UPPER VALVE FOR FILTER PRESS CONNECTION

PRESSURE VACUUM GAGE

MAGNETIC LIQUID LEVEL GAGE

TAP CHANGER

OIL PRESERVATION SYSTEM

INSTRUCTION NAMEPLATE

TANK GROUNDING PROVISION

COOLING SYSTEM

BASE

LOWER DRAIN VALVE, FILTER PRESS CONNECTION AND SAMPLING VALVE

PROVISIONS FOR JACKING

Fig. 3-10. Liquid-immersed self-cooled (OA) medium power transformer with external radiators. (Three-phase rating is 10,001-30,000 kVA; single-phase rating is 10,001-20,000 kVA.) (Westinghouse Electric Corp.)

the quantity of tubing required) is placed in the upper part of the tank and supported from either the cover or from the tank wall. Water circulated through this coil or coils effectively cools the oil. The cooling coils are made of copper, since iron coils are subject to corrosion. They are tested with pressures as high as 500 pounds per square inch as a safeguard against the development of leaks in service.

Oil-Immersed Forced-Oil Cooling

The one objection to water cooling is that in the event of a leak in the cooling coil the water, which is at a higher pressure than the oil, will enter the oil and greatly reduce its insulating quality. The water may reach certain live parts of the transformer and cause serious damage. The forced oil-cooled unit eliminates this hazard, since the oil is

Fig. 3-11. Complete radiator assembly. The headers in the complete radiator assembly may be welded directly to the tank or may be connected by flanges. (Westinghouse Electric Corp.)

tem. Any leaks will then cause the oil to enter the water of the external cooling system rather than water entering the oil. This type of cooling is used extensively in Europe but only occasionally in this country.

Oil-Immersed Self-Cooling with Air-Blast

With the addition of a blast of air against the radiators of large self-cooled units, additional heat may be dissipated. This permits increasing the rating of the transformer during the time that the air blast is supplied. The additional stream of cooling air is provided by fans mounted on each of the radiators or the radiators may be partially enclosed in a thin metal casing through which air is blown by means of a single large blower. The air-blast equipment is operated only when additional peak loads are required for short periods of time. The control for the equipment may be manual or automatic. With the automatic control, operation of the fans is started by a Mercoid control equipment whose bulb extends into the hot oil. When the oil reaches a predetermined temperature the Mercoid contacts close, starting the fans. When the oil temperature (due to increased radiation or because of a decrease in load) drops to a safe predetermined value, the control equipment again operates to shut off the fans. One complete control

circulated by means of pumps through an external water-cooled unit. The water pressure in this external cooling unit is less than the pressure in the oil circulating sys-

Fig. 3-12. Transformer equipped with fans to increase its capacity above the self-cooled rating. Fans are started automatically with controls sensitive to top oil and winding temperature. (McGraw-Edison, Power Systems Division)

equipment is usually supplied with each transformer. In some cases, if the individual units are small, one control is furnished per bank of transformers, to be operated by a Mercoid in the hot oil of one unit only. This arrangement simplifies and reduces the total cost of the control equipment so that it can be used economically with smaller transformers, Figs. 3-12 and 3-13.

Air-Blast Cooling

The air-blast type of transformer

161

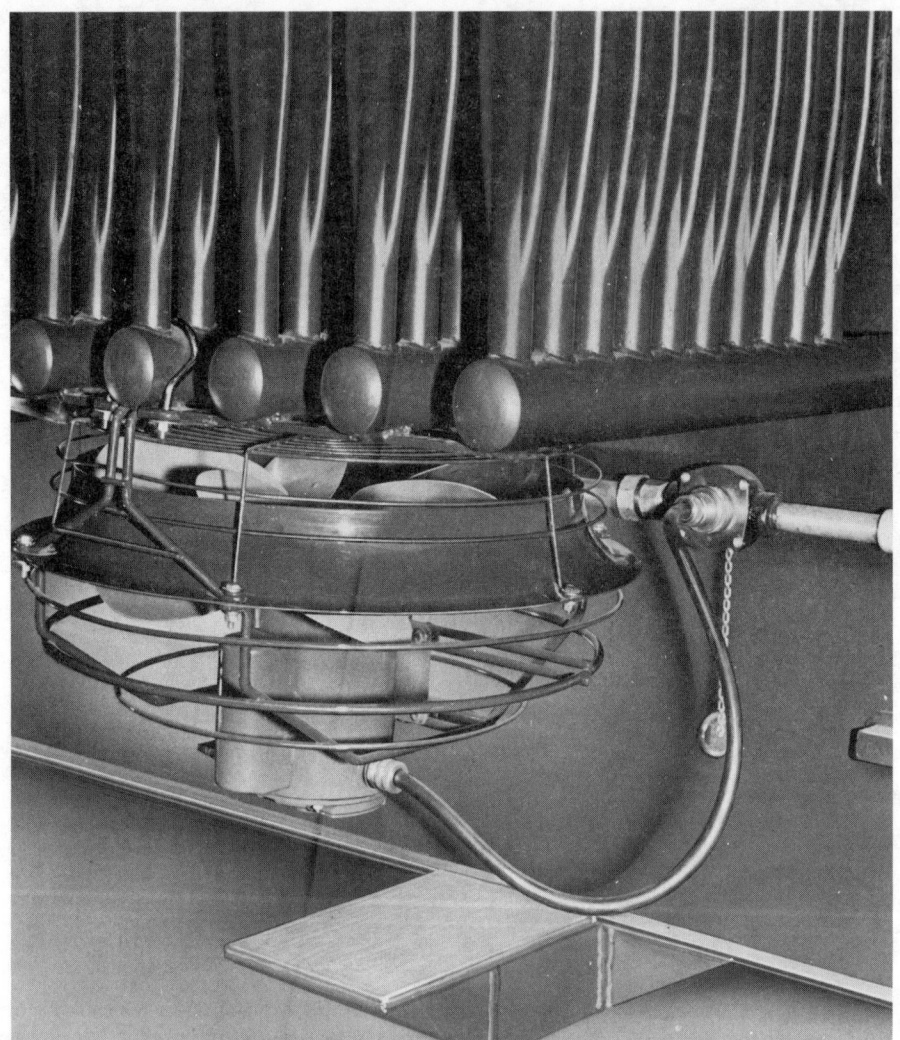

Fig. 3-13. Cooling fans are designed for quiet, trouble-free operation. Fan units include an orifice ring that guides the flow of air and reduces noise level. Electrical connections are water-tight. Motors have built-in thermal protection. (McGraw-Edison, Power Systems Division)

is a dry-type of transformer used where great economy of space and weight is desired, or where regulations prohibit the use of oil-insulated units because of the toxic fumes or fire hazard. These units are provided with ample air ducts between the coils, the coils and the core, and the coil and core assembly and the metal housing so the

Fig. 3-14. When cooling fans are used with a dry-type transformer, the kVA capacity is increased by 33 percent. (Distribution Transformer Dept., General Electric Co.)

air may be blown through them for the dissipation of the heat, Fig. 3-14. With the fans in operation the transformers have increased rating of 33 percent over the self-cooled ratings. The air-blast type of trans-former normally functions at 300 to 15,000 kVA at a voltage not exceeding 15 kV. Transformer manufacturers are producing air-blast transformers with higher kVA and voltage ratings.

Heating and Temperature Control

The heating of a transformer is caused by the iron and copper losses. Immediately after a self-cooled transformer is put into service, the heat generated by the losses is stored in the iron and copper and their temperature increases. After continued operation this heat is communicated to the oil and by convection is carried to the tank wall where it is radiated into the surrounding air. The temperature of the transformer will continue to rise until the rate at which the tank wall dissipates the heat is equal to the rate at which it is generated. When this condition is reached for a given steady load, the transformer is said to have reached a stable condition and it will operate continuously and indefinitely at this constant temperature and load, provided the maximum temperature reached is not high enough to injure the insulation.

Temperature Gradient

In an electric circuit having resistance, a difference of potential causes a current to flow through the circuit. Similarly, a difference of temperature must exist between the windings and the oil to cause the heat to flow from these windings into the oil and hence to the tank wall. This temperature difference is known as the temperature *gradient*. If a coil is relatively thick and has a large amount of insulation (which is not a good conductor of heat) it will require a larger gradient, or temperature difference between coil and oil, to cause the heat to flow into the oil, than if the coil were relatively thin with a small amount of insula-

tion and large area exposed directly to the oil. This gradient therefore is dependent upon the coil construction. Furthermore, this temperature gradient will vary as the total amount of heat generated in the windings varies, or as the copper loss varies (iron loss for a given voltage is constant). In terms of the load therefore, since the copper loss is proportional to the square of the load, the gradient will vary approximately as the square of the load.

This temperature gradient may be considered as made up of two parts. The first is from the point of maximum or "hot spot" temperature to the average temperature of the entire winding; the second is from the average temperature to that of the hot oil. The hot-spot temperature is seldom known for it can be measured only by means of thermocouples or resistance coils embedded in the windings during manufacture. This is rarely done because of the danger presented by the voltages of the windings. The average temperature can be determined by measuring the resistance of the transformer windings at a known temperature; for example, when the transformer is disconnected from the line and is at the same temperature as the surrounding air, known as the ambient temperature. A resistance measurement is again taken when the transformer is at load temperature. From these two resistances the temperature of the winding can be determined.

Temperature Limits

The limit of temperature for a transformer is established by the materials which are used in the manufacture of that transformer. Since most of the insulation used in the windings is of organic material and fibrous in nature (such as kraft paper, manila paper, kraftboard, and pressboard) this material will carbonize under excessive heating, and continued high temperature will accelerate such carbonizing action until a condition is reached where the insulating strength of the transformer is greatly impaired. Chemists are constantly developing new synthetic materials with intermediate properties that have not yet been fully evaluated. One unusual mixture that has been developed is a special varnish, pigmented with mica powder. This special mixture has a remarkable resistance to both oxidation and decomposition at temperatures well over 100°C. There are many other types of insulating materials used in transformers and the effective properties of these materials is largely dependent on the structure in which they are used. Dielectric strengths as measured by any of the standard tests have relatively little relation to the dielectric stresses that can be used in a conventional transformer design.

NEMA (National Electrical Manufacturers Association) and ANSI (American National Standards Institute) have established industry standards for dry-type transformers that stipulate the maximum temperature rise of the winding for various transformer insulating materials. There are currently four insulating classes as follows.

1. Class A. This class of insulating material is limited to use in transformers which are designed to have a continuous full load temperature rise not exceeding 55°C over a 40°C ambient. The standards allow for a hot spot 10° over the normal temperature rise. A Class A, 55°C rise transformer under full load, will have an average conductor temperature of 95° centigrade when operating in a 40°C ambient, and a maximum conductor hot spot of 105°C. Class A insulating material consists of: (1) Cotton, silk, paper, and similar organic material when either impregnated or immersed in a liquid dielectric; (2) Molded and laminated materials with cellulose filler, phenolic resins, and other resins of similar properties; (3) Films and sheets of cellulose acetate and other cellulose derivatives of similar properties; (4) Organic varnishes (enamel) as applied to conductors.

This class of insulating system is normally used in oil-filled transformers. Because of the low temperature rise this type of transformer insulation is normally limited to the small control transformer in the one kVA and below rating or in the larger transformer where a higher temperature would be a hazard such as in an area where combustible dust or fiber flyings might be present in the air.

2. Class B. This class of insulating material is limited to use in transformers having a maximum temperature rise under full load conditions not exceeding 80° centigrade, when the transformer is operating in a 40° centigrade ambient. The allowable hot spot temperature can be 30°C higher. In other words, the average conductor temperature in a Class B, 80°C rise transformer operating under full load conditions, in a 40°C ambient, will be 120° centigrade with the hot spot in the winding being no more than 150° centigrade.

Class B insulating materials consist primarily of materials or combinations of materials such as mica, glass fiber, or asbestos, with suitable bonding substances. Other materials or combinations of materials both organic and inorganic may be included in this class if by experience or accepted tests they can be shown to be capable of operation at 80°C rise temperatures.

In the past this was the most commonly used dry-type transformer and because of the lower inherent cost some manufacturers are still producing them through the 25 kVA rating. Some engineers still prefer the lower 80°C temperature rise because the transformer has a lower core loss, longer life, greater overload handling capabilities and greater safety where low flash-point combustible flyings are present in the atmosphere. If the above advantages are desired it would be advisable for the specification to call for Class H insulated, 80°C rise transformers rather than Class B insulated transformers.

3. Class F. This class of insulating material is limited to an average conductor temperature rise of 115°C under full load conditions when operating in a 40°C ambient. The allowable hot spot temperature is 30°C higher. Therefore, when operating under full load conditions in a 40°C ambient, a Class F, 115°C transformer will have an average conductor temperature of 155°C, and a hot spot temperature of 185° centigrade.

Class F insulating materials consist primarily of materials or combinations of materials such as mica, glass fiber or asbestos, with suitable bonding substances. Other materials, both organic and inorganic may be included in this class if by experience or accepted tests they can be shown to be capable of operation at 115°C average temperature rise.

The relative practicality of this insulation class for dry-type transformers is questionable. In order to legitimately qualify for this classification by NEMA and ANSI standards, the insulating materials must also qualify for the higher class H temperature rises, the only difference being the binding materials and impregnating materials. Many of the impregnating and binding materials that were once thought to be acceptable for 115°C operation have been found inadequate by NEMA and ANSI. Therefore it would appear that this insulation class will eventually fall into complete disuse as far as dry-type transformers are concerned, in favor of the Class H insulating materials, which use inorganic bonding materials that are primarily composed of silicon compounds or resins.

4. Class H. This class of insulating material allows a design temperature rise of 150° centigrade when a transformer is operating in a 40°C ambient. This rise is the average rise found in the conductor and an additional 30°C hot spot is allowed. Therefore the conductor temperature under full load conditions when the transformer is operating at a 40°C ambient will reach 190° centigrade, with the hottest spot not exceeding 220° centigrade.

Class H insulating materials primarily consist of materials or combinations of materials such as mica, glass fiber, asbestos, elastomer, and silicon, with suitable bonding substances such as appropriate silicon resins. Other materials or combinations of materials may be included in this class if by experience or by accepted tests they can be shown to be capable of operation at 150°C temperature rise. These other materials would have to be present in very limited quantities in order for the entire unit to qualify as a Class H insulated transformer.

Most of the dry-type transformers being manufactured today are of the Class H, 150°C temperature rise classification. Standard dry-type transformers, 30 kVA and larger are normally available from a manufacturer's stock. The higher temperature rise by this insulation class allows the transformer to be designed to a minimum size for convenience and by using a minimum amount of conductor and core permits it to be the smallest dimensionally and the most economical of all dry-type transformers.

NEMA and ANSI standards stipulate that a transformer constructed with Class H insulating material must withstand a 150°C temperature rise for 20,000 hours. If a transformer were designed to meet these exact specifications it would have an expected life of under three years which would obviously be unacceptable in most applications. To increase the life of the transformer it would be designed to operate at full load well below the maximum 150°C temperature rise, say in the 115°C range, but constructed with Class H insulation. The above transformer would not only have a longer life but would also withstand longer and greater overload periods and result in lower losses which will generally offset the higher purchase price within a relatively short time.

Ambient Temperature

Since the actual temperature of the transformer is the sum of the ambient temperature and the temperature rise, it is apparent that the ambient temperature very largely determines the load which can reasonably be carried in service.

Method for Approximating Ambient Temperatures and Their Influence on Loading. The following is an excerpt from NEMA Standards Publication No. TR 98-1964 "Guide For Loading Oil-Immersed Power Transformers With 65°C Average Winding Rise." This complete publication is available from NEMA (National Electrical Manufacturers Association), 155 East 44th Street, New York, N.Y. 10017. TR 98-1.05.*

---

*Although this publication was compiled by NEMA in 1964, it is, nevertheless, the most current publication and can be used as such.

TR 98-1.05 Method for Approximating Ambient Temperatures and Their Influence on Loading

The ambient temperature is an important factor in determining the load capability of a transformer since the temperature rise for any load must be added to the ambient to determine operating temperature. Wherever the actual ambient temperature can be measured, such ambient should be used in determination of winding hottest-spot temperature and the load capability of the transformer.

A. Approximating Ambient Temperature. It is often necessary to predict the load which a transformer can safely carry at some future time when the ambient temperature is unknown. The probable ambient temperature for any month may be approximated as follows from reports prepared by the Weather Bureau of the U.S. Department of Commerce which are available for various sections of the country:

1. Average temperature—use average daily temperature for the month involved averaged over a number of years.
2. Average of maximum daily temperatures—use average of the maximum daily temperatures for month involved, averaged over several years.

These ambients should be used as follows:

3. For loads with normal life expectancy use item 1 as the ambient for the month involved.
4. For short-time loads with moderate sacrifice of life expectancy use item 2 for the month involved.

During any one day the average or maximum temperatures may exceed the values derived from item 1 or 2. To be conservative, it is recommended that these temperatures be increased by 5 C since aging at higher than average temperatures is not offset by decreased aging at lower than average temperature. With this margin the approximated temperature will not be exceeded on more than a few days per month and, where it is exceeded, the additional loss of life will not be serious.

B. Influence of Ambient on Loading for Normal Life Expectancy. Average ambient temperatures should cover periods of time not exceeding 24 hours with maximum temperatures not more than 10 C greater than average temperatures for air and 5 C for water. Table I gives the increase or decrease in rated loads for other than average daily ambients of 30 C for air and 25 C for water. It is recommended that the 5 C margin described in par. A be used when applying the factors from the table. It should be pointed out that the increase or decrease obtained from Table I is more conservative than the corresponding values in Table III which are calculated from Table V, Assumed Transformer Characteristics, and therefore do not check these tabulations exactly. Table I is used for quick approximations.

Loading on the basis of ambient temperature with loads permitted by Table I will give approximately the same life expectancy as if transformers were operated at nameplate rating and standard ambient temperatures over the same period.

Table I covers a range in ambients of 0 C to 50 C for cooling air and up to 35 C for cooling water. A check should be made with the manufacturer before loading on the basis of cooling air ambient less than 0 C or before loading transformers in ambients greater than the limits given for Table I.

Since ambient temperature is an important factor in determining the load capability of a transformer, it should be controlled for indoor installations by adequate ventilation and should always be considered in outdoor installations.

Authorized Engineering Information 1-21-1964.

Since the evaluation of the cumulative effects of temperature and time in causing deterioration of transformer insulation is not thoroughly established, it is not possible to predict definitely the length of life of a transformer, even under constant or closely controlled conditions, much less under widely varying service conditions. However, experience and tests indicate that the rate of deterioration of transformer insulation approximately doubles

| | Percent of Rated Kva | |
Type of Cooling	Decrease Load for Each Degree C Higher Temperature	Increase Load for Each Degree C Lower Temperature
Self-cooled --OA	1.5	1.0
Water-cooled --OW	1.5	1.0
Forced-air-cooled --OA/FA, OA/FA/FA	1.0	0.75
Forced-oil-cooled --FOA, FOW or OA/FOA/FOA	1.0	0.75

TABLE I. LOADING ON BASIS OF AMBIENT TEMPERATURE

for each 8 to 10 degrees (Celsius) increase in temperature. Years of experience have shown that a transformer rated in accordance with the standards established by NEMA and ANSI will generally have a reasonably long life under usual service conditions.

Loading of Transformers

The loading of transformers is an important operating problem, as most units do not have a steady load but operate on load cycles which may have a decided peak at certain hours of the day.

The rated output of a transformer is fixed by its nameplate temperature-rise measured under specified test conditions. The output which it can deliver in service without causing undue deterioration of the insulation may be more or less than the rated output, depending upon the actual operating conditions.

To guide the operator in safely loading oil-immersed power transformers, a set of guides has been published by NEMA (National Electrical Manufacturers Association). The following excerpts are taken from NEMA, Standards Publication No. TR 98-1964 "Guide For Loading Oil-immersed Power Transformers With 65°C Average Temperature Rise." The complete publication is available from National Electrical Manufacturers Association, 155 East 44th Street, New York, N.Y. 10017. This guide covers general recommendations for loading all types of liquid-immersed power transformers specified to have an average winding rise and hottest-spot winding rise of not more than 65°C and 80°C, respectively, at rated load.

Part 1
GENERAL

TR 98-1.01 Limitations

A. It must be recognized that there may be limitations in the loads above ratings which a transformer may carry other than the capacity of the windings and cooling system. Among these limitations are: oil expansion, pressure in sealed units, bushings, leads, tap changers, and the thermal capability of associated equipment such as cables, reactors, circuit breakers, disconnecting switches and current transformers. Any of these items may limit the loading to less than the values suggested by this guide.

B. Transformers are sometimes installed in subsurface manholes and vaults of minimum size with natural ventilation through roof gratings. This type of installation results in a higher ambient temperature than the outdoor air. The amount of increase depends on the design of the manholes and vaults, net opening area of the roof gratings and the adjacent subsurface structures. Therefore, the increase in effective ambient temperature for expected transformer losses must be determined before loading limitations can be estimated.

C. Modern transformers with 65 C rise are generally designed to permit loading in line with this guide, but if there is any question as to the capability of the transformer, either old or new, to carry the desired load, the manufacturer should be asked for specific recommendations.

Authorized Engineering Information 1-21-1964.

TR 98-1.02 Transformer Life Expectancy

A. Recommendations in this guide are based on life expectancy of transformer insulation as affected by operating temperature and time.

B. Transformer life expectancy at various operating temperatures is not accurately known, but the information given regarding loss of life of insulation at elevated temperatures is considered to be conservative and the best that can be produced from present knowledge of the subject. The effects of temperature on insulation life are being investigated continuously, and new data may affect future revision of this guide. "Conservative" in the foregoing statement is used in the sense that the expected loss of insulation life for a single recommended overload will not be greater than the amount stated.

Authorized Engineering Information 1-21-1964.

TR 98-1.03 Transformer Rated Output

A. The terms "rated output" or "rated load" used in this guide refer to nameplate rating.

B. The temperature rise on which the rating of a transformer is based takes into consideration the experience of the industry regarding:

1. Insulation life as affected by operating temperature.
2. The ambient temperature assumed

to exist throughout the life of the transformer.

C. The actual output which a transformer can deliver at any time in service without undue deterioration of the insulation may be more or less than the nameplate kva rating, depending upon the ambient temperature and other attendant operating conditions.

Authorized Engineering Information 1-21-1964.

TR 98-1.04 Aging of Insulation

A. Aging or deterioration of insulation is a function of time and temperature. Since, in most apparatus, the temperature distribution is not uniform, that part which is operating at the highest temperature will ordinarily undergo the greatest deterioration. Therefore, in aging studies it is usual to consider the aging effects produced by the highest temperature.

B. Practically all of the data in reference to the aging of insulation at different temperatures has been obtained in laboratory and model tests in which the decrease in mechanical and electrical strength has been measured. The relation between life expectancy of insulation, as indicated by such tests, and actual life of a transformer is largely theoretical, so loading guides based on such information must be tempered by sound judgment based on experience of the industry.

C. Because the accumulated effects of temperature and time in causing deterioration of transformer insulation are not thoroughly established, it is not possible to predict with any great degree of accuracy the length of life of a transformer even under constant or closely controlled conditions, much less under widely varying service conditions.

D. The relation of insulation deterioration to changes in time and temperature are assumed to follow an adaptation of the arrhenius reaction rate theory which states that the logarithm of insulation life is a function of the reciprocal of absolute temperature:

Log life (hours) $= A + B/T$

where T is absolute temperature in degrees Kelvin (K) and A and B are constants.

E. The many variables mentioned and, particularly, the many varying conditions of load and ambient to which a transformer can be subjected in service make it impossible to give definite rules for the loading of transformers. It is possible to give only suggested loadings under specified conditions, and look to the user to make the best use of this information for his particular problem.

Authorized Engineering Information 1-21-1964.

BASIC LOADING FOR NORMAL LIFE EXPECTANCY

TR 98-2.01   Basic Conditions

A. The basic loading of a transformer for normal life expectancy is continuous loading at rated output when operated under normal service conditions as indicated in Sections 2.1.1(1), 2.1.1(2), and 2.1.1(3) of the "USA Standard General Requirements for Distribution, Power, and Regulating Transformers, and Shunt Reactors*." Section 2 General, C57.12.00. It is assumed that operation under these conditions is equivalent to operation at a continuous ambient temperature of 30 C for cooling air and 25 C for cooling water. The 5 C lower

---

*Copies are available from the American National Standards Institute, 1430 Broadway, New York, N.Y. 10018.

ambient for cooling water is to allow for possible reduction in efficiency of cooling due to deposits on cooling coil surfaces. Normal life expectancy will result from operating continuously with hottest-spot temperature of 110 C (equivalent with 120 C maximum) in any 24-hour period.

B. The hottest-spot winding temperature is the principal factor in determining life due to loading. This temperature cannot be directly measured on commercial designs because of the hazard in placing any temperature detector at the proper location because of voltage. The standard allowances have therefore been obtained from tests made in the laboratory.

C. The hottest-spot temperature at rated load is the sum of the average winding temperature and a 15 C allowance for hottest spot. For liquid-immersed transformers operating continuously under the foregoing conditions with normal life expectancy, this temperature has been assumed to be 110 C.

Authorized Engineering Information 1-21-1964.

TR 98-2.02   Loading for Normal Life Expectancy Under Specified Conditions

A. Loading by Hottest-spot Temperature Indicator. Hottest-spot temperature devices are supplied, when specified, which indicate or record the hottest-spot temperature. These devices may be used as a guide to limit loads.

Thermal relays, when supplied, indicate when predetermined time-temperature limits have been reached in the windings. These relays are calibrated for use with specific transformers and automatically take into account the hottest-spot temperature of the windings, the ambient temperature, and the pre-

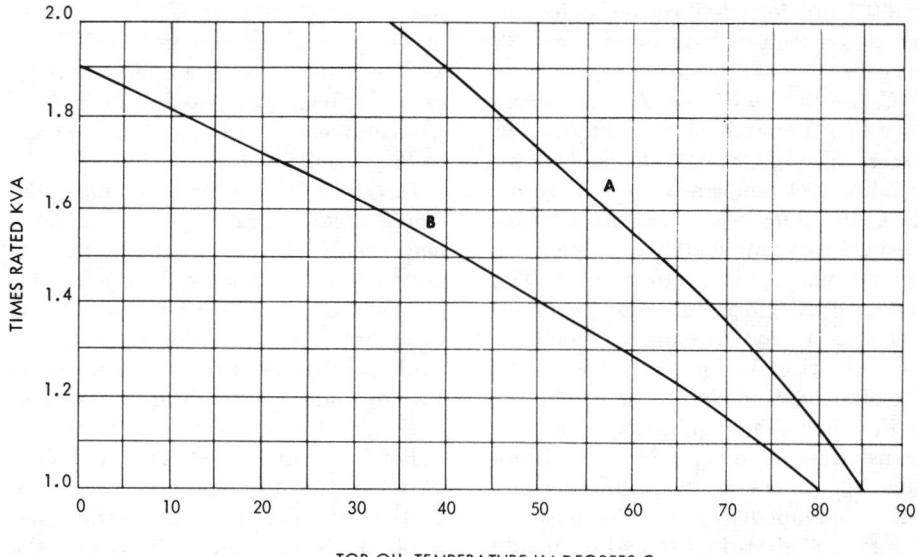

TOP OIL TEMPERATURE IN DEGREES C

Fig. 3-15. Approximate loading for normal life expectancy based on top oil temperature. A shows water-cooled, self-cooled and forced-air-cooled transformers rated 133 percent or less of the self-cooled rating. B shows forced-oil-cooled transformers or forced-air-cooled transformers rated over 133 percent of the self-cooled rating.

vious condition of loading. Higher loads are permitted for short periods than for long periods of time. When used as a guide for loading, the breaker trip or alarm indicates the maximum capability of the transformer.

B. Loading by Oil Temperature. Oil temperature alone should not be used as a guide for loading transformers. The hottest-spot winding rise over top oil temperature at full load should be determined from the factory tests and Equation 6-7* or, if unknown, a value should be assumed. The hottest-spot allowance should be corrected for the actual load carried by Equation 6-6 or Fig. 6-7.* This hottest-spot rise over top oil,

subtracted from 110 C will give the maximum permissible oil temperature for normal life expectancy. It should be recognized that, due to thermal lag in oil rise, time is required for a transformer to reach a stable temperature for any change in load. Therefore, higher peak loads can be carried for a short duration than for longer periods of time. This is reflected in the loading tables and the basic equations in par. C of TR 98-6.07.

If the transformer characteristics are not accurately known, the maximum oil temperature derived from Fig. 3-15 may be used as an approximate guide. Loading from this figure is based on a difference between the hottest-spot temperature and the oil temperature of 25 C for water-cooled, self-cooled, and forced-air-cooled transformers rated 133 percent or less of the self-cooled rating, and

---

*Refers to NEMA numbering. Not shown in this book.

173

of 30 C for forced-oil-cooled or forced-air-cooled transformers rated over 133 percent of the self-cooled rating.

C. Loading on Basis of Test Temperature Rise. For each degree centigrade in excess of 5 degrees that the average winding test temperature rise is below 65 C the transformer load may be increased above rated kilovolt-amperes by the percentages given in Table I. The 5-degree margin is taken to provide a tolerance in the measurement of temperature rise. When corrected by the factor given in the table, the new rating is the kilovolt-ampere rating which the transformer can carry at 65 C rise. Since this may increase the rating beyond that contemplated by the designer, the limitations given in TR 98-1.01 should be checked before taking full advantage of the increased rating due to low temperature rise.

Some transformers are designed to have the difference between the hottest-spot and average copper temperatures greater than the allowance of 15 C. This will result in an average winding temperature rise of less than 65 C, but the hottest-spot winding temperature rise may be at the limiting value of 80 C. Such transformers should not be loaded above their rating as outlined under this heading. The manufacturer should be consulted for information on the hottest-spot allowances used for these designs. This condition may be found in some self-cooled or water-cooled transformers where there are large differences in top and bottom oil temperatures and in some designs of forced-air-cooled and forced-oil-cooled transformers. Wherever possible, data on hottest-spot and oil temperatures obtained from factory temperature tests should be used when re-rating a transformer because of low test temperature rise, or when calculating temperatures for loads above rating by the formulae given in TR 98-6.07. Where auxiliary cooling is used (pumps or fans or both), the test data taken with this equipment operating should be used in the formulae.

D. Operation With Part or All of the Cooling Out of Service. Where auxiliary equipment, such as pumps or fans or both, is used to increase the cooling efficiency, the transformer may be required to operate for some time without this equipment functioning. The permissible loading under the conditions as specified is given in the following paragraphs.

For forced-air-cooled (OA/FA or OA/FA/FA) transformers with fans inoperative, use the self-cooled rating and apply loads on the same basis as if the transformer were self cooled. For forced-air, forced-oil-cooled transformers triple rated (OA/FA/FOA or OA/FOA/FOA) with all or part of the forced cooling inoperative, use the nameplate rating based on the cooling in operation and load on this basis.

For forced-oil-cooled (FOA or FOW) transformers with all pumps or fans or both inoperative, the following operating conditions are assumed to occur infrequently and without undue injury to the transformer:*

1. Rated load may be maintained for approximately 1 hour following normal operation at nameplate rating at 30 C ambient.

2. Rated load may be carried for approximately 2 hours if started with the windings and oil at 30 C ambient.

---

*The oil temperature for large units may exceed 110 C. Check with the manufacturer for limitations.

3. Rated voltage may be maintained for 6 hours at no load, following continuous operation at nameplate rating at 30 C ambient with cooling equipment in operation.
4. Rated voltage may be maintained for 12 hours at no load starting with the windings and oil at 30 C ambient.

For forced-oil-cooled transformer (FOA or FOW) ratings with part of the coolers in operation, use Table II.

TABLE II.

Percent of Total Coolers in Operation	Permissible Load in Percent of Nameplate Rating
100	100
80	90
60	78
50	70
40	60
33	50

The permissible loads in the table give approximately the same temperature rise as for full load with all cooling in operation.

E. Loading on Basis of Short-time Loads Above Rating. Transformers may be operated above 110 C average continuous temperature for short periods provided they are operated for much longer periods at temperatures below 110 C. This is due to the fact that thermal aging is a cumulative process. This permits loads above the rating to be safely carried under specified conditions without encroaching upon the normal life expectancy of the transformer. Suggested loadings for normal life expectancy are given in Table III. For conservative use of this table, it is suggested that the 5 C margin described in par. A

TABLE III. DAILY PEAK LOADS IN PER UNIT OF NAMEPLATE RATING TO GIVE NORMAL LIFE EXPECTANCY

Peak Load Time, Hours	Hottest Spot Temperature Reached, Degrees C	Continuous Equivalent Load in Percent of Rated Kva Preceding Peak Load																
		50 Percent Ambient, Degrees C						70 Percent Ambient, Degrees C						90 Percent Ambient, Degrees C					
		0	10	20	30	40	50	0	10	20	30	40	50	0	10	20	30	40	50
A. COOLING--SELF-COOLED OR WATER-COOLED (OA OR OW)*																			
1/2	149	2.00	2.00	2.00	1.97	1.84	1.70	2.00	2.00	2.00	1.87	1.73	1.58	2.00	2.00	1.87	1.73	1.59	1.43
1	142	2.00	1.95	1.84	1.72	1.60	1.47	1.99	1.88	1.76	1.64	1.51	1.38	1.90	1.78	1.66	1.54	1.40	1.26
2	134	1.76	1.67	1.56	1.46	1.35	1.23	1.72	1.62	1.52	1.41	1.29	1.17	1.66	1.56	1.45	1.34	1.22	1.10
4	127	1.52	1.44	1.34	1.25	1.15	1.05	1.50	1.41	1.32	1.23	1.13	1.02	1.48	1.39	1.29	1.20	1.09	0.99
8	120	1.37	1.29	1.20	1.11	1.02	0.92	1.36	1.28	1.20	1.11	1.01	0.91	1.36	1.27	1.19	1.10	1.00	0.90
24	110	1.26	1.18	1.09	1.00	0.90	0.80	1.26	1.18	1.09	1.00	0.90	0.80	1.26	1.18	1.09	1.00	0.90	0.80
B. COOLING--FORCED-AIR-COOLED RATED 133 PERCENT OR LESS OF SELF-COOLED RATING (OA/FA)†																			
1/2	149	2.00	2.00	1.89	1.78	1.67	1.55	2.00	1.93	1.82	1.70	1.58	1.46	1.94	1.83	1.72	1.60	1.48	1.35
1	142	1.85	1.76	1.66	1.56	1.46	1.35	1.80	1.71	1.61	1.50	1.40	1.28	1.74	1.64	1.54	1.43	1.32	1.20
2	134	1.61	1.52	1.43	1.34	1.25	1.15	1.58	1.49	1.40	1.31	1.21	1.11	1.54	1.46	1.37	1.27	1.17	1.07
4	127	1.43	1.35	1.27	1.19	1.10	1.01	1.42	1.34	1.26	1.18	1.09	0.99	1.41	1.33	1.25	1.16	1.07	0.98
8	120	1.33	1.25	1.17	1.09	1.00	0.91	1.33	1.25	1.17	1.09	1.00	0.91	1.32	1.25	1.17	1.09	1.00	0.91
24	110	1.25	1.17	1.09	1.00	0.91	0.81	1.25	1.17	1.09	1.00	0.91	0.81	1.25	1.17	1.09	1.00	0.91	0.81
C. COOLING--FORCED-AIR-COOLED RATED OVER 133 PERCENT OF SELF-COOLED RATING (OA/FA/FA) AND FORCED-OIL, FORCED-AIR OR FORCED-OIL, FORCED-WATER (FOA, FOW OR OA/FOA/FOA)*																			
1/2	149	1.69	1.63	1.56	1.49	1.42	1.34	1.65	1.58	1.52	1.45	1.37	1.29	1.60	1.53	1.46	1.39	1.31	1.23
1	142	1.53	1.47	1.41	1.34	1.27	1.20	1.50	1.44	1.38	1.31	1.24	1.17	1.47	1.41	1.34	1.27	1.20	1.13
2	134	1.38	1.32	1.27	1.21	1.14	1.07	1.37	1.31	1.25	1.19	1.13	1.06	1.36	1.30	1.24	1.18	1.11	1.04
4	127	1.29	1.24	1.18	1.12	1.06	0.99	1.29	1.23	1.18	1.12	1.05	0.98	1.29	1.23	1.17	1.11	1.05	0.98
8	120	1.24	1.19	1.13	1.07	1.00	0.93	1.24	1.19	1.13	1.07	1.00	0.93	1.24	1.19	1.13	1.07	1.00	0.93
24	110	1.19	1.13	1.07	1.00	0.93	0.85	1.19	1.13	1.07	1.00	0.93	0.85	1.19	1.13	1.07	1.00	0.93	0.85

* Subtract 5 C from each of the ambient column heading for water-cooled transformers. Minimum water temperature must be above zero C.

† The peak loads in this section of the table are calculated on the basis of all cooling in use during the period preceding the peak load. When operating without fans, use Section A of this table.

of TR 98-1.05 be used in determining the ambient.

F. Application of Tables for Normal Life Expectancy (For Assumed Characteristics See Table V). Assume a load cycle which resolves to a constant value of 50 percent followed by a 100 percent peak load for 2 hours (see method used in TR 98-6.06). Using Section A of Table III, if the ambient is 30 C, a self-cooled (OA) or a water-cooled (OW) transformer will carry 1.41 times nameplate rating for 2 hours following an equivalent continuous load up to 70 percent of nameplate rating. If the equivalent 2-hour peak load from the load cycle is 10,600 kva, the transformer which will carry this load on a daily basis without loss of life expectancy is: 10,600/1.41 = 7,500 kva. The constant equivalent load before the peak is 5,300 kva which is 70 percent of the nameplate rating of the transformer. Therefore, a 7,500-kva transformer is suitable for this daily load cycle.

Assume a 10,000-kva self-cooled (OA) transformer is installed in a substation. Will it carry a winter peak (10 C ambient) load at 15,000 kva for 2 hours following an equivalent continuous load of 8,000 kva before the peak? From Section A of Table III (a self-cooled transformer loaded to 90 percent of nameplate rating before a peak will carry 1.56 times nameplate rating for 2 hours with a 10 C ambient. The transformer required to carry this load is 15,000/1.56 = 9,620 kva and the previous equivalent load could be 8,660 kva. Therefore, a 10,000-kva transformer will carry this daily load cycle at this ambient without sacrifice of life expectancy.

Assume a 10,000-kva self-cooled (OA) transformer is installed in a substation. What peak load will it carry for

2 hours if the ratio of the peak load to the equivalent continuous load before the peak is 1.85 and the ambient is 0 C? From Section A of Table III, make an estimate of the peak kva. Assume a peak load of 15,000 kva; then the initial load in percent of nameplate is: (15,000/1.85)/10,000 = 0.81 or 81 percent. From Section A of Table III for a 2-hour peak at 0 C ambient, the per unit peak load is 1.72 and 1.66 for previous continuous loads of 70 and 50 percent, respectively, of nameplate rating. Interpolating for 81 percent, the per unit peak is 1.69, hence a peak of 16,900 kva. The assumed peak was therefore too low. Try 16,500 kva; the initial load in percent of nameplate is: (16,500/1.85)/10,000 = 0.89 or 89 percent. Interpolating from the table for 89 percent, the per unit peak load is 1.663 or 16,630 kva. This checks the assumption. Therefore, the transformer will carry up to 16,500 kva for 2 hours under these conditions, following an equivalent continuous load of 8,900 kva.

Authorized Engineering Information 1-21-1964.

SHORT-TIME LOADING WITH MODERATE SACRIFICE OF LIFE EXPECTANCY

TR 98-3.01 Aging of Insulation Due to Operation Above 110 C Average Hottest-Spot Temperature

A. When the aging effect of one overload cycle or the cumulative aging effect of a number of overload cycles is greater than the aging effect of continuous operation at rated load over a given period of time, the insulation deteriorates at a faster rate than normal. The rate of deterioration is a function of time and temperature and is commonly expressed

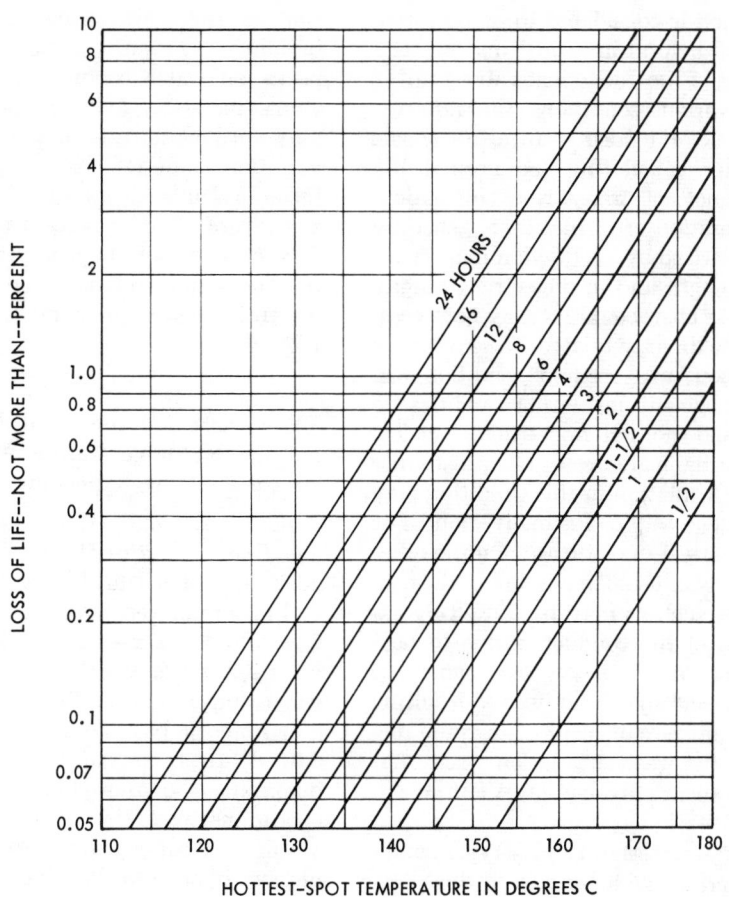

HOTTEST-SPOT TEMPERATURE IN DEGREES C

Time, Hours	Temperature in Degrees C to Use Not More Than the Following Life in Percent					
	0.1	0.25	0.50	1.00	2.00	4.00
1/2	160	171	180	...	...	...
1	152	163	171	180	...	...
2	145	155	163	171	180	...
4	137	147	155	163	171	180
8	130	139	147	155	163	171
24	119	128	135	142	150	158

For explanation of limitations and accuracy, see par. B and C of TR 98-3.01.

Fig. 3-16. Hottest-spot conductor temperature versus loss of life.

as a percentage loss of life. Charts and tables showing relative loss of life for various combinations of time and tem- perature are given in Fig. 3-16.

B. It should be clearly understood that, while the insulating aging rate in-

177

formation is considered to be conservative and helpful in estimating the relative loss of life due to loads above rating under various conditions, this information is not intended to furnish the sole basis for calculating the normal life expectancy of transformer insulation. Deterioration of insulation is generally characterized by a reduction in mechanical strength and in dielectric strength, but these characteristics may not necessarily be directly related. In some cases, insulation in a charred condition will have sufficient insulating qualities to withstand normal operating electrical and mechanical stresses. A transformer having insulation in this condition may continue in service for many months or even years, if undisturbed. On the other hand, any unusual movement of the conductors, such as may be caused by expansion of the conductors due to heat resulting from a heavy overload or to large electromagnetic forces resulting from short circuit, may disturb the mechanically weak insulation such that turn-to-turn or layer-to-layer failure will result.

C. The uncertainty of service conditions and the wide range in ratings covered are reasons why this loading guide is conservative in its suggested loading schedule. Some of the variables are: wide differences in ambient temperature between localities; differences in eleva-

tion; restricted air circulation caused by buildings, fire walls, etc.; mineral deposits on water cooling surfaces; previous emergency loading history which may not be known to the operator; and variations in design characteristics from those in Table V. As a guide, an average loss of life of 1 percent per year or 5 percent in any one emergency operation is considered reasonable.

Authorized Engineering Information 1-21-1964.

TR 98-3.02 Methods for Determining Loads Above Rating With Some Sacrifice of Life Expectancy

A. Transformers With Known Specific Characteristics. If the specific characteristics of a transformer are known and maximum recommended capability is required, the user may calculate the oil rise and hottest-spot temperature versus time data from the charts in Part 6 or from the basic formulae in TR 98-6.07. With these data, the user may determine the allowable load for his conditions by taking into account the ambient temperature and probable number of such loads during the life of the transformer, and the approximate percentage of life he is willing to sacrifice.

B. Transformers with Unknown Specific Characteristics or Conservative

	Following Initial Load of:			
	50%	70%	90%	100%
Type of Cooling	See Table	See Table	See Table	See Table
Self-cooled	3-3	3-4	3-5	3-6
Water-cooled	3-3	3-4	3-5	3-6
Forced-air-cooled*--OA/FA	3-7	3-8	3-9	3-10
Forced-air-cooled†--OA/FA/FA	3-11	3-12	3-13	3-14
Forced-oil-cooled--FOA or FOW or OA/FOA/FOA . . .	3-11	3-12	3-13	3-14

TABLE IV. TABLES OF LOADS WITH MODERATE SACRIFICE OF LIFE EXPECTANCY

* Forced-air-cooled rating is 133 percent or less of the self-cooled rating.
† Forced-air-cooled rating is over 133 percent of the self-cooled rating.

TABLE V. ASSUMED TRANSFORMER CHARACTERISTICS			
	OA/FA or OW (Self-cooled or Water-cooled)	OA/FA (Forced-air-cooled)*	FOA, FOW, OA/FOA/FOA or OA/FA/FA (Forced-oil with Forced Air Water Cooling or Forced-air-cooled) †
Hottest-spot rise, degrees C	80	80	80
Top-oil rise, degrees C	55	50	45
Time constant at full load, hours	3.0	2.0	1.25
Ratio of copper to iron losses	3.2 to 1	4.5 to 1	6.5 to 1
Ambient temperature, degrees C	30	30	30
"n" ‡	0.8	0.8	1.0

* Forced-air-cooled rating is 133 percent or less of the self-cooled rating.
† Forced-air-cooled rating is over 133 percent of the self-cooled rating.
‡ "n" is the exponential power of temperature rise versus loss.

Loads. Where specific characteristics of the transformer are not known or where conservatively recommended loads above the normal rating are satisfactory, refer to the capability tables listed in Table IV for tabulations of loads following 50, 70, 90, and 100 percent continuous loads with loss of life for each load cycle with peak durations from ½ hour through 24 hours.

1. Assumed Characteristics for Transformers Which Allow the Loading in Tables 3-3 Through 3-14.* Transformers vary widely in the characteristics which affect their short-time over-load capabilities. Any general guide which applies to all transformers of a given class must therefore be based on characteristics which give conservative overloads. The data given in this section for general use is calculated for transformers with the characteristics given in Table V.

2. Temperature and Load Limitations. The following temperature and load limitations are used in Table III:

---

*Refers to NEMA numbering. Tables 3-3 through 3-14 not shown in this book.

Maximum oil temperature
(unprotected) ....100 C
Maximum oil temperature
(protected) ....110 C
Maximum hot-spot temperature ....180 C
Maximum short-time loading
(one-half hour or
more) ....200 percent

LOADING OF TRANSFORMERS HAVING A NOMINAL RATING
TR 98-4.01 Determination of Loading

Nominal rating is defined in Section 12-81.014 of the "USA Standard Requirements, Terminology, and Test Code for Distribution, Power, and Regulating Transformers, and Reactors Other Than Current-Limiting Reactors," Section 12-80, Terminology, C57.12.80. The kva load which transformers so rated can carry continuously with a temperature rise of 65 C may be determined approximately by multiplying the nominal rated kva by the following conversion factors. The load so determined may be used as the basis for loading under desired conditions as covered by this guide. (See Table VI.)

TABLE VI	
Percent of Nominal Rated Kva for 2 Hours*	Conversion Factor
125	1.1
150	1.2

*Following 100 percent of nominal rated kva.

Authorized Engineering Information 1-21-1964.

SUPPLEMENTAL COOLING OF EXISTING SELF-COOLED TRANSFORMERS

TR 98-5.01 Increase in Loading

The load which can be carried on existing self-cooled transformers can usually be increased by adding auxiliary cooling equipment such as fans, external forced-oil coolers, or water spray equipment. The amount of additional loading varies widely, depending upon:
1. Design characteristics of the transformer.
2. Type of cooling equipment.
3. Permissible increase in voltage regulation.
4. Limitations in associated equipment.

No general rules can be given for such supplemental cooling, and each transformer should be considered individually.

Authorized Engineering Information 1-21-1964.

TR 98-6.07 Equations for Calculating Transient Heating of Oil-Immersed Transformers

A. General. The loading tables in this guide are based on average characteristics of a wide range of transformer ratings. Table V gives the characteristics used in calculating the loadings in the tables. When the characteristics of a particular transformer vary appreciably from those in the table and where more accurate loading capabilities are desired, the following basic equations may be used.

B. List of Symbols.

θ_a = ambient temperature (degrees centigrade)

θ_g = hottest-spot rise over top-oil temperature (degrees centigrade)

$\theta_g(fl)$ = hottest-spot rise over top-oil temperature at full load

θ_{hs} = hottest-spot temperature (degrees centigrade)

θ_o = top-oil rise over ambient temperature (degrees centigrade)

θ_{fl} = full-load top-oil rise (degrees centigrade)

θ_i = initial oil rise for $t = 0$ (degrees centigrade)

θ_u = ultimate oil rise for load L (degrees centigrade)

K = ratio of load L to rated load

L = load under consideration

R = ratio of load loss at rated load to no-load loss

e = 2.7128

t = duration of load, in hours

τ = thermal time constant of transformer for any load L and for any specific temperature differential between the ultimate oil rise and the initial oil rise, or $\tau = C\,(\theta_u - \theta_i)/P$

P = change in total loss due to change in load

τ_r = time constant for full load beginning with initial temperature rise of zero C (hours)

P_{fl} = total loss at full load (watts)

C = thermal capacity of transformer (watthours per degree centigrade)

C. Temperature Determination Equations (See par. C of TR 98-2.02).
Hottest-spot temperature

$$\theta_{hs} = \theta_a + \theta_o + \theta_g$$

(Equation 6-3)

Transient heating and cooling equation for top-oil rise over ambient temperature

$$\theta_o = (\theta_u - \theta_i)(1 - e - t/r) + \theta_i$$

(Equation 6-4)

Ultimate top-oil rise for load L

$$\theta_u = \theta_{fl}[(K^2R + 1)/(R + 1)]^n$$

(Equation 6-5)

Hottest-spot rise over top oil

$$\theta_g = \theta_{g(fl)}K^{2n}$$

(Equation 6-6)

$\theta_{g(fl)}$ = average conductor rise over top oil* + 15 C

(Equation 6-7)

$n = 0.8$ for OA, OW, and OA/FA transformers
$n = 1.0$ for FOA, FOW, OA/FOA, and OA/FA/FA transformers

Time-constant at rated kilovolt-amperes

$$\tau_r = \frac{C\theta_{fl}}{P_{fl}}$$

(Equation 6-8)

$C = 0.06$ (weight of core and coils) + 0.04 (weight of tank) + 1.33 (gallons of oil)

(Equation 6-9)

For FOA or FOW, use 0.06 (weight of tank) and 1.93 (gallons of oil) if $n = 1.0$ is used in Equation 6-6 and 6-7.

D. Equation Corrections. Theoretically, several corrections should be made when using the foregoing equations in

*At rated kilovolt-amperes from manufacturer's test.

calculating transient-oil rises, such as corrections for change in:

1. Time constant for loads above rating.
2. Ultimate conductor loss at end of load period.
3. Oil viscosity.

In making general calculations based on assumptions of transformer characteristics and maximum hot-spot temperatures which generally have a large factor of safety, results close enough for all practical purposes are obtained if all of these corrections are omitted, and the simpler formulae are used.

The overload values (not shown in this book) were calculated by the equations without corrections.

E. Time Constant. The time constant is the length of time which would be required for the temperature of the oil to change from the initial value to the ultimate value if the initial rate of change were continued until the ultimate temperature was reached.

The time constant may also be expressed as the length of time required for a specified percentage of the change in temperature to take place from initial value to ultimate value.

If n (the exponential power of temperature rise versus loss) equals unity, 63 percent of the temperature change occurs in a length of time equal to the time constant, regardless of the relationship of initial temperature and ultimate temperature. If n is not unity, the percentage varies and is a function of both initial temperature rise and ultimate temperature rise. In particular, if n equals 0.8, the percentage is 67 if the initial temperature rise is zero.

If the initial temperature rise is greater than zero, the percentage is lower than 67 and decreases as the initial temperature rise increases for a given ultimate temperature rise. If the initial temperature is approximately equal to the final temperature, whether just above or just below it, the percentage is approximately 63. If the initial temperature is greater than the ultimate temperature, the percentage is less than 63.

Since evaluation of the exact percentage for cases where n is not unity and the initial temperature rise is not zero becomes very laborious, it is frequently advisable to use the value of 63 percent as an approximation. In the more frequently encountered cases where n is approximately 0.8, the error resulting from this procedure is not large compared to the expected error in transient thermal calculations.

If $n = 1.0$, Equation 6-8 is correct for any load and any starting temperature. If n is less than 1.0, Equation 6-8 holds only for full-rating starting cold. If $n = 0.8$, the time constant for any load and for any starting temperature for either a heating cycle or cooling cycle is given in Equations 6-10 and 6-11.

$$\tau = \tau_r \; \frac{\left(\dfrac{\theta_u}{\theta_{fl}}\right) - \left(\dfrac{\theta_i}{\theta_{fl}}\right)}{\left(\dfrac{\theta_u}{\theta_{fl}}\right)^{1.25} - \left(\dfrac{\theta_i}{\theta_{fl}}\right)^{1.25}}$$

(Equation 6-10)

If starting cold (i.e.), $\theta_i = 0$, Equation 6-10 reduces to

$$\tau = \tau_r \left(\frac{\theta_{fl}}{\theta_u}\right)^{1.25}$$

(Equation 6-11)

F. *Winding Loss.* As the resistance of the winding for ultimate conditions is greater when the temperature is greater for loads above rating than for rated load conditions, a resistance correction

factor should be added to Equations 6-5 and 6-6 if this refinement is justified.

G. *Viscosity of Oil.* The ultimate temperature rise of oil for a constant loss decreases slightly as the temperature of the oil increases. This is due to a decrease in the viscosity of the oil. The viscosity correction tends to offset the effect of increased resistance.

Authorized Engineering Information 1-21-1964.

PROTECTIVE DEVICES

TR 98-7.01   Thermal Relays

A. Transformer thermal relays are devices, the operation of which indicates that predetermined time temperature limits in the transformer windings have been reached. They are calibrated for use with specific transformer apparatus, and the hottest-spot temperature of the windings, the ambient temperature, and previous conditions of loading are automatically taken into account. Higher loads are permitted for short periods of operation than for long periods of operation. (See Fig. 3-17.)

B. The relay can be adjusted to give indication at loads which will produce practically no sacrifice of life expectancy or some predetermined moderate sacrifice of such expectancy.

C. The device has one or more contacts which may be used for various functions, such as starting fans, giving a signal or alarm, or disconnecting of the transformers.

D. The application of such devices should be discussed with the manufacturers.

Authorized Engineering Information 1-21-1964.

TR 98-7.02   Overcurrent Protective Devices

For use in connection with applica-

TIME	TIMES RATED CURRENT
2 SECONDS	25.0
10 SECONDS	11.3
30 SECONDS	6.7
60 SECONDS	4.75
5 MINUTES	3.0
30 MINUTES	2.0

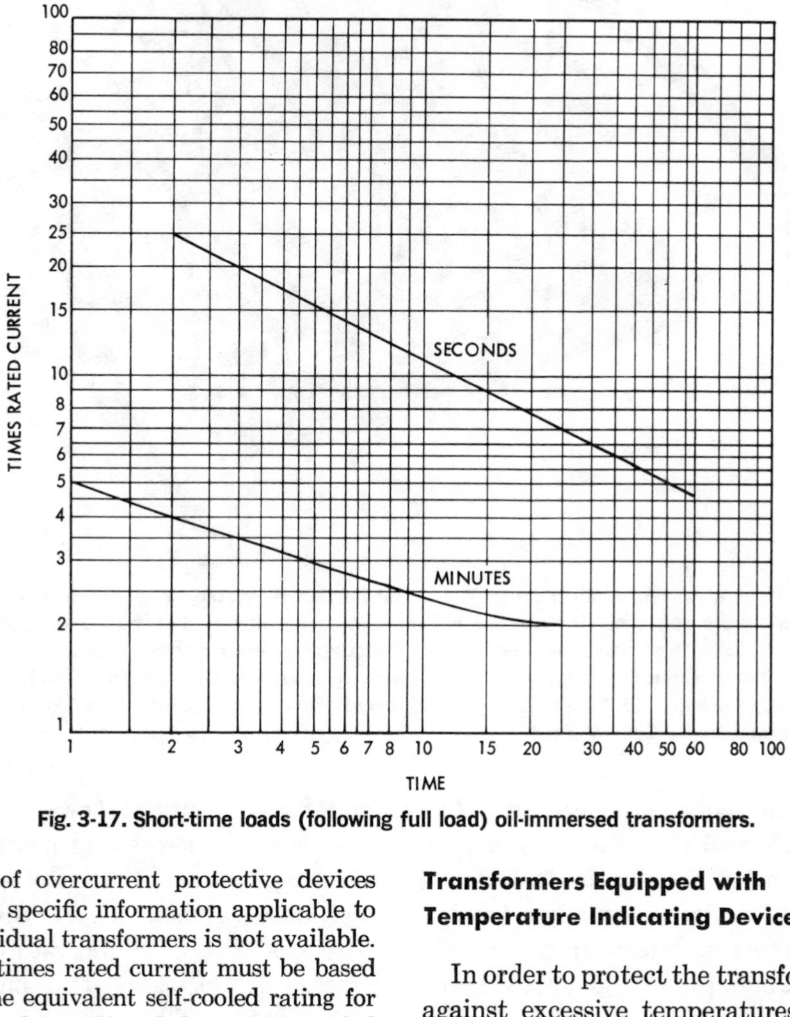

Fig. 3-17. Short-time loads (following full load) oil-immersed transformers.

tion of overcurrent protective devices when specific information applicable to individual transformers is not available. The times rated current must be based on the equivalent self-cooled rating for other than self-cooled or water-cooled transformers.

Authorized Engineering Information 1-21-1964.

Transformers Equipped with Temperature Indicating Devices

In order to protect the transformer against excessive temperatures and at the same time permit it to be loaded to its maximum safe operating temperature, many transformers

183

Fig. 3-18. Low-voltage circuit breaker. The circuit breaker is located below the oil level and is stripped by the deflection of bimetallic elements in series with the low-voltage leads on self-protected units. This breaker has been tested to successfully clear a bolted secondary fault five times without contact welding. A rigid bar gives simultaneous interruption of both breaker contacts. (See Fig. 1-39, Chapter 1, for actual installation in a transformer tank.) (Distribution Transformer Dept., General Electric Co.)

are now equipped with protective and thermal indicating devices. On the smaller distribution transformers this device is actually a breaker connected in series with the secondary windings and mounted inside the transformer case, Fig. 3-18. This breaker is actuated by a bimetal strip, comprising part of the breaker assembly, through which the secondary load current flows. As the load current increases or decreases, the temperature of this bimetal strip will likewise increase or decrease. Also, as the oil temperature increases or decreases, this bimetal temperature will be increased or decreased by the same amount, since it is mounted in the hot oil. Thus it can be seen that this bimetal can be made to simulate the actual temperature existing in the transformer winding. When the bimetal reaches a temperature corresponding to the maximum safe operating temperature of the winding, any additional temperature will cause the bimetal to release a latch and automatically trip the breaker, thus disconnecting the load from the transformer and preventing injury to the windings. Some of these breakers are provided with auxiliary latches which close a circuit in which an indicating lamp is connected. See Fig. 1-39, Chapter 1. The voltage for this auxiliary circuit is obtained from a small low-voltage winding built in the transformer for this special purpose. When the transformer reaches its safe operating temperature this auxiliary circuit latch trips and the indicating lamp, usually provided with a colored lens so that it can more readily be seen when burning, gives a visible indication of the load condition of the transformer.

If the load, and therefore the temperature, should be increased much above this safe operating temperature, as indicated by the signal light, the bimetal strips cause the main latches to release the tripping mechanism of the breaker and the load is thereby automatically disconnected. As the unit cools down, the breaker may be closed by hand and service restored. Such a device permits maximum safe loading of the transformer

at all times regardless of the value of short-time peak loads.

Power transformers are not subject to such fluctuating or large peak loads as are common on distribution transformers. For this reason they are generally equipped with devices which indicate only the temperature of the oil or the estimated hot-spot temperature of the winding. These devices are sometimes equipped with auxiliary contacts which close an alarm circuit actuated from an external source of voltage, causing a bell to ring or a light to burn when the temperature of the oil or of the winding reaches a predetermined value.

The simplest device used for indicating the oil temperature, from which the temperature of the winding may be estimated, consists of an alcohol thermometer and an operating mechanism similar to a Bourdon gage, Fig. 3-19. The alcohol thermometer is immersed in the hot oil and is connected to the operating mechanism by means of a flexible tube. An increase in temperature of the oil will heat the alcohol in the bulb of the thermometer, causing it to expand and exert a pressure which is transmitted through the flexible tube to the pointer of the operating mechanism. This pointer then moves over a scale graduated in degrees. As the temperature decreases, the pressure decreases and the pointer moves back to a lower

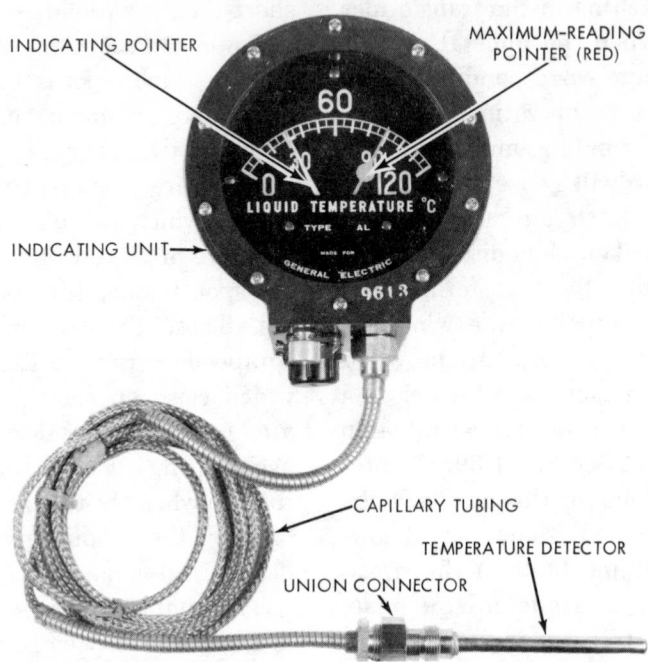

Fig. 3-19. Liquid temperature indicator. (Distribution Transformer Dept., General Electric Co.)

value on the scale. Some of these devices are equipped with two pointers, one indicating the maximum temperature attained and the other following the variations of temperature. This device may be designed to close contacts at some selected temperature (or temperatures). Those contacts in turn can be connected to start cooling fans or pumps, to sound an alarm, or to disconnect the transformer or load by operations of primary or secondary circuit breakers.

The thermometer-type hottest-spot indicator is applied chiefly to large power transformers where the money loss would be relatively large in case of damage due to excessive temperature in the windings. This device gives the operator an indication of the temperature of the winding rather than that of the oil. One type of hottest spot temperature indicator is an assembly utilizing a Bourdon gage, calibrated in degrees centigrade, connected to a bulb by a capillary tube, Fig. 3-20. The thermometer bulb and a heating coil are assembled in a well located in the hottest portion of the insulating oil near the top of the transformer. Fig.

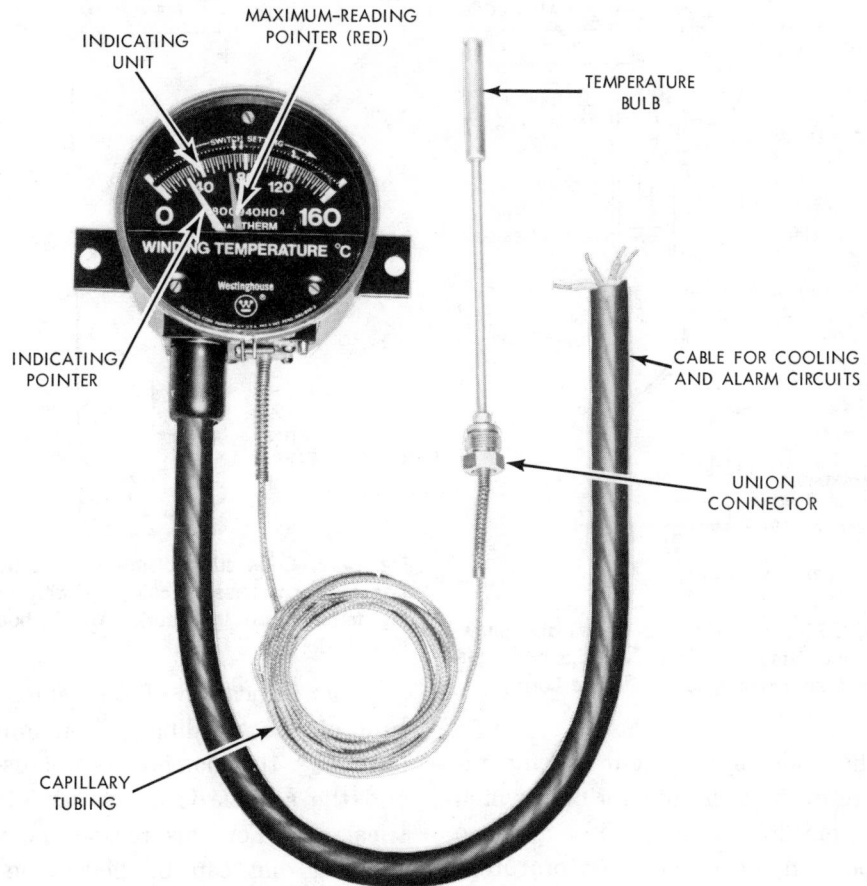

INDICATING
UNIT

MAXIMUM-READING
POINTER (RED)

TEMPERATURE
BULB

SWITCH SETTING

40 80 120

0 160

WINDING TEMPERATURE °C

Westinghouse

INDICATING
POINTER

CABLE FOR COOLING
AND ALARM CIRCUITS

UNION
CONNECTOR

CAPILLARY
TUBING

Fig. 3-20. Winding temperature indicator. (Westinghouse Electric Corp.)

3-21. In most cases the thermometer dial or indicator is mounted on the transformer case, but it can be placed a limited distance away from the transformer. The maximum distance is fixed by the practical length of the tube connecting the gage mechanism and the thermometer bulb. The current in the winding of the power transformer passes through the primary winding of a current transformer, as shown in Fig. 3-22, thus causing a current to flow in its secondary, proportional to that in the primary. This secondary current flows through a heating coil which encases the temperature indicator bulb, thus generating heat in the bulb proportional to that developed in the transformer winding.

187

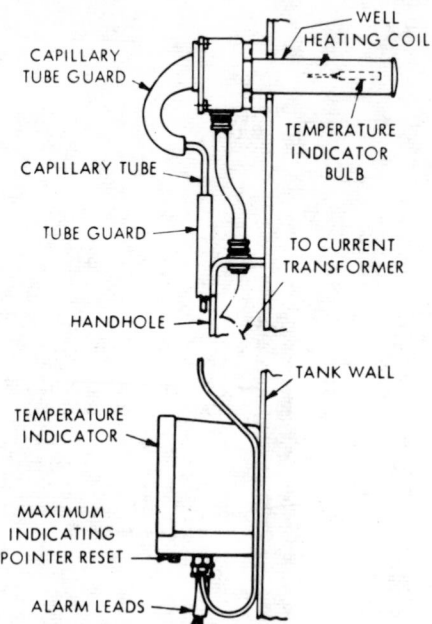

Fig. 3-21. Sectional view of transformer showing mounting of indicator with flexible capillary tube. (Westinghouse Electric Corp.)

This adds an increment of temperature to the thermometer bulb which is equal to the winding hot spot rise above the hottest oil temperature.

The temperature of the bulb is therefore at all times proportional to the temperature of the main transformer winding, therefore the instrument will indicate the hottest spot temperature of the windings.

Switches are normally provided in this instrument to operate forced-cooling control and alarm circuits.

Another type of hottest-spot indicator used only on power transformers is known as the bridge-type hottest-spot indicator. This also

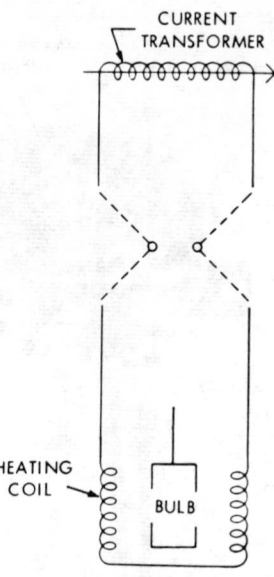

Fig. 3-22. Connection diagram for current transformer and heating coil (used with winding temperature indicator). (Westinghouse Electric Corp.)

gives an indication of the temperature of the windings of a transformer. A thermometer is not used and the scheme is purely an electrical one. For this reason the indicating dial can be placed on a switchboard at any reasonable distance from the transformer.

The operation of this device is as follows: A current transformer is energized from the current in the main transformer winding. A heat-insulated coil is included in the secondary circuit of this current transformer. The amount of heat generated in this coil is proportional to the heat generated in the main transformer windings. A noninductive resistance embedded in the

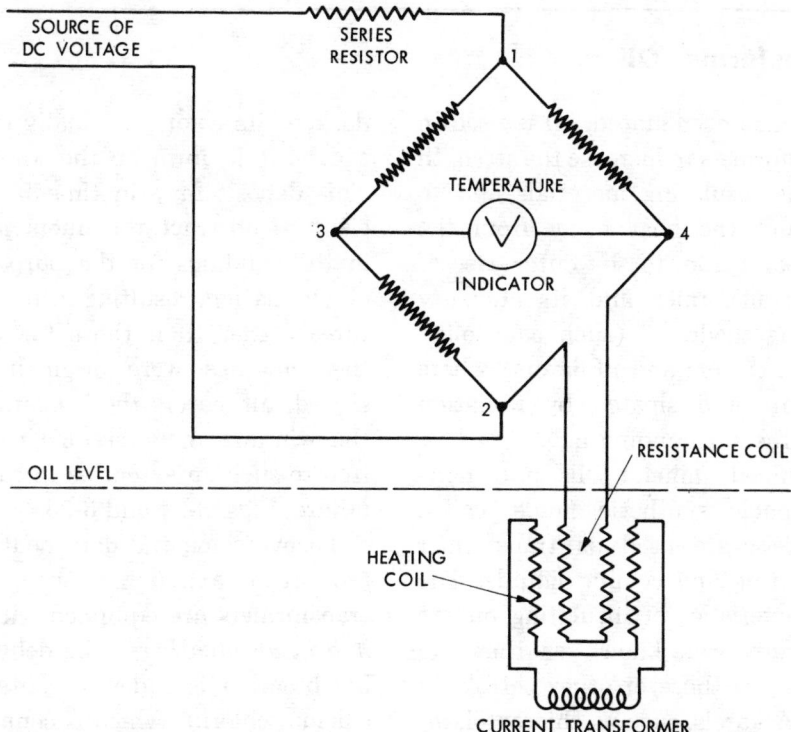

SOURCE OF
DC VOLTAGE

SERIES
RESISTOR

TEMPERATURE

INDICATOR

3

4

1

2

OIL LEVEL

RESISTANCE COIL

HEATING
COIL

CURRENT TRANSFORMER

Fig. 3-23. Schematic diagram of bridge-type hottest-spot temperature indicator.

heating coil forms the fourth leg of a Wheatstone bridge. See Fig. 3-23. The bridge, completed by means of resistances outside of the transformer, derives its voltage from a direct-current source. A voltmeter is connected between two points, normally of equal potential.

When a current flows in the power transformer winding, a current proportional to the power current flows in the current transformer. This latter current raises the temperature of the heating coil, which, in turn, increases the temperature of the noninductive resistance. The bridge therefore becomes unbalanced and the amount of unbalance is indicated by the voltmeter. There is a definite relation between this current and the variation of temperature, and the voltmeter scale is calibrated to read the temperature of the heating coil directly in degrees Celsius. The Wheatstone bridge is balanced when normal current is flowing in the transformer. Any unbalance of the bridge will be an indication of the temperature of the heating coil, which in turn is proportional to the temperature of the main winding.

Transformer Oil

As has been stated, oil is used in transformers to increase the strength of the insulating materials and to conduct the heat away from the coils and iron, to the outer case of the transformer and its auxiliary cooling mediums (such as cooling tubes, coolers and radiators) where it can be dissipated by radiation into the surrounding air.

Refined mineral oils and non-flammable synthetic fluids, called askarels, are used as transformer insulating and cooling liquids. The characteristics of insulating oil are generally well known as they are similar to those of other petroleum oils. Askarels, such as chlorextol, inerteen, and pyranol, are essentially composed of two synthetic chemicals with the addition of a fractional amount of a third compound.

When the oil of an ordinary transformer expands due to an increase in temperature, the air above the oil is forced out of the case; conversely, when the oil contracts due to decreased temperature, air is drawn into the case. This "breathing" action is undesirable, for the interchange of air brings oxygen and moisture from the atmosphere into contact with the oil. The moisture weakens the dielectric strength of the oil, and the oxygen combines with the oil to form a sludge which

darkens its color and finally causes a deposit to form on the windings. This deposit may in time be sufficient to obstruct the ducts placed in the windings for the purpose of oil circulation, resulting in temperatures higher than those for which the windings were originally designed; ultimately the insulation on the windings may become carbonized to such an extent as to cause a failure, Figs. 3-24 and 3-25.

To overcome the detrimental effect of breathed moisture, some transformers are equipped with *dehydrating breathers*. The dehydrating breather is a device containing calcium chloride which is connected to a pipe entering the tank above the oil level. Breathing takes place through this device and all moisture

Fig. 3-24. Coil distortion caused by short-circuit current flow. (Kemper Insurance Co.)

Fig. 3-25. Proper maintenance would have prevented this high-voltage coil damage laid to short-circuit coil fault. (Kemper Insurance Co.)

The expansion tank is connected by a pipe to the main tank, which is completely filled with oil. With increase in temperature, the oil due to its expansion will flow into the expansion tank which has a relatively small surface exposed to the air. The minimum oil level in the expansion tank is such that the main tank is always filled with oil. Breathing then occurs in the expansion tank only and, since it offers a much smaller area of oil surface exposed to the air than does the oil surface in the main transformer, the amount of oxygen absorbed will be considerably lessened.

The expansion tank may be provided with a dehydrating breather to absorb moisture contained in the ingoing air. The expansion tank is usually provided with a sump from which any accumulated moisture may be drained off.

The maintenance of oil-insulated transformers is largely a matter of preserving the qualities of the oil, since the solid insulation of the transformer will usually remain in good condition if the oil is properly taken care of. The enemies of transformer oil are oxygen and moisture, and any device that will operate to keep one or both from contact with the oil is a valuable addition to the transformer equipment.

Some power transformers are equipped with what is known as "The Inertaire Equipment" which

in the air is absorbed by the calcium chloride. A marker and scale indicate when the dehydrating agent has taken up a predetermined amount of moisture and should be replaced. Check valves are provided to prevent deterioration of the dehydrating material by contact with the atmosphere.

To prevent the main body of oil from coming in contact with the oxygen of the atmosphere, some larger distribution and power transformers are equipped with expansion conservator tanks mounted near the top of the main tanks, Fig. 3-26.

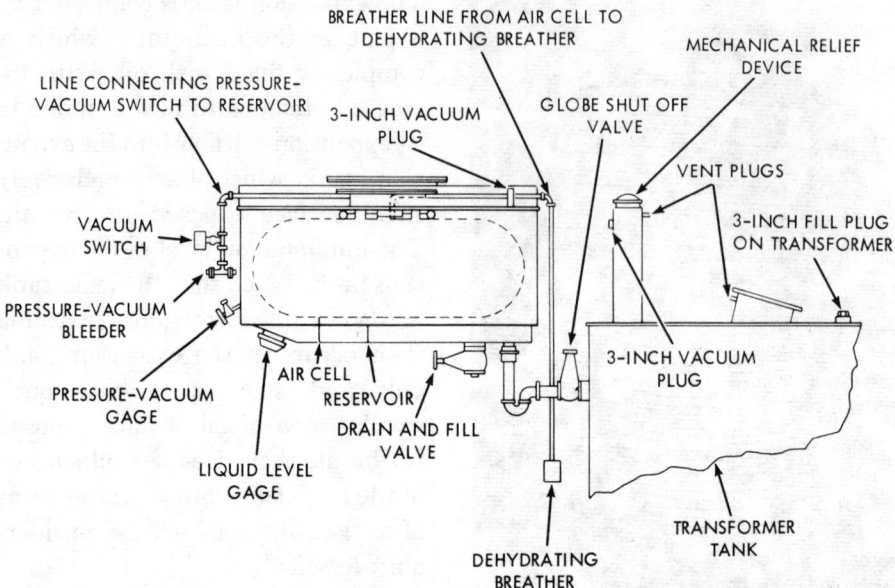

LINE CONNECTING PRESSURE-
VACUUM SWITCH TO RESERVOIR

BREATHER LINE FROM AIR CELL TO
DEHYDRATING BREATHER

MECHANICAL RELIEF
DEVICE

3-INCH VACUUM
PLUG

GLOBE SHUT OFF
VALVE

VENT PLUGS

VACUUM
SWITCH

3-INCH FILL PLUG
ON TRANSFORMER

PRESSURE-VACUUM
BLEEDER

3-INCH VACUUM
PLUG

PRESSURE-VACUUM
GAGE

AIR CELL

RESERVOIR

DRAIN AND FILL
VALVE

LIQUID LEVEL
GAGE

TRANSFORMER
TANK

DEHYDRATING
BREATHER

Fig. 3-26. Constant oil pressure oil preservation system (COPS) maintains a constant pressure on the surface of the oil in the transformer. The tank and the oil are sealed from the atmosphere, preventing exposure of the oil to oxygen and moisture. The oil expansion in the transformer, caused by thermal cycling, is absorbed into a steel reservoir mounted above the main tank. Contact between the oil and the atmosphere is prohibited by a flexible nitrile air cell in the expansion reservoir. The air cell is vented to the atmosphere through a dehydrating breather and inflates or deflates as the oil volume in the transformer changes. Connection between the main tank and the reservoir is made through a globe valve, thus permitting isolation of the main tank. (Westinghouse Electric Corp.)

effectively eliminates both moisture and oxygen from the transformer, Fig. 3-27. The transformer with such equipment has the air above the oil space in the tank replaced with pure dry nitrogen. The Inertaire Equipment consists of a supply tank of pure nitrogen connected to the space above the oil level at the top of the tank. The supply of nitrogen is automatically controlled by a reducing valve and is conserved in the gas by means of a mer-cury regulator. The mercury regulator permits the gas to escape if, due to the temperature increase, the pressure in the space above the oil exceeds a predetermined value. As the temperature of the oil decreases and the internal pressure falls below a predetermined value, the mercury regulator functions to permit the release of additional gas from the nitrogen supply tank through the reducing valve into the gas space above the oil.

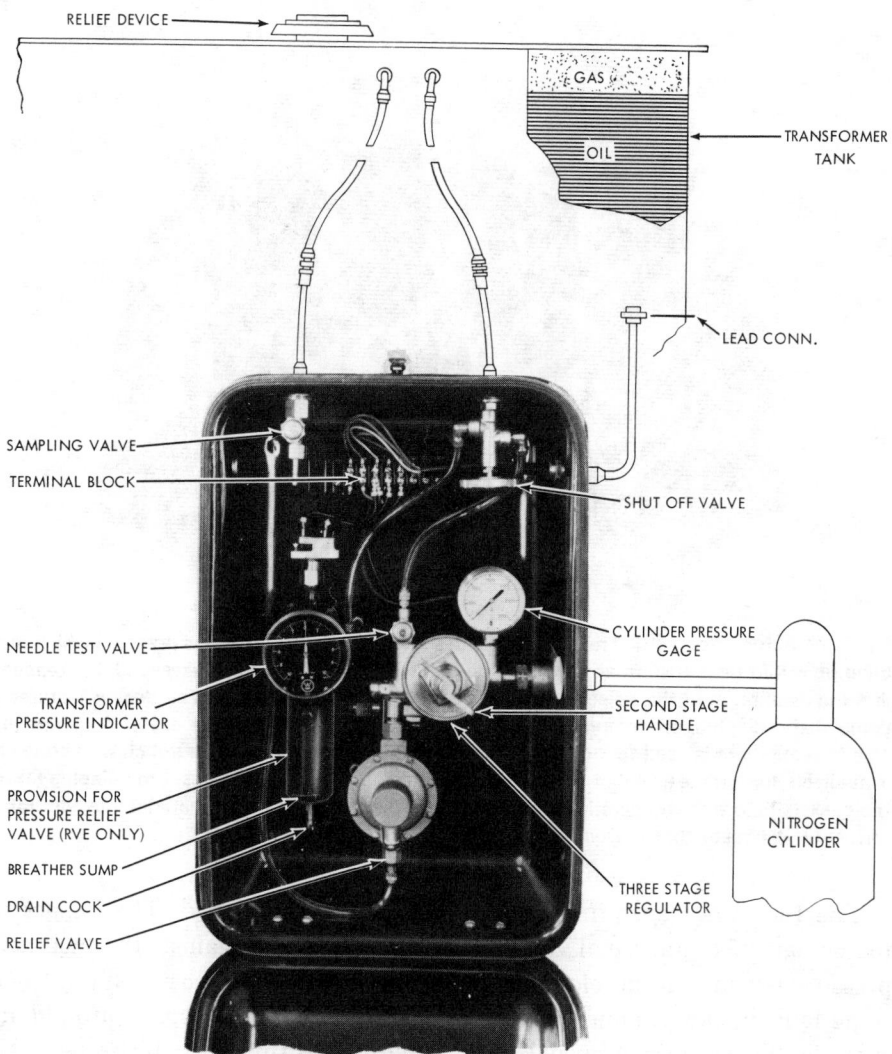

RELIEF DEVICE

GAS

OIL

TRANSFORMER TANK

LEAD CONN.

SAMPLING VALVE

TERMINAL BLOCK

SHUT OFF VALVE

NEEDLE TEST VALVE

CYLINDER PRESSURE GAGE

TRANSFORMER PRESSURE INDICATOR

SECOND STAGE HANDLE

PROVISION FOR PRESSURE RELIEF VALVE (RVE ONLY)

NITROGEN CYLINDER

BREATHER SUMP

DRAIN COCK

THREE STAGE REGULATOR

RELIEF VALVE

Fig. 3-27. Inertaire® equipment—type RBE—assures long insulation life and negligible oil deterioration by maintaining a cushion of dry nitrogen above the oil of the transformer. The nitrogen is supplied from a steel cylinder and is automatically fed into the transformer through a reducing valve whenever the internal pressure in the tank falls below ½ pound per square inch. A relief valve incorporated into the controls conserves nitrogen in the gas space by permitting it to escape into the atmosphere only when the pressure in the transformer reaches a predetermined value. A valve connected to the gas space provides a means of sampling the gas for oxygen content. A compound pressure gage indicates internal tank pressure or vacuum. All equipment except the nitrogen cylinder is enclosed in a weatherproof cabinet with padlocking provisions. (Westinghouse Electric Corp.)

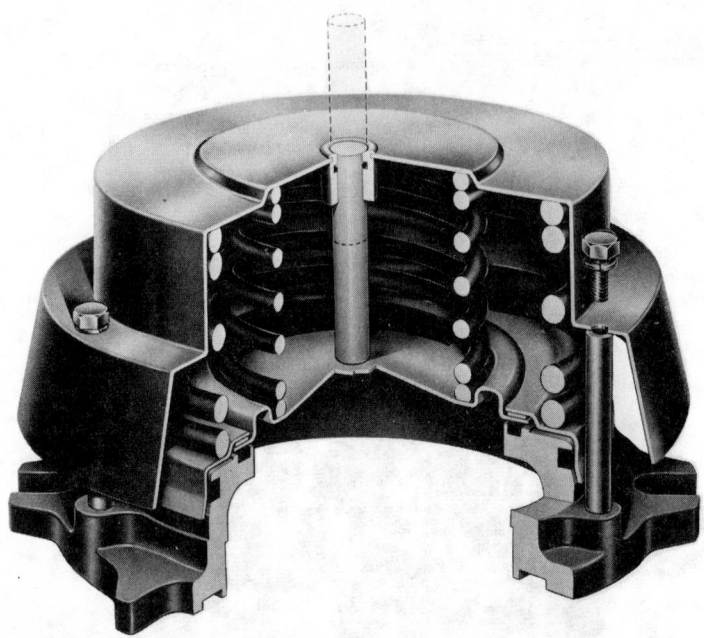

Fig. 3-28. Automatic resetting relief device is designed to relieve dangerous pressure which may build up within the transformer tank. When a predetermined pressure is exceeded, the reaction lifts the diaphragm of the relief device and vents the transformer tank. The design features a dome-shape diaphragm, compression springs, gaskets and a protective cover. A lightweight plastic semaphore is used to indicate an operation of the relay. An alarm switch with contacts is available for remote signal of an operation. The compression springs reset the diaphragm on the gaskets following an operation thus preventing the entry of foreign material into the transformer tank. (Westinghouse Electric Corp.)

The Inertaire transformer is protected against abnormal internal pressure which may develop due to some fault in the winding, although such faults are very unusual. For this purpose a thin diaphragm of some suitable material such as micarta (which can be ruptured by such pressures) is used in combination with the manhole opening. The manhole cover holds the diaphragm in position. If an abnormal pressure develops inside the tank, the diaphragm is ruptured. The impact of the pressure against the manhole cover raises it almost instantly and forms an annular space around its periphery through which the pressure is relieved. This cover is fastened to the diaphragm supporting ring by a set of heavy springs, Fig. 3-28. As soon as the pressure is relieved, the springs return the cover, thus preventing the escape of additional gas or the entrance of air and water into the gas space.

Mechanical Stresses

Whenever a current flows in a conductor, a magnetic field is produced in the region about this con- ductor. If another conductor, also carrying a current, is placed near the first, the resultant magnetic

Fig. 3-29. Typical three-phase coil-and-core assembly. (McGraw-Edison, Power Systems Division)

The force action between the two conductors in this case amounts to an attraction, that is, the two conductors will tend to be drawn towards each other. When the currents are in opposite direction, the field intensity will be increased in the space between the two conductors. The force action between the two conductors is then a force of repulsion or a force tending to push the two conductors farther apart.

When a current is flowing through a conductor which has been wound into a coil of several turns, the various turns attract each other at all points, since the current is flowing in the same direction in all turns.

When the primary and secondary coils of a transformer are assembled together, the currents in these two windings are opposite in direction. The force between them is then a force of repulsion at all times. Under normal conditions of operation the mechanical stresses which are developed due to these forces of repulsion are very small. On short circuit, however, these stresses are multiplied many times. If they are of sufficient magnitude, the coils are braced to prevent their injury due to the stresses when a short circuit occurs. Figs. 3-29 and 3-30 show a core and coil assembly of a three-phase core-type transformer in which the coil columns are supported at both ends by heavy insulating rings and steel plates so mounted on the

Fig. 3-30. Typical shell-type core-and-coil structure of a small single-phase transformer. The permanently welded core clamp cradles the core and minimizes stress. It is rigidly built to protect edges and to withstand any operating conditions. Steel channels brace the coils at the top and bottom to prevent movement under short-circuit conditions. (Distribution Transformer Dept., General Electric Co.)

field at each point surrounding the two conductors is the vector sum of the fields due to each of the currents taken separately. If the currents are both in the same direction, the field intensity will be decreased in the space between the conductors and increased in the region outside.

end frame structure which supports the entire core and coil assembly, that pressure is maintained against the ends to counteract the forces set up by heavy overload or short-circuit stresses.

Fig. 3-30 shows a core and coil assembly of a small single-phase shell-type transformer. In the shell type of construction the magnetic circuit tends to support the coils against any displacement due to the short-circuit stresses on that portion of the windings enclosed by the magnetic circuit. On large shell-type units, wound with the pancake type of winding, the portions of the windings extending beyond the magnetic circuit are braced with heavy steel frames and supporting plates.

Under short-circuit conditions the mechanical stresses depend upon several factors, the most important of which are the frequency, size, voltage, and reactance of the transformer. The conditions of maximum stress would be represented by a large transformer, operated on a low-frequency circuit, of moderate voltage and low impedance, and fed from a power house of large capacity compared to the rating of the transformer.

While low reactance is desirable from the standpoint of good regulation, a transformer having low reactance may not be able to withstand the mechanical stresses set up during a short circuit.

Transformers must be protected from short circuits or grounds. The National Electrical Code NFPA (National Fire Protection Association) No. 70-1971; ANSI (American National Standards Institute) CI 1971 sets specific requirements as to the rating of the required primary overcurrent protection. The rating of the required primary overcurrent protective device is smaller for dry-type transformers than for askarel- or oil-insulated transformers. The rating of the primary overcurrent device may be increased if the secondary is properly protected by an overcurrent device or is equipped with a coordinated thermal overload protection by the manufacturer as previously described in this chapter.

Transformer Accessories and Fittings

There are several accessories found on power transformers with which the reader should be acquainted.

Explosion Vent

When an electrical fault occurs under oil, high pressures are possible.

These pressures could burst the sheet steel tank if some means were not used to guard against this. The explosion pipe (4 inches in diameter or greater) extends a few feet above the cover of the transformer and is curved in the direction of the ground at the outlet end of the pipe. Fitted at the curved end of this length of pipe is a diaphragm that will break at a relatively low pressure and will release the forces from within the transformer. This diaphragm may be of glass or very thin (0.16 inch) phenolic sheeting.

Gas Relay

It has been noted that an electrical arc or fault can result in a pressure wave in the liquid. Should such a fault develop, it is desirable to remove the transformer from service as quickly as possible to prevent extensive damage. A thin diaphragm in the gas relay moves when acted upon by this wave, and a mercury switch connected to this diaphragm energizes relays to switch the transformer off the load. However, gradual overheating of any part, such as a hot joint, while not causing a pressure wave can ultimately result in failure of the transformer. This local overheating will decompose or crack the oil, forming gases which rise to the top of the tank. These are accumulated in a dome-like section of the relay in which a float is riding in the oil. The gas displaces the oil, dropping the surface of the oil, with the float following the surface of the oil down. The float operates a mercury switch in an alarm circuit. On receiving the alarm, the condition may be investigated before extensive damage results.

Level Gages

To ensure that the correct liquid level exists in the transformer, a gage is fitted. This will be on the main tank if it is of the smaller variety with no expansion tank or on the expansion tank if one is fitted. This gage will have a mark at the correct level for 25-degree-C temperature of oil. This ensures that should the temperature rise, the unit will not overflow nor will a dangerously low level be reached at very low temperature conditions. Contacts are provided on the larger transformer to signal an alarm when the oil level drops to a dangerous level.

Thermometer

The thermometer is fitted to measure the temperature of the top oil; in small transformers this may be in the form of a liquid-filled thermometer mounted at the top of the tank; while in large units, it is usually of the gas-filled type with the bulb fitted into a well in the cover and the indicator at the bottom of the tank at eye level. Contacts are provided to signal an alarm for high temperature.

Bushings

The electrical power circuits must be insulated where they enter the tank. In addition, this entrance must be oil-tight and weatherproof. A bushing is used for this purpose. It is usually composed of an outer porcelain body; and at higher voltages, additional insulation in the form of oil and molded paper is used within this porcelain.

The four types of bushings used on transformers as main lead entrances are as follows:

1. Solid porcelain bushings.
2. Cable terminators or potheads.
3. Oil-filled bushings.
4. Condenser bushings.

Low-voltage transformers with separate leads generally have solid porcelain bushings. These bushings consist of high grade porcelain cylinders through which the connections pass. The outside surface may be plain or have a series of corrugations or skirts to increase the surface leakage path to the metal case. When the conductors are brought to the transformers in lead covered cables, the leads are often brought in through cable terminators. These terminators are similar to the ordinary potheads and are attached to the transformer with the bushings inside the case. High voltage bush-

ings are either the oil-filled type or the condenser type. The oil-filled type has a central conducting rod or tube through which the conductor passes. Around this is a series of insulating barriers held apart by spacers. The barriers and spacers are enclosed in a skirted porcelain shell which is filled with oil. The condenser type is similar except that the central rod is wound with alternating layers of insulation and tin foil. This results in a path from the conductor to the case consisting of a series of condensers. The layers

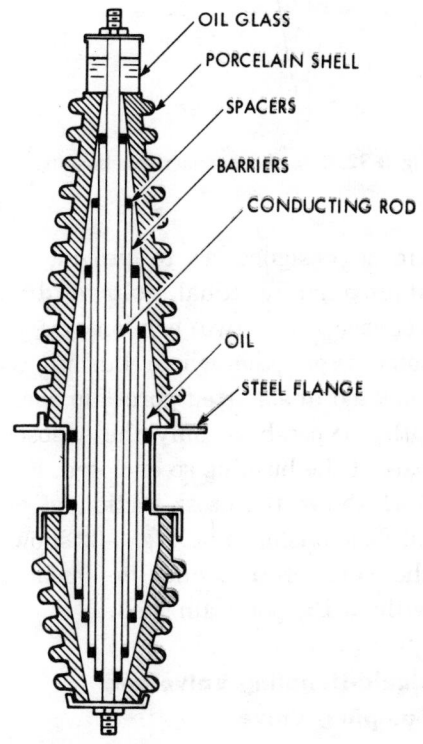

Fig. 3-31. Oil-filled bushing.

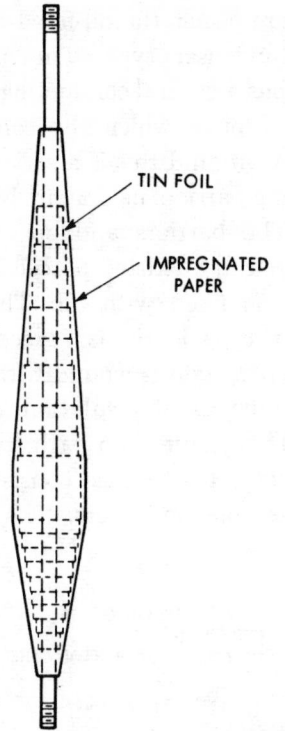

TIN FOIL

IMPREGNATED PAPER

Fig. 3-32. Core of condenser type bushing.

are so designed as to provide an approximately equal voltage drop between each two condensers. In some types the whole bushing is enclosed in a skirted porcelain shell. Other types have only the exposed part of the bushing so enclosed. Fig. 3-31 shows the cross section of an oil-filled bushing and Fig. 3-32 shows the core of a condenser bushing without the porcelain shell.

Liquid-Handling Valve and Sampling Valve

In order to conveniently admit or remove liquid from the tank, a valve is fitted at the bottom of the tank. It is usually connected to a sump to ensure that all liquid will be removed. Adjacent to this is a ½-inch needle valve also leading to the sump which is used for drawing off a sample of liquid for test purposes. By taking liquid from the lowest point in the tank, any free water should appear in the sample and give warning of contamination of the liquid.

Transformer Nameplates

A transformer, like all other electrical equipment, is designed and manufactured to function in an electrical system at a specified voltage, frequency, load etc. Broadly, the information contained on the nameplate, Fig. 3-33, can be subdivided into five sections.

1. **General.** Information is given of the transformer serial number, type, rating, nominal voltages, impedance, tap changer range (if fitted) together with oil capacity, weights, etc.

2. **Physical Arrangement of Terminals.** In one form or another, usually as a block drawing, the physical locations of all terminals are given with reference to some obvious feature of outside construction. This enables all connections to be positively identified.

3. **Schematic Diagram.** A complete winding schematic diagram shows all internal connections, coil

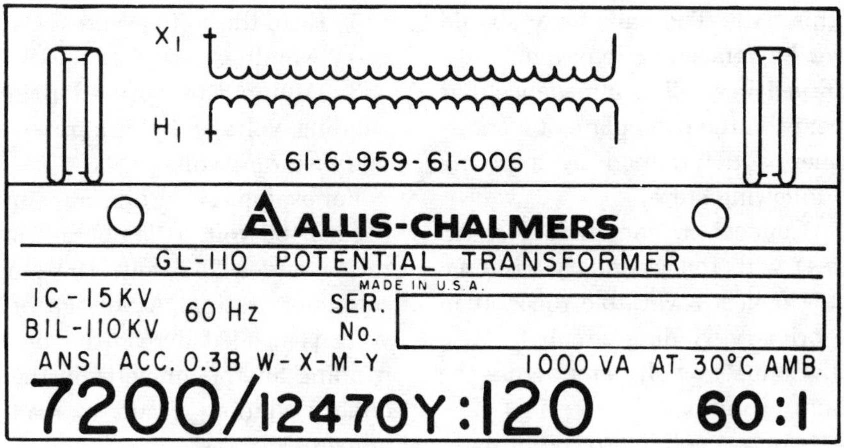

Fig. 3-33. Potential transformer nameplate. (Allis Chalmers)

tap numbers, internal selector switches and main terminal markings. Thus the combined use of the physical arrangement drawing offered above and the schematic, enables the user to determine exactly how each bushing is connected internally. Also included on the schematic are all auxiliary components such as current and potential transformers showing their actual electrical positions and usually the ratios.

4. Vector Reference. On all three-phase units, primary and secondary voltage reactors are drawn to indicate the phase angle between primary and secondary voltages. The vector diagrams are labeled with the terminal designations used on the schematic.

5. Voltage Ratio Table. One or more tables were given for transformers with some form of tap changer. This table lists the tap indicator number, the actual coil numbers and how they are connected, using the numbers shown on the schematic diagram. The voltage ratio is given for each tap position and, in some instances, the full load current at that tap position. On many nameplates, to avoid the use of voltage ratios which would mean small decimal numbers, the nominal voltage of the winding without the tap is taken and the voltage required to produce this nominal value is listed for each tap. Both methods give the same information.

It is important that the installer carefully check the nameplate data to be sure that the transformer is installed and maintained according

201

to this data. The nameplate should never be defaced or removed.

Impedance. The impedance that appears on the nameplate of a transformer is determined by applying the following steps:

1. Connect an ammeter across (in series) with the secondary winding.

2. Connect a variable rheostat in the primary winding circuit.

3. Connect a voltmeter across the primary winding.

4. Apply a voltage to the primary circuit and by adjusting the rheostat increase the voltage across the primary winding until the secondary winding current reaches its rated current.

5. Read the voltage across the primary winding.

6. Divide the applied primary winding voltage by the rated primary winding voltage.

For example, a certain transformer has a 2400 volt, 10 ampere primary rating and a 240 volt, 100 ampere secondary rating. A voltage of 120 volts is needed across the primary winding for the ammeter in the secondary winding circuit to read 100 amperes.

By using step number 6 in the above procedure, we divide 120 by 2400 and arrive at a figure of 0.05 or 5 percent impedance. (See Fig. 3-34).

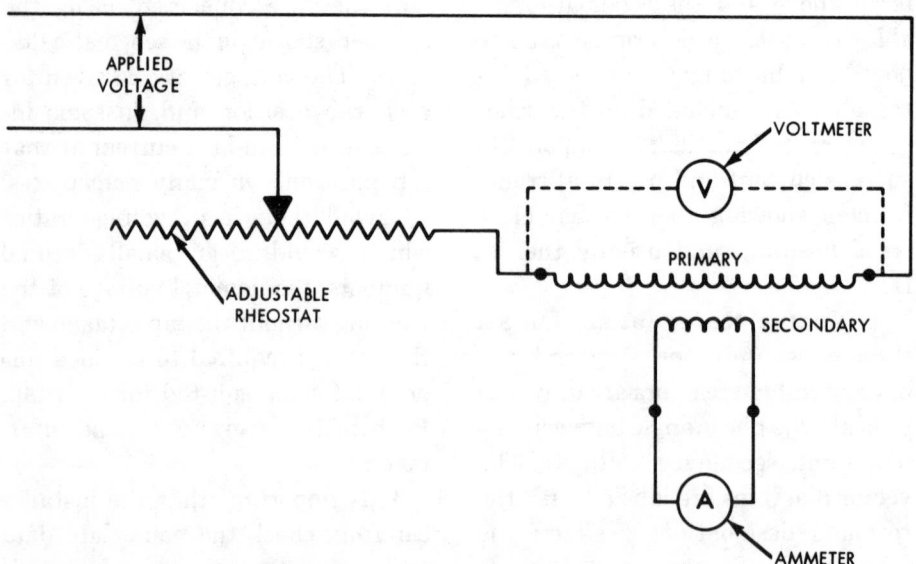

Fig. 3-34. Connecting an adjustable rheostat, a voltmeter and an ammeter in the primary and secondary windings to determine the impedance of a transformer.

Sound Levels of Transformers

Sound Levels

All transformers make some sound when they are energized. This is due to the vibrations generated within the laminated steel core structure and is heard as a hum. This hum has a fundamental frequency of twice that of the applied frequency. The volume is determined by the transformer design, construction characteristics and the methods used in their installation. The amount of sound created by a transformer is measured by a unit known as the decibel (db).

Test procedures have been established so that transformer manufacturers can publish the sound level ratings of their transformer. See Table VII.

Normally when transformers are installed in substations, vaults or out-of-doors, sound is no problem but there are specific critical locations where sound is a very important factor. Fortunately the noises in and around most indoor locations, known as the "ambient sound level,"

TABLE VIII	
Area	Sound level, decibels (db)
Residence (incl. apt. bldgs.)	25–45
Retail store	45–60
Office areas	45–75
Manufacturing plants	75–95

usually mask transformer sounds. (See Table VIII for the average ambient sound levels in areas where transformer noise could be a problem.) As an example, if we were to install a 150 kVA distribution transformer with a 50 db rating, in a factory that has an ambient sound level of 85 db the sound of the transformer would not be heard above the ambient sound, but if this same transformer were installed in an apartment building where the ambient sound level was 30 db the transformer sound would be noticeable and would be considered objectionable. In other areas such as schools, churches and hospitals where the ambient sound level is very low, special precautions must be taken to select a transformer with a low sound rating and to locate and install the transformer in such a manner so as to keep the sound level at a minimum.

Low Sound Level Installations

An interesting observation could be made which is relative to locating more than one transformer in the same area.

TABLE VII	
KVA	ANSI std. in decibels
0–5	40
6–9	40
10–25	45
26–50	45
51–150	50
151–225	55
226–300	55
301–500	60

When two transformers with equal sound levels are installed side-by-side their combined sound level is only 3 db higher than their individual level. The combined level of three similar transformers will be 4.8 db higher than their individual level. For example if three 50 kVA transformers with an individual rating of 45 db are installed in the same area their combined sound level will be 49.8 db, not 135 db. Although the foregoing could be an important factor, other installation precautions must be taken when installing transformers if the sound level is to be kept at a minimum.

1. Select a transformer that has a low sound level rating. The sound level rating of the transformer should be below the ambient sound level of the area where it is to be installed. In some critical, low sound level areas it may be necessary to factory order a special quiet-type transformer. Such an area could be in a hospital, where the National Electrical Code requires that isolating transformers be installed to feed special circuits in the operating rooms.

2. Locate the transformer as far away as possible from areas where sound is objectionable.

3. Locate the transformer where the sound level is not amplified by sound reflection from the walls or ceiling. Tests have shown that the *least* desirable location for a trans-former is in a corner near the ceiling as the walls and ceiling tend to act as a megaphone. If the transformer is free-standing and the room where it is to be installed is near a low sound level area, the only accurate way to determine the lowest sound level location is to make a temporary connection and actually move the transformer to different locations in the room. The walls and ceiling of the room could be covered with acoustical tile or fiberglass which will dampen the sound on the high harmonics of the transformer but will have little effect on the fundamental hum generated by the transformer.

4. Mount the transformer so that the mechanical vibrations of the transformer are not transmitted to the connected raceway system or to the structural parts of the building.

Rigid mounting on a heavy reinforced concrete wall or floor is generally suitable for a small transformer since its mass is comparatively small. When mounting transformers on a structural frame, wall, ceiling or column, a method referred to as "flexible-mounting" is normally necessary. This method requires the transformer to be installed on flexible mounts which isolate the transformer vibrations from the structural part of the building. These mounts must be properly selected and loaded and are normally furnished and installed by the installer. Mounts must be installed so there is no metal to metal contact between the transformer and the structural part of the building. On specially constructed low sound level transformers internal vibration pads which isolate the core and coil assembly from the enclosure are installed at the factory by the manufacturer. When small transformers not having built-in vibration isolators are installed, the conduits or wiring method must be flexible so as to eliminate the conducting of the transformer vibrations to the structural part of the building.

Frequency and the Transformer

In this country the standard frequencies now in common use are 25 and 60 hertz. The frequency of 60 hertz is used for lighting and power installations while the 25 hertz is often employed for railroad electrification. A frequency of 125 hertz has been used for aircraft electrification. In Europe frequencies of 50 and $16\frac{2}{3}$ hertz are in common use. The 50 hertz is used for lighting and power installations and the $16\frac{2}{3}$ hertz is used for railroad electrification.

The question may arise as to why 25 hertz electricity is not used for lighting and power systems because the 25 hertz system would have a lower impedance due to the lower inductive reactance. This type of system would therefore also have a better voltage regulation. There are other advantages of the 25-hertz system but there is one basic problem and that is, when this system is used for lighting, the filament of the lamps cool down between cycles at this lower frequency thus causing a noticeable flickering of the lamps. At the frequency of 60 hertz the filament of the lamp is subjected to the low current for a much shorter time and the filament does not really have time to cool off, therefore the flickering is not noticeable.

When selecting or installing a transformer it is important that the transformer is rated to operate at the same frequency as that of the system.

A detailed study on the effect a change in frequency will have on a transformer is made in the chapter "Design of Small Transformers" but for the present it should be kept in mind that if the voltage applied to a transformer is kept constant, the current will vary inversely as the frequency. This is shown by the following:

The current of a transformer is determined by:

$$I = \frac{E}{Z}$$

Where $Z = \sqrt{R^2 + X_L{}^2}$

Where $X_L = 2\pi f L$

Substituting $2\pi f L$ for X_L

$$Z = \sqrt{R^2 + (2\pi f L)^2}$$

Substituting $\sqrt{R^2 + (2\pi f l)^2}$ for Z

$$I = \frac{E}{\sqrt{R^2 + (2\pi f L)^2}}$$

As the value of L (henries) increases with the number of turns it is also evident that the number of turns of the primary may be decreased as the frequency of the primary voltage is increased.

From the foregoing it is evident that if a 60-hertz transformer were connected to a 25-hertz system, the current of the primary would be more than twice that of the current rating of the primary therefore serious damage could be done to the primary winding of the transformer.

Tap Changers

Taps are frequently placed in one or both of the windings of a transformer in order to vary its voltage ratio. The purpose in changing the voltage ratio depends upon the service to which the transformer is subjected. For example, a distribution transformer connected near a substation may have a higher voltage on its primary side than one connected farther out on the line. In the case of power transformers used for transmitting large amounts of power, the primary voltage may be constant but the secondary voltage will vary, under varying loads, due to line drop. With distribution transformers it is usually sufficient to change the voltage ratio occa-sionally by means of taps as the load changes due to natural growth. Such changes are therefore infrequent. With power transformers, however, it is necessary to change the voltage ratio frequently to take care of daily and seasonal changes of load. This change in voltage ratio is made by changing connections of taps brought to a tap changer within the transformer. Tap changers are also used with auto-transformers to form a regulator or booster for varying the voltage in small steps on distribution feeder lines. They are also used for adjusting power interchange and for phase-angle shifting on interconnected transmission systems.

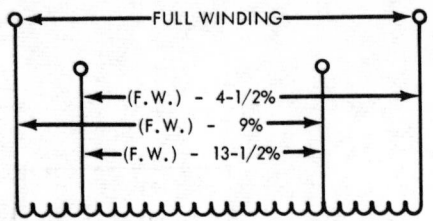

Fig. 4-1. Elementary primary winding on a transformer.

Figure 4-1 is an example of an elementary primary winding on a transformer. This is used to illustrate the taps taken off a primary coil. Most of the transformers in use today have the primary winding in two parts with the taps taken off between the two primary coils.

It may be seen from Fig. 4-1 that when we speak of the $4\frac{1}{2}$ percent, 9 percent, or $13\frac{1}{2}$ percent tap, we mean we are actually cutting out this percentage of primary winding. In this way the ratio of primary turns to secondary turns may be changed.

Table I shows what the primary voltage setting will be in order to maintain a constant secondary voltage of 120 volts.

From this table, we can see that if the primary line voltage were to drop to approximately 2,292 volts, we could put our transformer on $4\frac{1}{2}$ percent tap (full winding minus $4\frac{1}{2}$ percent) and maintain the secondary voltage at 120-240 volts.

Taps on a transformer may be changed only after considering the voltage at peak load and, also during off-peak, load must be known. A recording voltmeter must be placed on the secondary circuit for at least twenty-four hours to determine the lowest and highest voltage. If we were to set the tap setting for peak-load voltage, we may have too high a voltage during off-peak load. Therefore, we must try to have a satisfactory voltage at all times. See Table II. A satisfactory voltage is between 117-122 volts at the customer's service entrance.

Voltage should be kept at satisfactory voltage as much as possible and only in the tolerable low and high voltage ranges for short periods of time to determine the primary line voltage:

$$\frac{\text{Primary}}{\text{voltage}} = \text{ratio} \times \text{secondary voltage.}$$

TABLE I			
TAP WINDING	PRIMARY VOLTAGE	SECONDARY VOLTAGE	RATIO
FULL WINDING	2,400	120 (240)	20-1 (10-1)
4-1/2%	2,292	120 (240)	19-1 (9.5-1)
9%	2,184	120 (240)	18-1 (9-1)
13-1/2%	2,076	120 (240)	17-1 (8.5-1)

TABLE II			
	TOLERABLE LOW VOLTAGE	SATISFACTORY VOLTAGE	TOLERABLE HIGH VOLTAGE
LIGHTING, MOTORS, ETC. STOVES, MOTORS, ETC.	112 230	117 235	122–127 245–250

$$\text{Ratio} = \frac{\text{primary tap setting}}{\text{secondary rating}}$$

For example, in a 2,400—120/240-volt transformer, the secondary voltage is 226 volts (should be 240 v). On checking the tap setting, it is found to be on full winding—2,400 volts.

$$\text{Ratio} = \frac{2,400}{240} = \frac{10}{1} = 10{:}1$$

Primary voltage $= 10 \times 226 = 2,260$

Looking at our Table I we can see by putting the tap setting on 4½ percent, we would have approximately 240 volts which would be satisfactory.

No-Load Tap Changer

Terminal Board

If the change in voltage ratio is infrequent, the taps may be brought to a terminal board, Fig. 4-2, inside the transformer tank. Taps are changed after the transformer is disconnected from the line by changing the position of one or more connectors which connect the studs on the terminal board. The taps from the winding are in turn connected to these studs. This change in connections is made according to instructions found on the nameplate or diagram of connections supplied by the manufacturer of the transformer. Such a method of changing taps requires the removal of the main cover or handhole cover and working at times in hot oil, and it is therefore inconvenient. Aside from the inconvenience and the time required to make such a change, the operator may accidentally drop a tool or loose metal part, such as a washer or nut, from the terminal board into the transformer winding, or a careless operator may make the wrong connection and cause serious trouble within the transformer or to apparatus connected to it. However, this method of tap changing is used for distribution and small power transformers inasmuch as it is the cheapest method of tap changing from the standpoint of first cost.

Fig. 4-2. Tap terminal board (dry-type transformer). The taps can be reached from the front or back by removing a panel which also protects the taps from tampering. The taps are rigidly supported on a terminal board centrally located on the coils. Taps are changed by moving the terminal strips from one connecting point to the other. To simplify these changes the connections are clearly identified. (Westinghouse Electric Corp.)

Manually-operated Switch

To safeguard the transformer from improper connections and at the same time decrease the time required to make a change of taps, the taps from the winding are brought to a switch instead of a terminal board having studs and connectors. This switch is operated by some form of handwheel or handle, which may be on the underside of the cover for internal operation, or which may be connected to a shaft extending through the cover or side of the tank wall for external operation. In either case the transformer must be disconnected from the line before the change in taps is made. Such a method of changing taps eliminates the possibility of (a) making the wrong connections, (b) dropping loose parts into the winding, (c) having high resistance in contacts due to improper tightening of connectors to studs on the terminal board; and when the handwheel is on the outside of the transformer tank, it eliminates the necessity of having to remove the cover.

The method of connecting a tap-changing switch in the transformer winding is shown in Fig. 4-3. For the position shown in Fig. 4-3, namely Position *1*, the full high-voltage winding is connected to the line. As the insulated handle is ro-

tated to the right or in a clockwise direction to positions *2, 3, 4, 5,* alternate left and right sections of the winding are removed from the circuit. In Position *1*, the winding taps *4* and *5* are connected together by means of a current-carrying segment large enough to span two adjacent studs to which these taps are connected. This segment is part of the operating handle but is insulated from it. As the handle is turned to Position *3*, for example, this segment connects winding taps

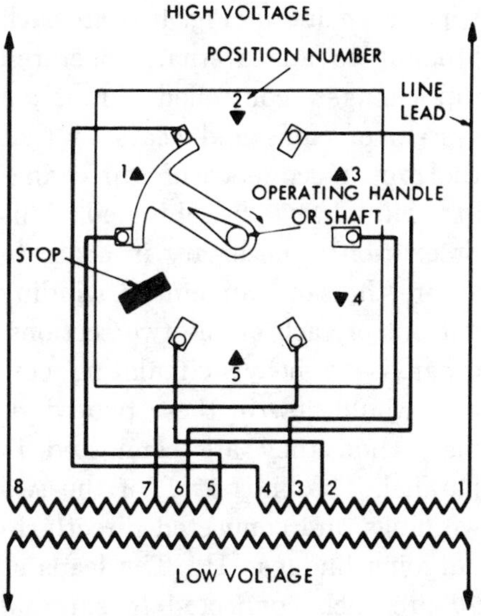

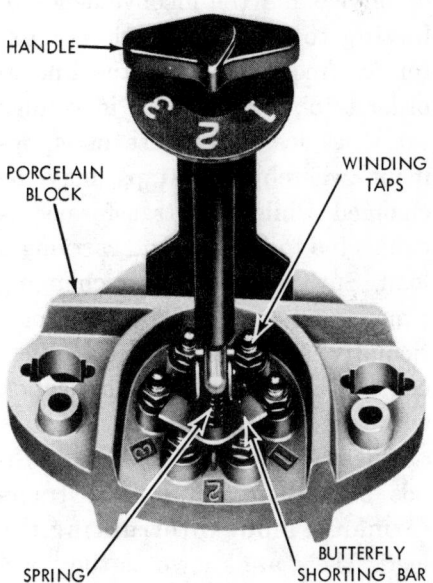

Fig. 4-4. No load tap changer. A tap changer handle, located above the oil, links a butterfly shorting bar under spring tension to assure clean wiping and positive positioning of the tap changer. All positions are clearly identified by numerals located on a collar around the tap changer handle shaft. The high-voltage taps are terminated in a special porcelain block. (Westinghouse Electric Corp.)

Fig. 4-3. Connections of a tap-changing switch.

3 and *6*, thereby omitting from the circuit all turns between taps *3* and *4* and between taps *5* and *6*. Fig. 4-4 illustrates one form of tap-changing switch for use in small distribution transformers.

Tap Changing Under Load

With the expansion and interconnection of power systems it often becomes necessary to change the transformer taps several times daily to obtain the required voltage variation. The demand for continuity of service and the inconvenience of having to disconnect the transformer frequently from the line in order to obtain a change in voltage ratio has led to the use of equipment whereby the taps may be changed while the transformer is connected to the line and carrying a load. Such equipment for changing transformer taps under load is particularly adaptable where voltage control is required in connection with large blocks of power.

There are two fundamental methods of changing taps on a transformer without interrupting the load. These are known as the "parallel-winding" scheme and the "single-winding" scheme. In the parallel-winding scheme, as shown in Fig. 4-5, one winding of the transformer is arranged with parallel circuits. Each of the parallel windings is provided with taps which are changed only when their respective winding is opened by a circuit breaker. In the normal operation of the transformer, the two windings are operated in parallel and therefore each winding is only half the capacity of the third winding. However, each of the parallel windings must be capable of carrying the full-load current of the transformer while the taps are being changed on the winding which is opened.

The operation of this method is as follows: taps *2* to *9* of parallel windings *a* and *b* (Fig. 4-5) are each brought to an externally operated tap changer, controlled through a system of cams and gears so that the proper sequence of tap changing will always be obtained. This precaution is necessary in order to assure the same amount of winding cut out of each of the two sections, otherwise a heavy circulating current would flow in these two windings when they are connected in parallel. The ends at *1*, of the two windings, are connected directly to outgoing line lead H_1. The leads at *10* are each connected to external circuit breakers *A* and *B*, which in turn are connected in parallel on the

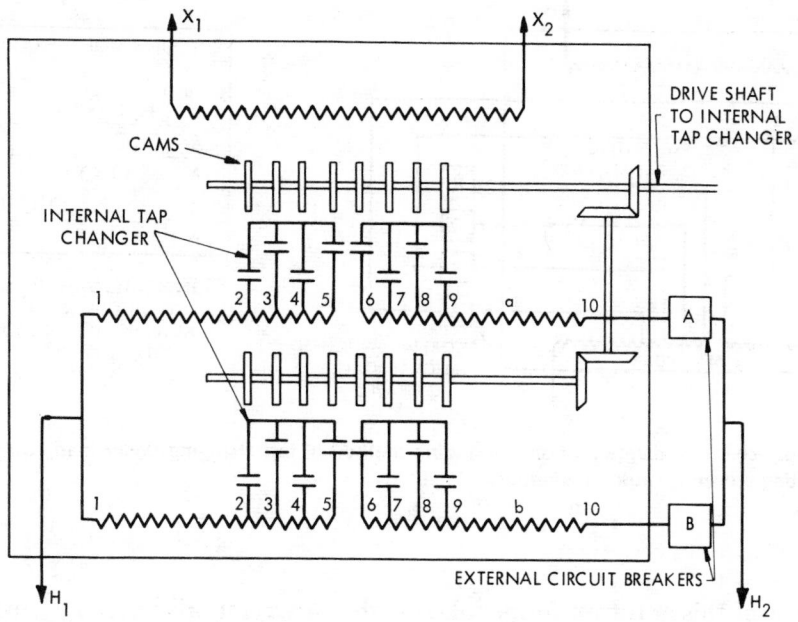

Fig. 4-5. Schematic diagram of parallel winding method for tap changing under load.

outgoing side to line lead H_2. The internal tap changer mechanism is operated through a drive shaft extending through the transformer case.

The sequence of operation for each change in voltage ratio is then as follows:

(1) Open circuit breaker A
(2) Change taps in winding a
(3) Close circuit breaker A
(4) Open circuit breaker B
(5) Change taps in winding b same as in winding a
(6) Close circuit breaker B.

The foregoing cycle of operations is repeated for each change of taps. As circuit breaker A is opened,

winding a is disconnected from the line and all the load must be carried during this period by winding b. When A is reclosed and B is not yet opened, the load is again divided between the two windings a and b. As B is opened, the load is transferred to winding a. Upon reclosing circuit breaker B, windings a and b again each carry half of the total load.

The foregoing method of tap changing has been replaced to a large extent by the single-winding method. In the single-winding method of tap changing under load, a reactor or preventive autotransformer is used to bridge across taps on the main transformer winding.

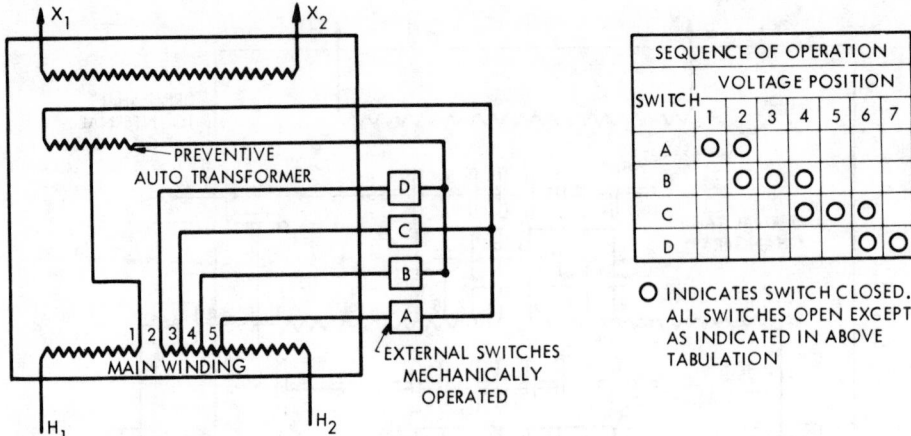

Fig. 4-6. Schematic diagram of single-winding method of tap changing under load, with self-protecting preventive autotransformer.

The use of this autotransformer also gives an additional voltage step between each adjacent tap. For example, if the transformer has four 5 percent taps, an additional 2½ percent step is obtained during each tap-changing cycle when the autotransformer is connected across two adjacent taps. This is possible because one end of the main transformer winding is connected to a mid-point of the autotransformer. The switching cycle is as given in Fig. 4-6.

With the autotransformer connected as in Position *1* with only switch *A* closed, the transformer is connected across the line with all the taps out of the circuit and half of the autotransformer winding in series with the main winding. This portion of the autotransformer is then carrying all the load current. In Position 2, with switches *A* and *B* closed, the autotransformer is connected across taps *4* and *5* and half the load current flows through each half of the autotransformer winding. The voltage across H_1–H_2 is increased by one half of the voltage between taps *4* and *5*. In Position *3*, with only switch *B* closed, half the autotransformer winding is again in series with the main winding, but the voltage has been stepped up across H_1–H_2 by an amount equal to the voltage between taps *4* and *5*. This sequence of operation can be continued as indicated in the table of Fig. 4-6 until all the tap sections of the main transformer are inserted in the circuit. It will be noted from this table that at all times one or more of the breakers are always closed, therefore the variations of voltage ratio are obtained without interrupting the load connected to the transformer.

Various modifications of this circuit may be employed. For example, in the circuit just described the voltage taps are not equal because of the reactance drop of the autotransformer when it is connected to one tap only, as in positions *1, 3, 5* and *7*. When full-load current is passed through one half of the autotransformer with the other half open, the full-load current of the main transformer becomes the exciting current of the autotransformer. Under this condition there are no neutralizing ampere turns from the other half, so the autotransformer becomes a reactor. Air gaps are provided in the core to prevent saturation and to give low impedance when operating in this manner. This of course makes the exciting current relatively high when the autotransformer is connected across taps.

To make the voltage steps exactly equal for each position, a circuit breaker is sometimes added to this circuit so that after the switching cycles to Positions *1, 3, 5* and *7* are completed, the circuit breaker short-circuits the autotransformer and thereby removes it electrically from the circuit. This breaker is mechanically interlocked with the externally-operated switches so that the

alternate positions *2, 4,* and *6* cannot be made without first opening the circuit breaker, thus again placing the autotransformer in the circuit during the switching operation.

Location of Tap Changers

There is no fixed rule for the location of the tap changer with respect to primary or secondary windings. In general, it is desirable to locate the tap changer in the high-voltage winding because of its smaller current. Its position may be fixed by the application of the transformer which may require parallel operation on the one winding or the regulation of power flowing in a given direction. If the voltage of the high-voltage winding becomes so great as to make it difficult to insulate the tap-changer parts, the tap changer is sometimes put in the low-voltage winding. If the current in the low-voltage winding is too large for the capacity of the tap-changing switch, or the voltage of either winding is too high, a series transformer is required.

In a wye-connected transformer with the neutral solidly grounded, the taps and tap-changing switches are placed in the grounded neutral end of the winding. Although the winding may have a very high voltage at its line end, the maximum voltage to which the tap-changer switches are subjected is equal only to the voltage existing between the neutral and the highest tap position. The nearer the taps are placed to the neutral end, the lower will be this voltage.

Automatic Control. In applications where the variation in voltage would be considerable throughout the day, it is customary to have this voltage drop compensated for automatically by using a motor-driven tap changer.

If the transformer is unattended, voltage drop control may be achieved by pulsing the tap changer's drive motor with a voltage regulating relay. Operation of this relay will cause the motor to rotate in such a direction that taps are changed which will maintain the voltage at approximately a constant value. As some voltages are self-correcting, a time-delay relay is used in conjunction with a voltage regulating relay. This allows a certain length of time to elapse between the moment the

voltage variation is picked up and the motor is set in operation. Thus, this device will eliminate a large number of unnecessary operations and greatly increase the life of the apparatus.

Remote Manual Control. In applications where it is necessary to correct only for voltage drop or phase angle during peak load periods, a remote hand switch or supervisory control is used. Both of these devices will actuate the tap changer's drive motor from a remote location and are used in place of the voltage regulation relay.

Hand Crank. For maintenance or emergencies, a hand wheel or crank is provided so that the mechanism can be operated at the transformer. In order that the motor cannot be operated while the crank is being used, the crank is usually the means by which a safety switch in the motor circuit is closed.

Phase-Angle Control

With the rapid increase in the use of electricity and the demand by consumers for uninterrupted service, together with the desire of the power companies to be able to supply such service, it may become necessary to transmit power to a given load center over two parallel transmission lines. Also, to generate the

electricity more economically, it has often become advisable to interconnect several generating stations so that, as the load becomes too large for one station, additional stations may be placed on the line or, as the load center changes, the load may be shifted from one station to another. In order to do this, it is

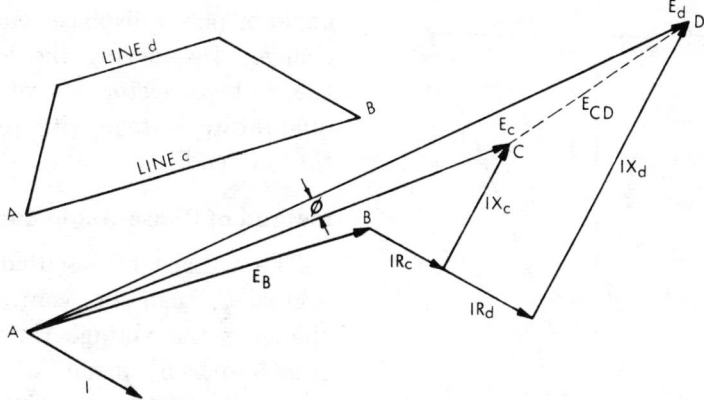

Fig. 4-7. Vector diagram showing phase displacement between voltages of two lines of different impedances supplying equal loads.

necessary to regulate the phase displacement between the two parts of the circuit while carrying load. This phase-angle control is obtained by adding to the line a suitable voltage at some angle with the line voltage.

To see just how this is accomplished, let it be assumed that two systems are to supply equal loads to a given load center and that these two systems or lines are of unequal length having unequal resistances and reactances. Such a system is shown in Fig. 4-7 where B represents the load center with load supplied over the two lines c and d connected in parallel. IR_c and IX_c represent the resistance and reactance drops in the line c due to the load of I amperes in line c. Assuming that a voltage E_B is required at the load center, then by adding IR_c and IX_c vectorially to E_B gives voltage E_c required at A

to send this load I amperes to B. Assuming now that line d has greater resistance and reactance than line c, and that it is to carry the same amount of current, the resistance drop IR_d and the reactance drop IX_d will be greater than in line c. The resultant voltage E_d required at A to send this current to B at a fixed voltage of E_B will be $E_B + IR_d + IX_d$ added vectorially. It will be seen from the diagram that this voltage E_d required at A is greater than E_c required at A and these two voltages are out of phase with each other.

If the two lines are to carry equal power load and equal wattless power in each line when connected together as shown in Fig. 4-7, it is necessary either to increase the impedance of the line c to that of line d so that the impedance drops in both lines are the same, or to add

217

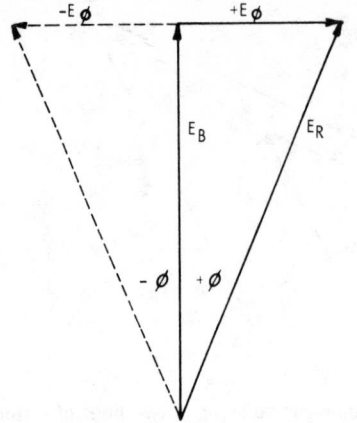

Fig. 4-8. Vector diagram showing principle of phase-angle control.

to the circuit a voltage E_{CD}, which is the difference in voltage between E_c and E_d. Since the two voltages E_c and E_d are not in phase with each other it is obvious that it is not sufficient simply to add voltage to circuit c by changing the voltage ratio, but that it is necessary to add a voltage which is at some angle with the reference voltage.

Fig. 4-8 shows the fundamental principles of the phase-angle control equipment. E_B represents vectorially the reference voltage, and $+E$ or $-E$, the voltage which is to be added at right angles to voltage E_B in order to obtain a resultant voltage E_R having a phase displacement of ϕ degrees with respect to the reference voltage E_B. Voltage control is obtained by varying the length of the reference voltage vector E_B.

The angle ϕ which represents the

angle of phase displacement can be changed by varying the length of the voltage vector E which is a quadrature voltage with respect to the voltage E_B.

Method of Phase-Angle Control

The voltage represented by the vector E_B can be controlled by changing the voltage ratio of the transformer by means of the tap-changing-under-load equipment previously explained. The quadrature voltage vector E can be controlled similarly by combining suitable variable quadrature components of voltage with the normal line voltage.

Fig. 4-9 shows a schematic diagram of a three-phase regulator unit for phase-angle control. For the sake of explanation let it be assumed that approximately plus or minus 6 degrees of phase control is desired in eight 1½ percent steps. The equipment consists of an exciting transformer excited from the line which is to be controlled, with its secondary connected to a series transformer inserted in the same line. Each secondary winding of the exciting transformer is divided into four equal parts so that, with the use of the preventive autotransformer, eight steps in voltage are obtained. The principle of operation of this regulating transformer secondary and autotransformer is the same as explained for the single-winding method of voltage control;

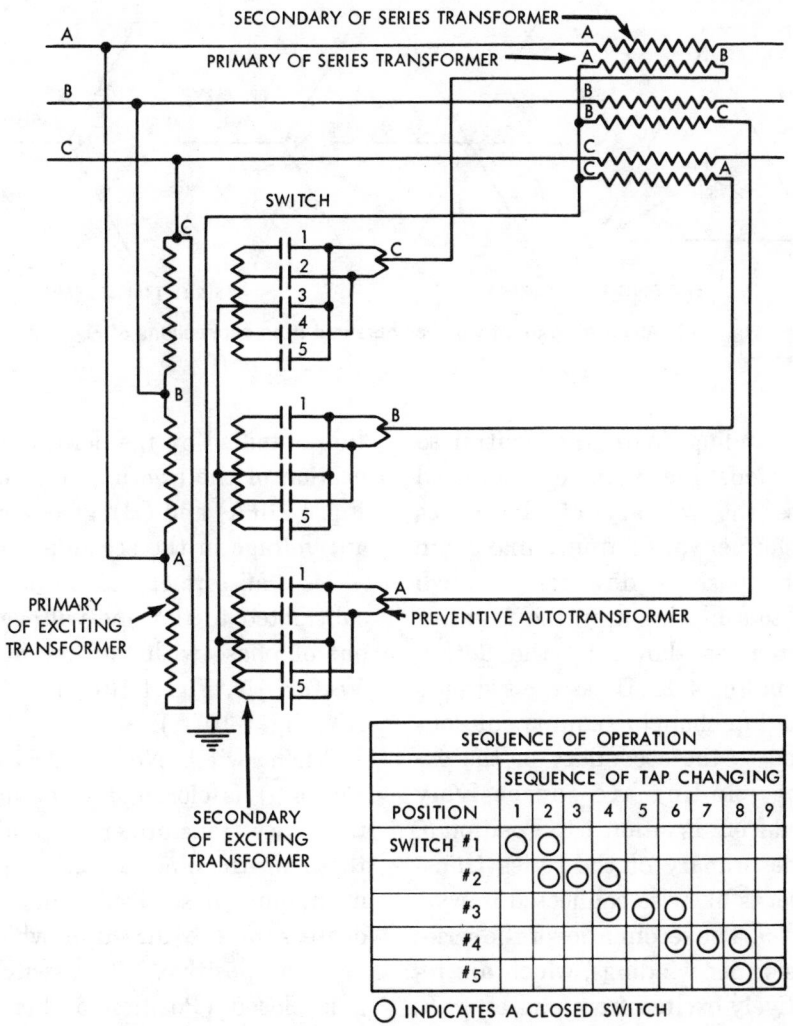

Fig. 4-9. Schematic diagram of three-phase regulator unit with phase-angle control.

in fact, the same type of equipment is used for both.

The object of the equipment is to obtain a voltage in the secondary of the series transformer which will be at right angles to the normal line-to-neutral voltages, *AN*, *BN* and *CN*, as indicated in Fig. 4-10. The primary of the exciting transformer is connected in delta, Fig. 4-9. The secondary is connected in wye, with the neutral made in the middle of

219

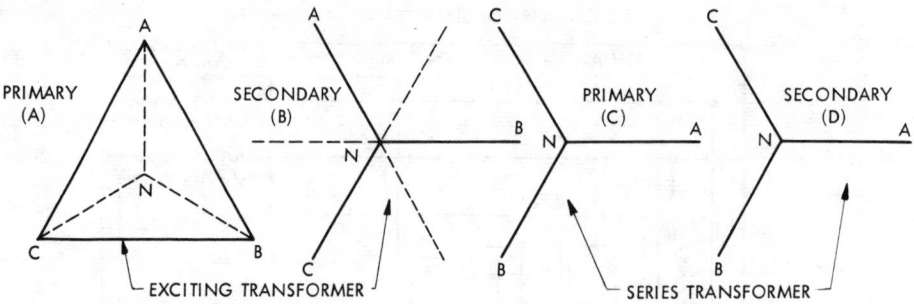

Fig. 4-10. Vector diagram of voltages obtained from connections of Fig. 4-9.

each winding. With the neutral so connected, the voltage impressed across the primary of the series transformer varies, from a maximum in the positive direction through zero, to a maximum in the negative direction as shown by the dotted lines in Fig. 4-10(B) as the switches of the tap-changing equipment connected to the secondary of the exciting winding are successively closed from Position *1* to Position *5.*

The primary of the series transformer is likewise connected in wye, Fig. 4-9. The secondaries of the series transformer windings, which are respectively excited from phases *A, B, C,* of the exciting transformer, are respectively connected in series with phases *B, C* and *A* of the line in which phase-angle control is desired. This 120° displacement between the primary of the series transformer, *AN,* Fig. 4-10 (C), and the secondary of the exciting transformer, *AN,* Fig. 4-10(B), together with the 30-degree displacement in the opposite direc-

tion obtained by the delta-wye connection of the exciting transformer, Fig. 4-10(A) and (B), gives a resultant voltage in the secondary of the series transformer (to be added or subtracted to the line voltage) 90° out of phase with the line voltage. Vecto*r AN,* Fig. 4-10(D) is 90° to *AN,* Fig. 4-10(A).

When switch No. *1* (Position *1,* Fig. 4-9) is closed, the voltages of the series transformer are added to those of the line so as to give a maximum phase displacement of 6 degrees in the direction which we may call positive. When switch No. *3* is closed (Position *5,* Fig. 4-9), the voltages impressed across the series transformer are zero, therefore there is no displacement of voltage in the line. When switch No. *5* is closed (Position *9,* Fig. 4-9), the voltages of the series transformer are added to those of the line so as to give a maximum phase displacement of 6 degrees in the opposite direction.

220

In our example we have obtained a maximum phase control from plus 6 degrees to minus 6 degrees, or a total of 12 degrees. The actual transitions between these two limits are made in 8 equal steps of approximately 1½ degrees per step, in accordance with the sequence shown in Fig. 4-9, which is similar to that shown in Fig. 4-6 for the single-winding method of tap changing under load.

Referring to Fig. 4-8, the voltage E is the secondary voltage of the series transformer 90° out of phase with the line voltage represented by E_B, and ϕ for maximum phase displacement as assumed in the foregoing example is 6 degrees. ϕ is positive $(+\phi)$ for Positions *1* to *5*, and negative $(-\phi)$ for Positions *5* to *9*. The actual value of the voltage E of the secondary of the series transformer for 6-degree phase displacement is 10.5 percent of the line voltage. Therefore, a regulating unit for plus or minus 6 degrees phase-angle control corresponds very nearly to a regulating unit of 10 percent buck or boost for voltage control.

By changing the primary connection of the series transformer from wye to delta in the proper sequence with respect to the secondary of the exciting transformer, the voltages in the secondary of the series transformer will be in phase with the line-to-neutral voltages, which is the condition required for voltage control. Thus it is possible to use the same equipment for either voltage control or phase-angle control by proper connections of the series transformer. Frequently both voltage and phase-angle control are combined in one regulating unit. This requires an additional set of tap-changing equipment, including the preventive autotransformer and another series transformer. The same exciting winding may be used for both circuits however. When both voltage and phase-angle control are desired for new transformers, it is more economical to obtain the control directly on the main power transformer.

Transformer Connections

Practically all the electrical power in use today is generated by three-phase alternators. Likewise the transmission of this power and the distribution on the primary side are based on the three-phase system. In the early development of the alternating current circuit many two-phase generators were installed in generating stations, and this power was transmitted on the primary side over a two-phase circuit. The secondary distribution circuit therefore provided energy for two-phase motors and single-phase lighting and motor loads. Because of the greater economy effected by the transmission of power over a three-phase system, many of these two-phase generators have been replaced with three-phase installations. With the use of the two-phase to three-phase transformer connection, many of these two-phase generators are still in use, but the power is transmitted on the primary side over a three-phase system. On the distribution end of this system, three-phase power is thus available for three-phase motors. To permit the continued use of the two-phase motors which were originally connected to the two-phase distribution system, the three-phase to two-phase transformer connection is employed.

Although the single-phase circuit is not used in the transmission of power nor on the primary side of the distribution circuit, it must not be inferred that single-phase transformers are few in number or have little application. The majority of the three-phase transmission circuits in this country are made with three single-phase transformers connected in various ways to obtain the three-phase power. Thus it is seen that transformers are used not only for voltage transformation, that is for stepping up or stepping down voltage whether it be single-phase, two-phase, or three-phase, but they are also used to obtain various phase transformations.

Among the many transformer connections that are thus possible in power and distribution service for both voltage and phase transformation, the following are the most common:

1. Voltage transformation
   (a) Single-phase
   (b) Two-phase
   (c) Three-phase
2. Phase transformation
   (a) Three-phase to single-phase
   (b) Three-phase to two-phase or two-phase to three-phase
   (c) Three-phase to two-phase and three-phase
   (d) Two-phase to six-phase
   (e) Three-phase to six-phase

Voltage Transformation

Single-Phase Circuits

The low-voltage winding of a standard distribution transformer connected to a single-phase circuit is ordinarily made in two equal sections. These sections are arranged so that they may be connected in series or in parallel. This arrangement permits current to be delivered at two voltages, one of which is double the other. Figs. 5-1, 5-2, and

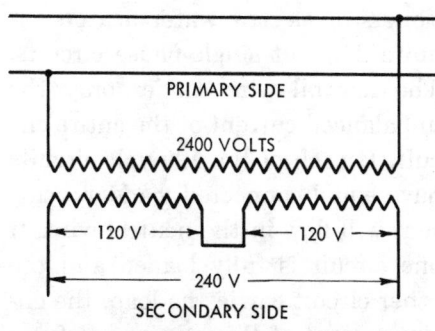

Fig. 5-1. Low-voltage sections connected in series, forming two-wire circuit.

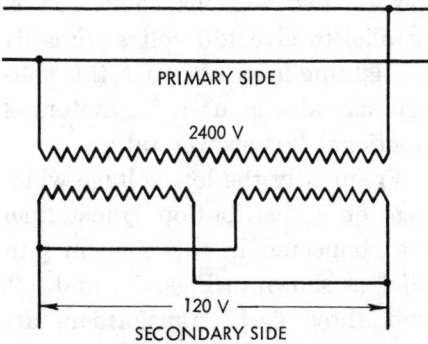

Fig. 5-2. Low-voltage sections connected in parallel—two-wire circuit.

223

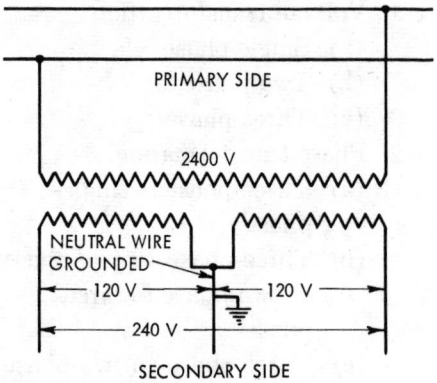

PRIMARY SIDE

2400 V

NEUTRAL WIRE GROUNDED

120 V 120 V

240 V

SECONDARY SIDE

Fig. 5-3. Low-voltage sections connected for three-wire service — three-wire, single-phase circuit.

5-3 show a distribution transformer with a single-winding primary and a two-section secondary, connected to a 2400-volt primary (supply) circuit.

In Fig. 5-1 the secondary winding has its two sections (each of 120 volts) connected in series to give a load voltage of 240 volts. Such a connection is employed where a motor load only is required.

In Fig. 5-2 the secondary winding has its two sections connected in parallel to give 120 volts, primarily for lighting loads although this voltage may also be used for motors of fractional horsepower rating.

Frequently the low-voltage windings of a distribution transformer are connected in series or in parallel, as shown in Figs. 5-1 and 5-2, and three such transformers are then connected into a three-phase bank giving three-phase service for

motor loads. It is more common, however, to connect the two sections of the low-voltage winding in series for three-wire service. The third wire of this three-wire circuit is connected to the secondary winding at the point where the series connection of the sections is made. This third wire is commonly known as the *neutral wire*. Such a connection is shown in Fig. 5-3. This permits the use of the 240-volt circuit for motor load and at the same time two 120-volt circuits for lighting.

Today the majority of new homes are wired for three-wire service. Electric ranges, dryers, central climate control systems are designed for three-wire operation, therefore it is necessary to provide three-wire service to homes which have these appliances. In the home the three-wire service entrance conductors terminate at the service equipment panel. Most of the individual branch circuits carried through the house are at 120 volts. Some circuits which feed larger loads such as air conditioners or electric water heaters require 240-volt single-phase circuits. The neutral wire carries only the unbalanced current of the entire circuit. If each of the 120-volt circuits have equal connected loads, no current will flow in the neutral wire. If one circuit is fully loaded and the other circuit carries no load, the entire current of this one circuit flows through the neutral wire. If the second circuit is carrying a load less in value than that of the first circuit, the neutral wire will carry only the difference in current of these two loads. The 240-volt motor load is carried by the two outside wires. The neutral wire therefore seldom carries as much current as either of the two outside wires. However, it should be made large enough to safely carry whatever current would be caused to flow through the circuit due to a fault such as a short circuit in either one of the two sides of this three-wire circuit. Each of the two outside wires should be fused or provided with circuit breakers which will operate to protect both the transformer winding and the connected load against undue overloads, short circuits, or grounds.

The neutral wire of a three-wire, single-phase circuit is required to be grounded. Circuits are grounded to limit excessive voltages from lightning, line surges, or unintentional contact with higher voltages from lines and to limit the voltage to ground during normal operation. If the neutral were not grounded and one of the outside wires came in contact with the ground, the other side of the circuit would rise to a voltage above ground equal to the series voltage of the two winding sections. Also in case of a puncture of the insulation between the primary and the secondary windings of the transformer, the secondary winding would

have a potential above ground equal to that of the primary winding with which it is in contact. If the neutral is grounded, the low-voltage winding cannot be raised to a potential above ground any higher than the voltage induced in each section.

The four leads of the low-voltage sections, as shown in Figs. 5-1 to 5-3 inclusive, are brought directly to three or four bushing leads mounted on the transformer case. If the current-carrying capacity of the bushing lead is sufficient to carry the current of the two sections when connected in parallel, only three bushing leads are required, although in many instances all four leads are brought out of the case. When the capacity of the transformer is such that the bushing lead is not large enough to carry the current of the two sections connected in parallel, all four leads are brought out of the case. When only three bushing leads are provided, the transformer is usually connected for three-wire service by the manufacturer. For two-wire service as shown in Fig. 5-1, it is then necessary to make this connection on the inside of the case. With all four leads brought out of the case all connections can be made externally as shown in Figs. 5-1 to 5-3 inclusive and it is not necessary to remove the cover to see that the transformer is properly connected.

Primary Winding Taps. Taps usually are provided on the primary

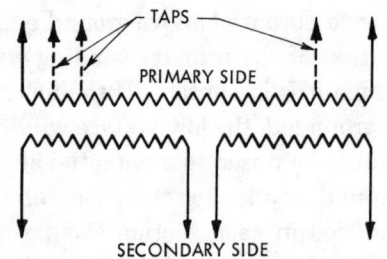

Fig. 5-4. Schematic diagram of transformer with end taps in the primary winding.

windings of higher-voltage classes of transformers, i.e., voltages 6,900 volts and above. The main purpose of these taps is to obtain normal secondary voltages when the transformer is used on the line at points where the primary voltage is low due to line drop. These taps may be located either at the two ends of the primary winding, as shown in Fig. 5-4, or in the middle of the winding, as shown in Fig. 5-5. End taps usually are provided on relatively low-voltage windings, with all leads from the winding terminating at a terminal board within the transformer

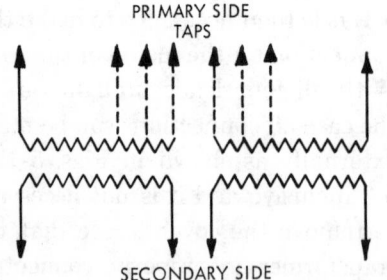

Fig. 5-5. Schematic diagram of a transformer with mid-taps in the primary winding.

case. Mid-taps, or taps located near the center of the winding, as shown in Fig. 5-5, are provided on those windings in which the voltage ratio is to be changed by means of a tap changer, or where the voltage of the winding is relatively high. If end taps were used on windings of relatively high voltages, that is, above 13,200 volts, it would be necessary to place additional insulation in the winding on each side of the taps to protect the winding from surges on the line.

When end taps are used the voltage ratio of the transformer is changed by connecting the line leads directly to the various tap leads. On transformers provided with mid-taps, as in Fig. 5-5, the line leads are permanently connected to the two ends of the winding, which are properly insulated for voltage surges on the line.

These taps whether placed near the ends or the middle of the winding are often spoken of as "full-capacity" or "reduced-capacity" taps. Full capacity means that the normal rating of the transformer may be secured when operating on taps. In order that normal rating may be obtained at these tap voltages the input current must be increased because of the reduced voltage at which this current is delivered. With reduced-capacity taps, the normal rating of the transformer decreases by the same percentage by which the tap reduces the voltage of the winding. The input current therefore never exceeds the normal current rating. The taps referred to above are called reduced-voltage taps because they are used on a reduced voltage to maintain constant rated voltage on the secondary winding. At times the winding is extended beyond the normal voltage rating of the transformer and taps are placed in this extended winding. Such taps are known as over-voltage taps and are used for voltages above the normal line voltage in order to maintain constant rated voltage on the secondary winding.

Polarity: Single-Phase Transformers

By common usage, polarity refers to the voltage vector relations of transformer leads as brought outside the tank, both high- and low-voltage leads being taken in the same order (from left to right or right to left) facing the same side of the transformer in both cases, and is, therefore, independent of the arrangement of the windings on the magnetic circuit. The polarity of the transformer can be changed by merely interchanging the position of the two leads of any one winding as brought out of the tank.

The voltage-vector relations can best be understood if the induced voltages only in the high-voltage and low-voltage windings are considered, regardless of which winding is the primary. Since the induced voltages in both coils are induced by the same flux, they must be in the same direction in each turn of both windings. This is indicated by the arrows on each turn of the windings of Figs. 5-6 to 5-9. If it is assumed that the induced voltage in the high-voltage winding H_1–H_2 is in the same direction as the order of lettering of the leads H_1–H_2 then the induced voltage in the low-voltage winding will be in the direction X_1–X_2 in each of the four figures.

The A.N.S.I. (American National Standards Institute) standardization rules state that high-voltage leads brought outside the case are to be marked H_1, H_2, etc., and the low-voltage leads X_1, X_2, etc., the order being such that when H_1 and X_1 are connected and voltage is applied to the transformer, the voltage between the highest numbered H lead and the highest numbered X lead shall be less than the voltage of the full high-voltage winding. When leads are thus marked, the polarity of the transformer is subtractive when H_1 and X_1 are adjacent and additive when H_1 is diagonally located with respect to X_1.

Referring to Figs. 5-6 to 5-9 inclusive, let it be assumed that a 1-to-1 ratio of voltage transformation exists. If H_1 lead is connected to X_1 lead in Figs. 5-6 and 5-9 and voltage applied to the high-voltage winding H_1–H_2, a voltmeter connected across the other two leads H_2 and X_2 will read the difference of these two voltages, since they are then in opposition, as is readily discernible from an inspection of these

figures. The voltmeter reading from the foregoing assumption will be zero. Therefore the polarity of the transformers of Figs. 5-6 and 5-9 is subtractive. If H_1 lead is connected to its adjacent X_2 lead in Figs. 5-7 and 5-8 and voltage applied to the high-voltage winding H_1–H_2, the voltmeter connected across the other two leads H_2 and X_1 will read the sum of the two voltages, since the two voltages are now in the same direction. The voltmeter reading from the foregoing

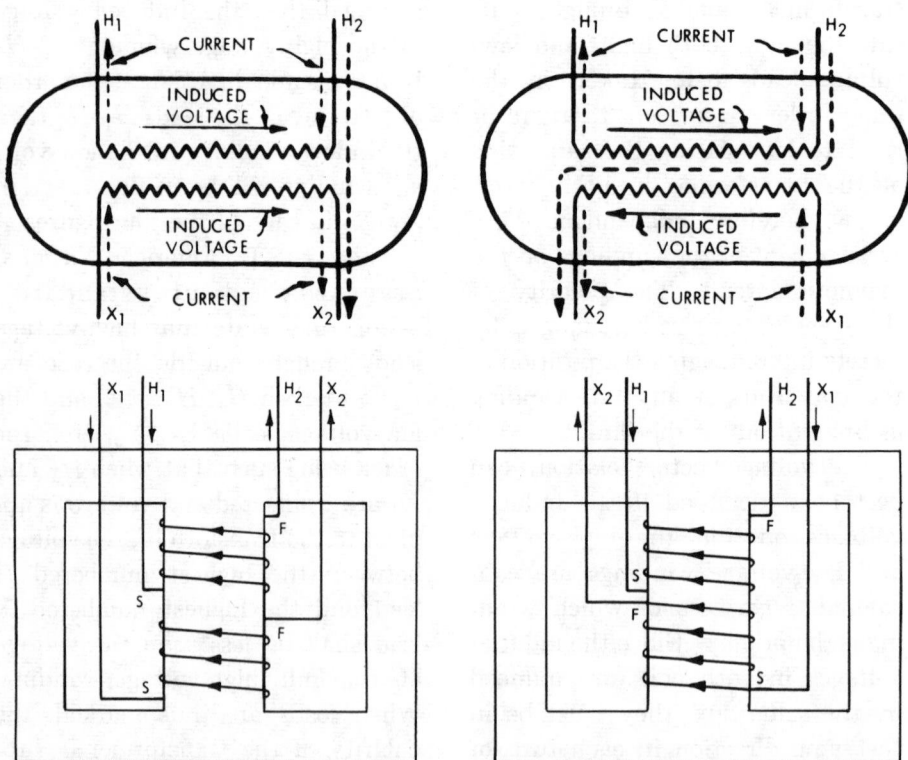

Fig. 5-6. Subtractive polarity.

Fig. 5-7. Additive polarity.

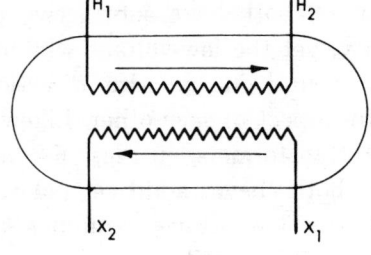 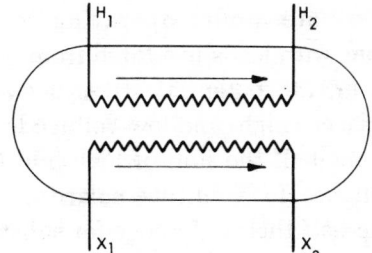

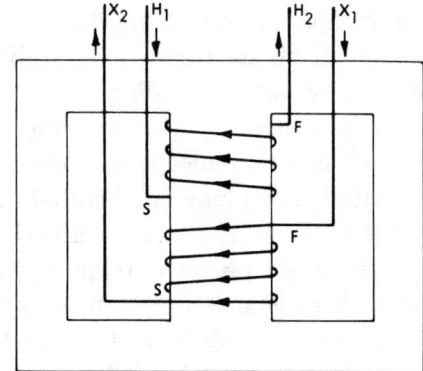

Fig. 5-8. Additive polarity.

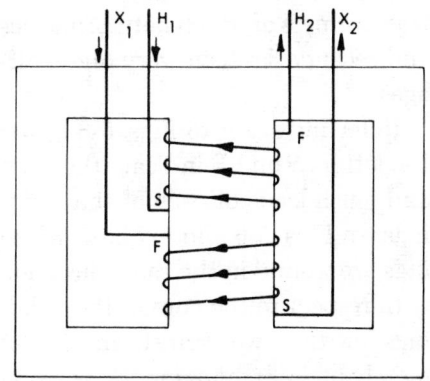

Fig. 5-9. Subtractive polarity.

assumption will be twice the voltage applied to the high-voltage winding. The polarities of the transformers of Figs. 5-7 and 5-8 are therefore additive. Therefore the lead lettering in each of the four figures agrees with the above rule.

To simplify the work of connecting transformers in parallel it is recommended that the H_1 lead be brought out on the right-hand side of the case facing the high-voltage side.

The polarity of a transformer is no indication of the direction of the turns of the high-voltage and low-voltage windings about the core, nor does it indicate the voltage stresses that exist between the turns of these two windings. However, subtractive polarity has a small advantage over additive polarity in the matter of the voltage stresses between external leads. That is, if two adjacent high- and low-voltage leads should accidentally come in contact, the voltage across the other leads would be the sum of high and low voltages for additive polarity and their differences for subtractive polarity. Fur-

229

thermore, under operating conditions with leads insulated from each other, the potential stress between adjacent high- and low-voltage leads is one-half the sum of high and low voltages for additive polarity, and one-half their difference for subtractive polarity. This advantage of subtractive polarity, which is ordinarily negligible, becomes appreciable for transformers of which both primaries and secondaries have very high voltages.

Referring again to Figs. 5-6 to 5-9 the letters S and F indicate the start and finish leads of each of the windings. In Figs. 5-6 and 5-7 both windings are wound in the same direction with respect to the core of the windings of the two transformers, yet merely by interchanging the position of the two X leads as indicated in Fig. 5-7 the polarity has been changed from subtractive to additive.

In Figs. 5-8 and 5-9 both windings are wound in opposite directions with respect to the core, and the same voltage stresses exist between similar parts of the windings of the two transformers although they are not the same as those which exist between the windings of Figs. 5-6 and 5-7. Again merely by interchanging the position of the two X leads as indicated in Fig. 5-9 the polarity has been changed from additive to subtractive.

The transformers of Figs. 5-6 and 5-9 both have subtractive polarity, yet the low-voltage windings are wound in opposite directions with respect to each other. Likewise the transformers of Figs. 5-7 and 5-8 both have additive polarity, yet the low-voltage windings are wound in opposite directions with respect to each other. The same polarity in each of these examples is obtained by interchanging the numbering of the start and finish leads of the low-voltage windings.

Transformers having leads marked in accordance with the rule as stated above may be operated in parallel by connecting similarly marked leads provided their ratio, voltages, resistances, reactances and frequencies are such as to permit parallel operation. Also, if the transformers are equipped with tap changers, they must be set on identical taps in each transformer.

If an attempt is made to parallel two transformers of opposite polarities but with similar voltage ratios, without due regard to the bus connection, the voltages cancel each other, the voltage drops to zero, and the circulating current rises to a very high value with a possible chance of burning out both transformers or blowing their fuses. It might be said that the two transformers short-circuit one another.

In general, distribution transformers of 200 kilovolt-ampere rating and less, having voltages of 7,500 volts and less, have additive polarity. All other distribution transformers have subtractive polarity. Some power transformers have additive polarity, although the majority have subtractive polarity.

The terms *additive* and *subtractive* generally are not applicable to instrument transformers. In connecting instrument transformers to wattmeters, watthour meters, power factor meters, etc., it is necessary to know the relative instantaneous direction of currents in the leads. The lead polarity of a transformer may also be considered as a description of the relative instantaneous direction of currents in its leads. Primary and secondary leads are said to have the same polarity when, at a given instant, the current enters the primary lead in question and leaves the secondary lead in question in the same direction, as though the two leads formed a continuous circuit. For example, referring to Fig. 5-6, the broken line arrows represent the instantaneous direction of current flow. The current is flowing in at the primary lead H_2 and out of the secondary lead X_2 as though the two leads H_2 and X_2 formed a continuous circuit as indicated by the dotted arrow. Similarly, in Fig. 5-7, the primary lead H_2 and the secondary lead X_2 although brought out at diagonal corners of the case are of the same polarity.

To indicate the relative instantaneous direction of currents in the instrument transformer one primary and one secondary lead of the same polarity as defined above, are each marked with a white polarity marker.

Referring again to Figs. 5-6 and 5-7, if these two figures represented instrument transformers the two leads marked H_1 and X_1 would instead, be marked with a white polarity marker while the other two leads would be left unmarked.

Test of Polarity

Three methods commonly are employed in testing for polarity and checking the lead markings. The method which is most convenient may be chosen.

Method One. When a given transformer of known polarity and correct lead marking, having the same ratio as the one to be tested, is available, the polarity can be checked by comparison as follows: Connect the high-voltage windings of both transformers in parallel by connecting the respective H_1 and H_2 leads. If the leads of the unit to be tested are not marked, assume that similarly located high-voltage leads on the two transformers are of the same marking. Also connect the left-hand side low-voltage leads (facing low-voltage side) of both transformers, leaving the right-hand side leads free. With these connections completed apply a reduced value of voltage to the high-voltage windings and measure the voltage between the two free leads on the low-voltage side. A zero reading of the voltmeter will indicate the relative polarity of both transformers to be identical, and the lead lettering of the low-voltage winding of the transformer under test should be the same as that of the known transformer. If, however, the voltmeter reads the sum of the low-voltages of both transformers, the polarities of the two units are relatively opposite, and the lead lettering of the low-voltage winding of the transformer under test should be reversed with respect to that of the known transformer. It is recommended that as a precautionary measure a light fuse or sufficient lamps be connected into the circuit before connecting in the voltmeter.

Method Two. The polarity may be determined by the use of direct current as follows: With direct current passing through the high-voltage winding, connect a high-voltage, direct-current voltmeter across the outlet terminals of the same winding so as to get a small positive deflection of the voltmeter needle. Then transfer the two voltmeter leads directly across the transformer, i.e., the lead from the right-hand high-voltage terminal being placed on the adjacent low-voltage terminal and the other lead to the other low-voltage terminal. The direct-current excitation is broken thereby, inducing a voltage in the low-voltage winding which will cause a deflection in the voltmeter. If the needle swings in the same direction as before, the polarity is additive. If the needle swings in the opposite direction, the polarity is subtractive.

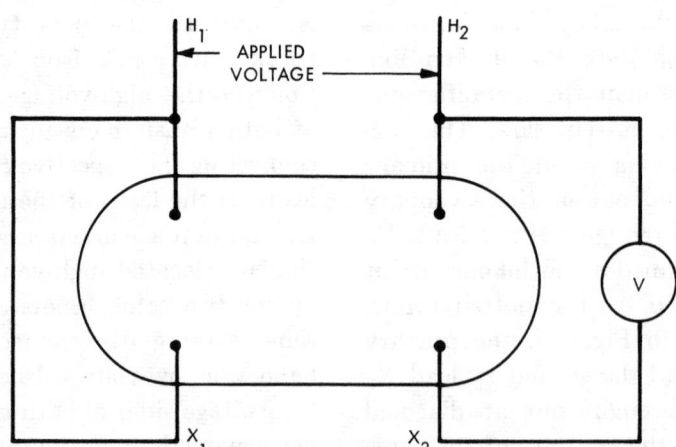

Fig. 5-10. Voltmeter method of checking polarity and lead markings.

Method Three. Connect the adjacent left-hand high-voltage and low-voltage outlet leads facing the low-voltage side of the transformer, such as H_1 and X_1 of Fig. 5-10. Apply any convenient value of alternating current voltage to the high-voltage winding, and take readings of the applied voltage and the voltage between the right-hand adjacent high-voltage and low-voltage leads, H_2 and X_2. If the latter reading is less than the former, indicating the difference in voltage between that of the high-voltage and low-voltage windings, the polarity is subtractive. If the latter reading is greater than the former the polarity is additive. This method is practically limited to transformers in which the ratio of transformation is 30 to 1 or less.

Single-Phase Transformer Connections

From the previous discussion on polarity, it was shown that since the same flux is responsible for both primary and secondary induced voltages, if both windings are wound in the same direction, the directions of the induced voltages must also be in the same direction. With such an arrangement, adjacent terminals of both windings will have similar numbers; thus at the instant that the induced voltage is in the direction H_2 to H_1 in the primary winding, its direction in the secondary is also X_2 to X_1. If the secondary winding is wound opposite of that to the primary, then the induced voltage will also be in the opposite direction. However, this also reverses the subscripts of the secondary winding, so that regardless of whether the transformer is of additive or subtractive polarity, the induced voltages will always be in the same direction with respect to their terminals.

Since in both cases, the H_2 and H_1 and X_2 to X_1 voltages are in phase, we may then represent both additive and subtractive transformers by voltage vectors as shown below.

$$H_1 \leftarrow H_2 \qquad X_1 \leftarrow X_2$$

Note that the first high-voltage vector may be drawn in any direction, but thereafter the direction and sense of all other vectors must agree with the direction and sense of the first high voltage vector. For instance, $H_2 \rightarrow H_1$ is another way of drawing $H_1 \leftarrow H_2$ and merely means that the vector has been rotated physically through an angle of 180 degrees without any change of phase in the voltage it represents.

If the vector which represents

233

voltage H_2 to H_1 is drawn $H_2 \rightarrow H_1$, the voltage vector H_1 to H_2, which is 180 degrees out of phase, must be drawn $H_2 \leftarrow H_1$; and similarly if voltage H_2 to H_1 is drawn $H_1 \leftarrow H_2$, the voltage vector for voltage H_1 to H_2 must be drawn $H_1 \rightarrow H_2$.

It must be noted that these vectors bear no relationship to the physical arrangement of the terminals on the transformers.

Let us now proceed to connect the two single-phase transformers to a single-phase supply. Since the H_1 terminal is always on the left when facing the low voltage, we will connect similarly marked terminals to the same line conductor as shown in Fig. 5-11.

If both single-phase transformers have the same polarity, subtractive for example, the low-voltage terminals will be marked as shown in Fig. 5-12.

To parallel these two single-phase transformers on the low voltage, we will start connecting the X_1 terminals together as shown in the previous figure.

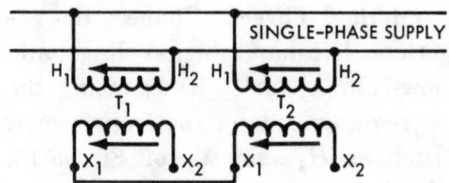

Fig. 5-12. Current direction of two single-phase transformers connected to a single-phase supply.

We can now predict what the voltage between the two X_2 terminals will be when the two X_1 terminals are connected together by adding the two low-voltage vectors. That is, voltage X_2 of T_1 to X_2 of T_2 will be the vector sum of X_2 to X_1 of T_1 and X_2 of T_2.

Assuming as before that the H_2 to H_1 vector is $H_1 \leftarrow H_2$, then regardless of whether the transformers have additive or subtractive polarity (which determines the physical location of the terminals only) the X_2 to X_1 electrical vectors will be $X_1 \leftarrow X_2$. (A voltage X_1 to X_2 which is 180 degrees out of phase with X_2 to X_1 may be represented by a vector of the same length but opposite in direction to X_2 *to* X_1, namely by $X_1 \rightarrow X_2$). Therefore, voltage X_1 to X_2 of T_2 would be shown as $X_1 \rightarrow X_2$. Voltage X_2 to X_1 of T_1 plus voltage X_1 to X_2 of T_2 equals:

$$X_1 \overset{T_1}{\longleftarrow} X_2 + X_1 \overset{T_2}{\longrightarrow} X_2 =$$

$$X_1 \overset{T_1}{\longleftarrow} X_2 = 0$$

$$\downarrow$$

$$X_1 \longrightarrow X_2$$

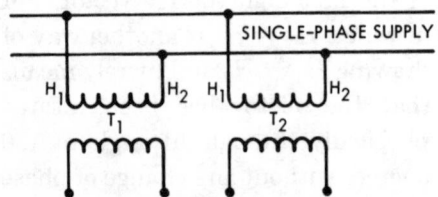

Fig. 5-11. Two single-phase transformers connected to a single-phase supply.

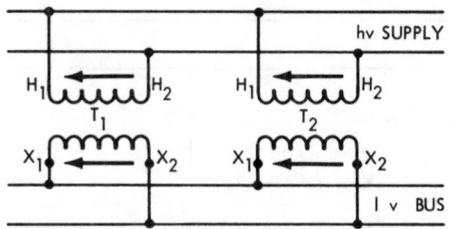

Fig. 5-13. Secondary connections of two single-phase transformers.

These vectors indicate no voltage difference between terminals X_2 of T_1 and X_2 of T_2; therefore, they may be safely connected as shown in Fig. 5-13.

If, however T_1 is of opposite polarity to that of T_2, the low-voltage terminals would be arranged as shown in Fig. 5-14.

We can then proceed as before by connecting similarly marked high-voltage terminals to the same line conductors and connect the low-voltage X_1 terminals together as shown in Fig. 5-14.

It should be noted that the determination of voltage X_2 of T_1 to X_2 of T_2 will be exactly the same as the previous example, since the low-volt-

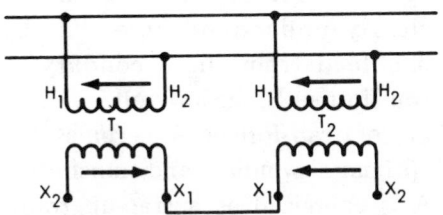

Fig. 5-14. Incorrect secondary connection.

age vectors will not change for additive or subtractive transformers having the same high-voltage reference voltage vector.

From the preceding discussion it is apparent that to properly parallel single-phase transformers, similarly marked high-voltage terminals are connected to the same high-voltage conductor, and similarly marked low-voltage terminals are connected to the same low-voltage conductor.

If an attempt is made to replace the T_1 transformer with one of opposite polarity to T_2 as shown in Fig. 5-14 without making the required alterations in the bus connection or retaining the same bus connections as shown in Fig. 5-12, a serious condition would result.

Two types of connections which may be used when connecting two transformers together for single-phase lighting and heating are the double-parallel or banked connection and the parallel-series secondary connection.

Figure 5-15 illustrates the banked connection which is used only for temporary or emergency conditions as a single transformer is more economical for the same output.

As shown previously, the similarly marked terminals are connected to the same line conductors with the addition of the neutral connection. Since the H_1 terminals are connected to the A phase, the line conductor to which the X_1 terminals

235

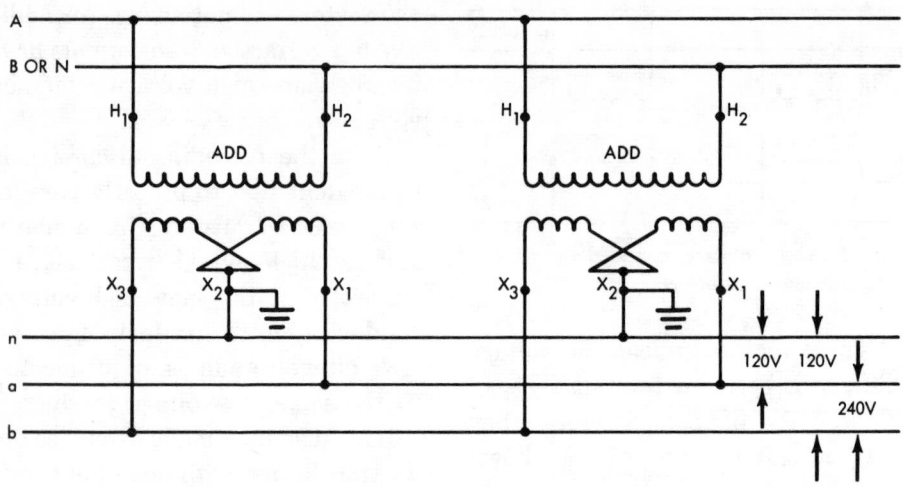

Fig. 5-15. Paralleling two single-phase transformers.

are attached will also be the same phase.

When connecting the secondaries, the voltage between terminals to be connected should be read with a voltmeter before the secondary connections of the second transformer are made to complete the parallel connection. The voltmeter should read zero with the primaries energized since similar terminals will be at the same potential if the proposed connection is correct. If there is a voltage reading, these are the wrong terminals to connect together and the connection must be changed.

There is always a great danger of *backfeed* when paralleling transformers; thus it is a good practice to always consider the transformer primary *HOT*, even if the fuse is blown.

It is, however, a common industry practice to backfeed a transformer put into service to avoid having an outage.

Figure 5-16 illustrates the backfeed on "banked" parallel connected transformers. Transformer B steps the primary voltage down to 240 volts and transformer A acts as a step-up transformer through the low-voltage winding at 240 volts. The high-voltage winding of transformer A would be energized at the rated voltage of the high-voltage winding even though its high-voltage fuses are blown. It should be clearly understood here that by backfeed from the secondary of transformer B, the low-voltage winding of transformer A becomes the "primary" winding, and transformer A is energized as a step-up transformer.

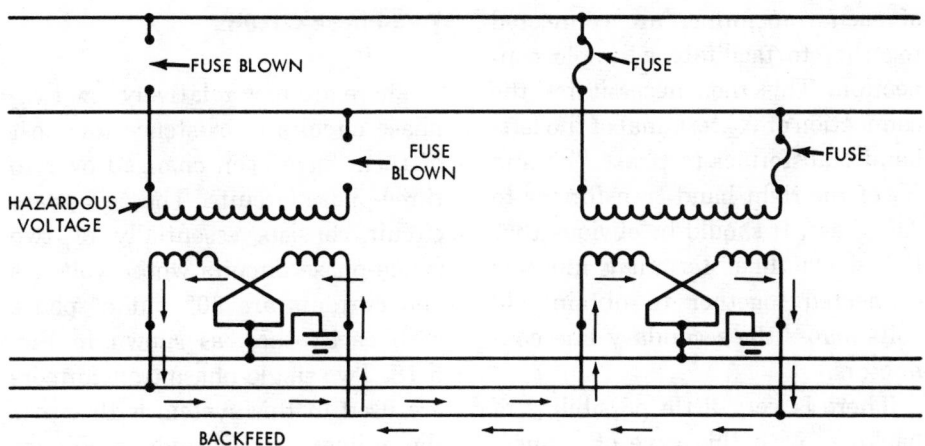

Fig. 5-16. Possible backfeed in connecting two single-phase transformers.

The parallel series secondary connection is again used only for temporary or emergency conditions as a single transformer is much more economical.

In this connection, each transformer is supplying service to only one side of the secondary. When transformers are connected parallel series, one transformer will not pick up the load from the other transformer when a fuse is blown. This connection would be better to use when transformers are paralleled without breakers in the secondary between them.

As can be seen from Figure 5-17, the two adjacent "unlike" terminals

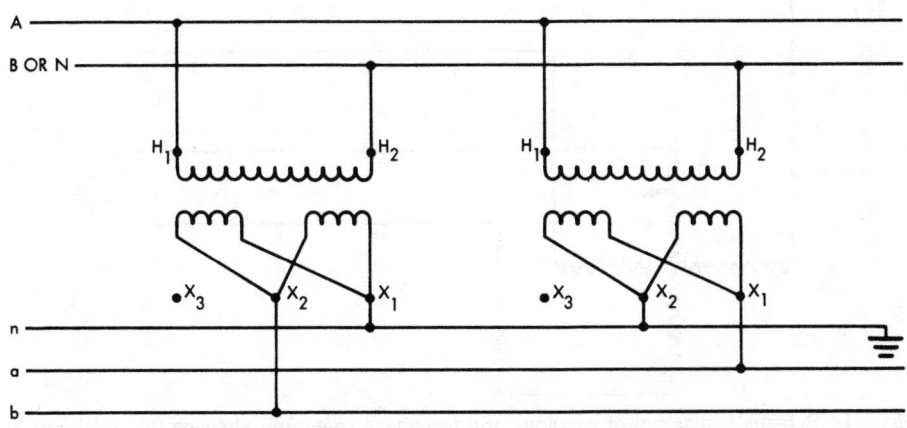

Fig. 5-17. Two single-phase transformers paralleled to supply.

of each transformer are connected together to facilitate a simple connection. This then necessitates the connection of X_2 terminal of the left-hand transformer to phase "b" and X_1 of the right-hand transformer to "a" phase. It should be obvious that the two "unlike" terminals must be connected together to obtain 240 volts across the secondary line conductors.

There is very little possibility of backfeed with this type of connection unless there is a load connected across the hot leads. The disadvantage of using this connection is that when a fuse blows, there is an interruption to a customer's service. This connection can be used when there is not a transformer of sufficient capacity to carry a load, yet two smaller transformers are available.

Two-Phase Circuits

There are now relatively few two-phase circuits in existence for most of them have been changed over to three-phase circuits. The two-phase circuit consists essentially of two single-phase circuits whose voltages and currents are 90° out of phase with each other, as shown in Fig. 5-18. Two single-phase transformers are used in this system with either the primary or secondary side or both sides connected for three-wire or four-wire circuits. Which of these arrangements is to be used depends on whether or not the two phases are interconnected within the generator or motor windings. For example, if two phases of the generator are connected at their neutral points the three-wire circuit on the

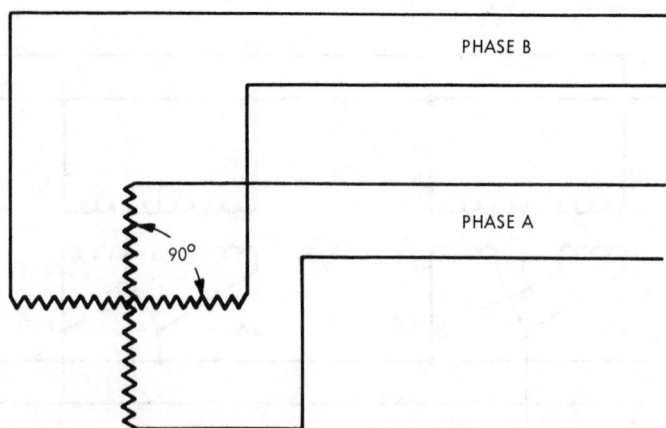

Fig. 5-18. Schematic diagram of windings of a two-phase generator, showing 90° relationship of voltages and currents.

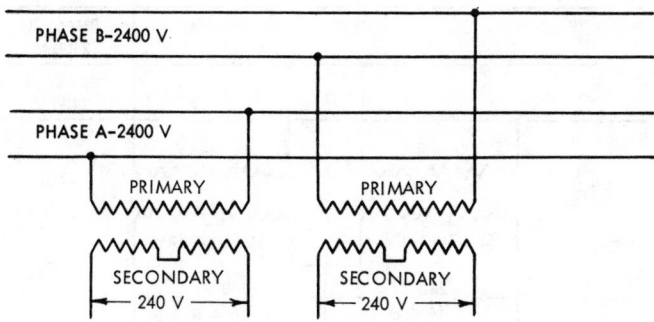

Fig. 5-19. Two-phase, four-wire circuit.

primary side will not be possible. If the phases are entirely free within the generator or motor, any external interconnection desired may be made. Fig. 5-19 shows the most commonly used two-phase connection, consisting of two single-phase transformers connected for two-phase, four-wire circuits on both the primary and secondary sides. The voltage-vector diagram of this connection is shown in Fig. 5-20. When a two-phase, three-wire circuit is desired the two single-phase transformers are connected as shown in Fig. 5-21, with the corresponding voltage-vector diagram as in Fig.

5-22. With this connection the voltage across the outside wires is $\sqrt{2}$, or 1.41 times the voltage across the winding of each transformer, and the middle wire carries the vector sum of the currents of the two windings. If the load on each of the two phases is the same, the middle, or "neutral," wire will carry $\sqrt{2}$, or 1.41 times the current of the outside wires. If one of the outside wires of the three-wire connection shown in Fig. 5-21 should become grounded, the voltage to ground becomes 1.41 times the voltage of either phase. If, however, the neutral wire is also grounded the maximum voltage can-

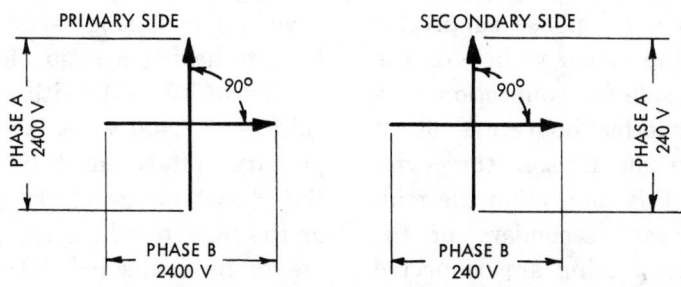

Fig. 5-20. Voltage-vector diagram of Fig. 5-19.

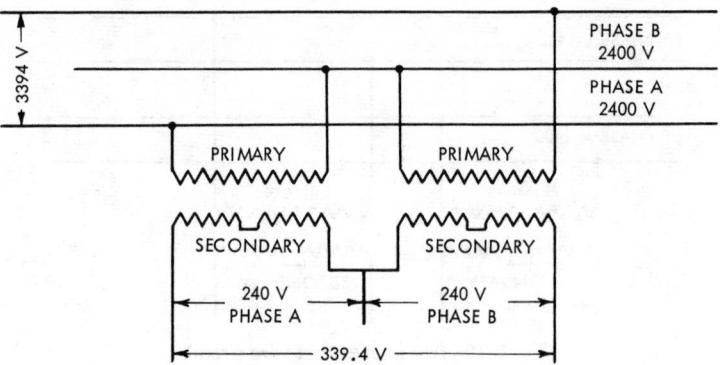

Fig. 5-21. Two-phase, three-wire circuit.

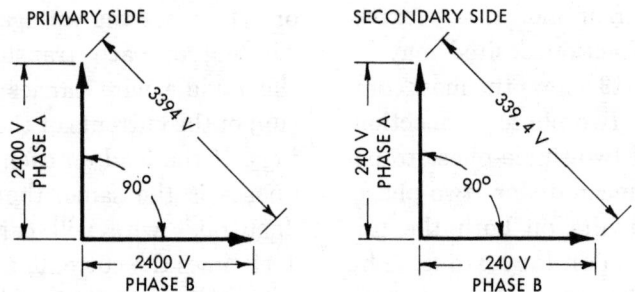

Fig. 5-22. Voltage-vector diagram of Fig. 5-21.

not exceed the phase voltage of either phase.

Figs. 5-23 and 5-24 show the connections of two single-phase transformers to a two-phase supply for obtaining the proper voltage on the secondary side for connection to a rotary converter delivering direct current to an Edison three-wire circuit. In this connection the midpoints of each secondary of the transformer winding are connected together and grounded, thereby providing the neutral of the Edison three-wire system, which usually is grounded, Fig. 5-23.

In all the foregoing diagrams we have assumed single-phase transformers having a ratio of transformation of 10 to 1, with a primary voltage of 2,400 volts. Actually the primary voltage must be equal to the phase voltage of the generator or the lines to which the primaries are to be connected. The ratio of transformation depends upon the

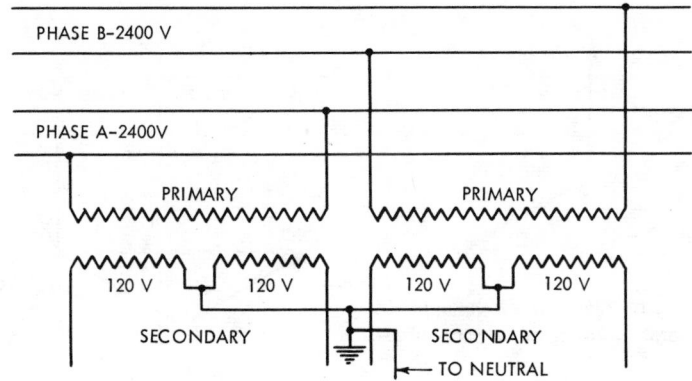

Fig. 5-23. Two-phase, five-wire circuit for rotary converter service.

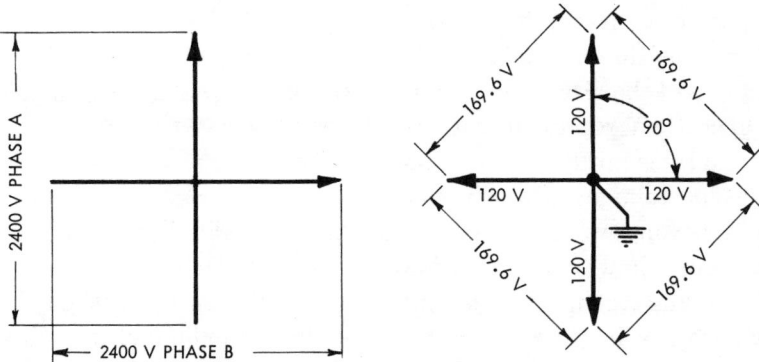

Fig. 5-24. Voltage-vector diagram of Fig. 5-23.

application, that is, it depends upon the desired secondary load voltage and the primary line voltage.

Three-Phase Circuits

Three-Phase Wye and Delta Connections. A three-phase system originates from three separate windings of an alternating-current generator or alternator. These three windings are located within the alternator so as to give three separate voltages which are out of phase with each other by 120 electrical degrees.

If these three circuits were kept separate from the generator to the apparatus using the currents, from these windings a total of six line leads would be required. This, of course, is impractical and therefore common wires are used. In forming the common wires, the individual phases of the alternator are connected in either of two methods, which are known as the wye and delta connections. For example let us assume that the three lines in Fig. 5-25 represent the vector rela-

241

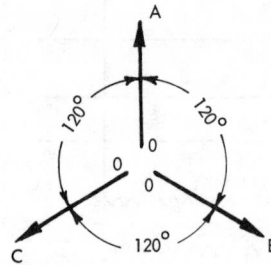

Fig. 5-25. Vector diagram of voltages induced in separate windings of a three-phase alternator.

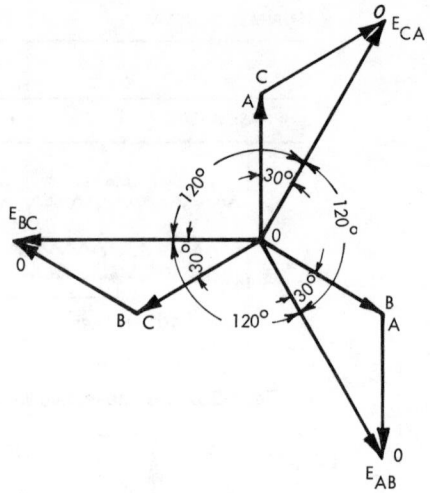

Fig. 5-26. Vector diagram of three-phase voltages connected in wye.

tionship of the three voltages generated in the three windings OA, OB, and OC of the alternator. These three lines then represent the direction and magnitude of the instantaneous voltages of each coil. In the following we shall assume that the current which would flow in each winding if connected to some external load is in phase with the voltage generated in each of the respective coils. If now we connect the three ends O together and bring out the remaining three ends, A, B, and C, as line leads we have a three-phase wye connection. The common point O of the three windings forms the neutral of the three-phase system. It is called the neutral because it is symmetrical with regard to the three coils and may be grounded without creating an unsymmetrical condition. With the three windings connected thus in wye the voltage E_{AB} between the two line leads A and B, for example, is the vector sum of the coil volt-

ages E_{AO} and E_{OB}, as shown in Fig. 5-26. Similarly the voltage E_{BC} between the two line leads B and C is the vector sum of the coil voltages E_{BO} and E_{OC}, and the voltage E_{CA} between the two line leads C and A is the vector sum of the coil voltages E_{CO} and E_{OA}. This vector sum gives the three-phase line voltages of the system equal to $\sqrt{3}$ times the voltage of one coil and 120° out of phase with each other. The "Y" (Wye) system is also known as the star system.

If instead of connecting these coils in wye we arrange them in a three-coil closed circuit and connect a line lead at each of the three junction points thus formed we have a delta system, as shown in Fig. 5-27. The three-phase line voltages and their phase relations are shown in Fig.

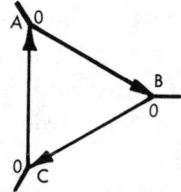

Fig. 5-27. Three generator coils connected in delta.

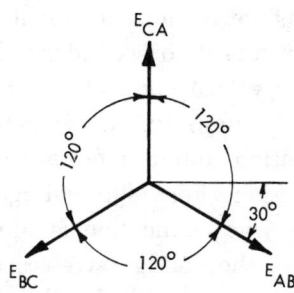

Fig. 5-28. Vector diagram of three-phase voltages connected in delta.

5-28. With this connection the line voltages are equal to the individual coil voltages. Although the three coils are connected to form a closed circuit, no current will flow in this circuit at no load, because the vector sum of three equal voltages 120° out of phase with each other is at all times equal to zero. With no voltage impressed across the entire circuit there can be no current flow. This connection is also known as a Δ or mesh system, although the latter term is seldom used in this country.

An inspection of Fig. 5-26 will

show that the line current of the three-phase system must be equal to the current in each of the individual coils. In Fig. 5-27, however, the line current as it reaches the junction point of the two windings has two paths through which it flows. Therefore the current in the windings of the delta-connected alternator is less than the line current. Actually the winding current is 57.7 percent of the line current, or is equal to the line current divided by $\sqrt{3}$.

To sum up the foregoing, it should be noted that in a wye-connected alternator the line voltage is equal to the individual coil voltage multiplied by $\sqrt{3}$, and the line current is equal to the coil current. In the delta-connected alternator the line voltage is equal to the individual coil voltage, and the line current is equal to the winding current multiplied by $\sqrt{3}$. It is also to be noted that the line voltages of the wye-connected alternator are 30° out of phase with the line voltages of the delta-connected alternator.

The windings of three single-phase transformers or one three-phase transformer are connected in the same manner as are separate windings of the three-phase alternator in order to make a three-phase group or a single three-phase transformer. The primary windings may be either wye- or delta-connected and the secondary windings may be so connected also. This gives four possible combinations of the transformer windings as follows:

1. Wye to wye
2. Wye to delta
3. Delta to delta
4. Delta to wye

Some of these combinations are approved and in common use and some have objections which limit their use.

Three-Phase Wye-to-Wye Connection. The wye-to-wye connection has its advantages and its disadvantages, but the disadvantages predominate and have restricted its use. One advantage is that the voltage of each single-phase unit connected to form a wye-to-wye connection is only 57.7 percent of the line voltage. This tends to make the connection suitable for power transmission, as the individual transformers are wound for a relatively low voltage, and nearly double this voltage (173 percent) is obtained on the line. The wye connection also permits the grounding of the neutral on either or both primary and secondary windings. If the primary neutral is carried back to the neutral of the alternator, unequal loads may be taken off the secondary windings between the secondary neutral and any of the three line leads. The unbalanced current will flow through the neutral back to the alternator. This permits lighting loads to be taken off between the neutral and any of the three-phase wires and, at the same time, power load from the three-phase leads.

The wye-to-wye connection is used when tying together two high-voltage transmission systems of unequal voltage. In this case it is necessary that the connection used be such as not to cause a shift in phase from primary to secondary. Therefore only two connections are applicable, namely, the wye-to-wye connection and the delta-to-delta connection. For high voltages the wye-to-wye connection is preferred because the voltage stresses of the windings to ground with the neutral grounded are only 58 percent of the voltage stress to ground of similar units connected delta-to-delta. When the wye-to-wye connection is used for this purpose a tertiary winding is provided to suppress the third harmonic voltages which would otherwise appear on the system. The tertiary is a third winding and is often used in power transformers to provide station power requirements or tie with synchronous condensers.

In many cases, unless a tertiary winding is added to provide a path for the circulating currents to flow through, interference with telephone circuits may result. When using single-phase transformers, this trouble may be avoided by grounding the primary and secondary neutrals.

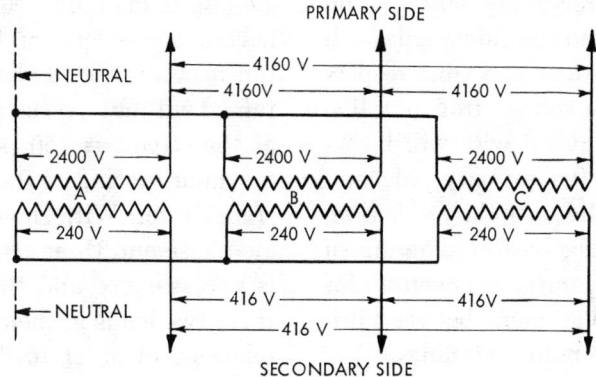

Fig. 5-29. Three single-phase transformers connected wye-wye.

The fact that the wye-to-wye connection is subject to disturbances from harmonic voltages and currents constitutes one of the objections to its use. Another objection to its use is that unbalanced loads cannot be carried on the secondary side unless the primary neutral or fourth wire is provided. Another objection to the use of this connection arises from the fact that it is practically impossible to construct three single-phase units, or even a three-phase unit, in which the magnetizing currents of each of the three windings are exactly the same. This, then, inherently makes it impossible to have perfectly balanced voltages in each of the three windings of the wye-to-wye connection.

Fig. 5-29 shows three single-phase transformers connected wye-to-wye, with the neutral or fourth wire indicated by a broken line. If the single-phase voltage rating of each transformer is 2,400 volts to 240

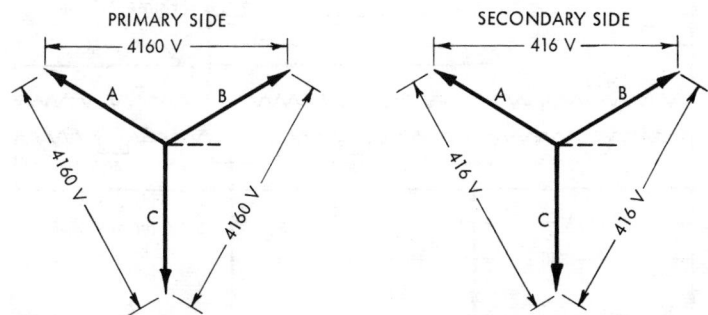

Fig. 5-30. Voltage-vector diagram of Y-Y connection with 0° phase displacement.

volts, the three-phase voltages on the primary and secondary sides will be 4,160 volts and 416 volts respectively, and the voltage from any line lead to the neutral wire will be 2,-400 volts on the primary side and 240 volts on the secondary side. Fig. 5-30 is a voltage-vector diagram of these three units connected for 0° phase displacement between primary and secondary windings.

Three-Phase Wye-to-Delta Connection. The wye-to-delta connection obtained with three single-phase transformers or one three-phase unit is used to a great extent for power transmission and distribution. It permits single-phase and three-phase loads to be drawn simultaneously from the delta-connected secondary at the same voltage.

Assuming that the wye-to-delta connection is made with three single-phase transformers as shown in Fig. 5-31 and assuming that a single-phase load only is to be connected across one of these phases, the maximum load that can be obtained without overloading any part of the circuit is 150 percent of the maximum rating of the single-phase transformer. With simultaneous single-phase and three-phase loading it is necessary to add the currents of these two loads in their proper phase relations in order to determine the maximum loads that can be drawn from this connection. The resultant current in any one of the three transformers due to this loading must not continuously exceed the maximum current rating of the winding. The voltage-vector diagram for wye-delta connected transformers is shown in Fig. 5-32.

With the wye-delta connection of three single-phase transformers one unit may be disconnected from the circuit and service maintained with the secondary operating in open-delta at 57.7 percent of normal bank

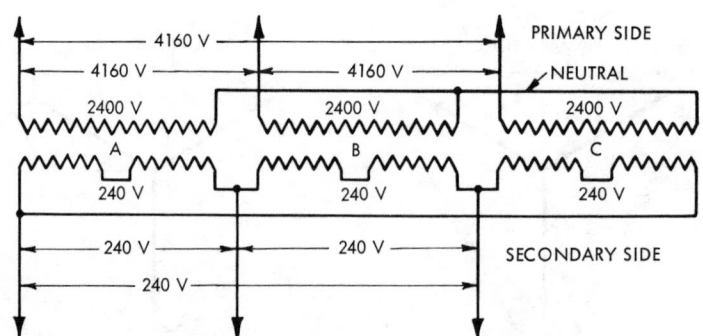

Fig. 5-31. Three single-phase transformers connected wye-delta.

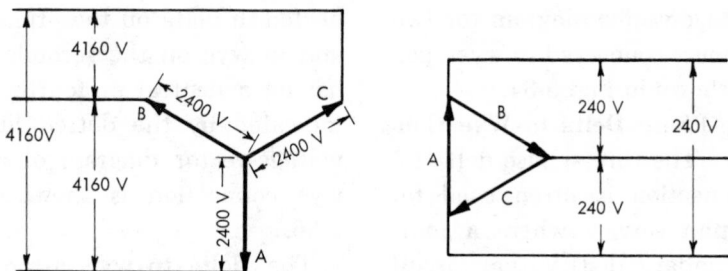

Fig. 5-32. Voltage vector diagram of Fig. 5-31.

capacity, provided the neutral on the primary side and the neutral of the source of supply for the primary are grounded. Such an emergency connection is shown in Fig. 5-33. The system will be unbalanced and considerable telephone interference may result from such a connection.

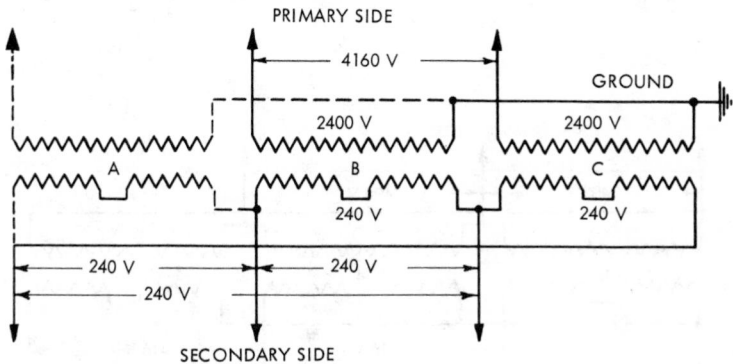

Fig. 5-33. Emergency connections of a wye-to-open delta transformer bank with the disconnected transformer shown dotted.

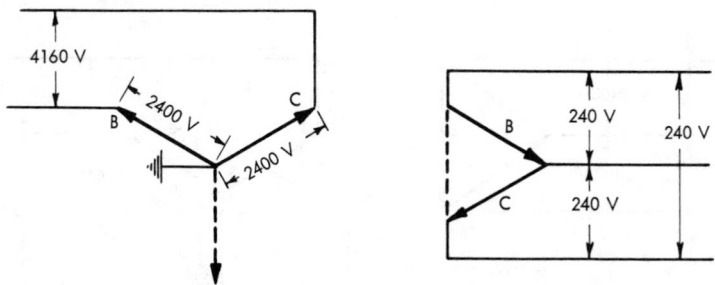

Fig. 5-34. Voltage-vector diagram of Fig. 5-33. Same as Fig. 5-32 except *HV* neutral is grounded and transformer *A* is not connected in the circuit.

The voltage-vector diagram for two transformers connected in wye-open delta is shown in Fig. 5-34.

Three-Phase Delta-to-Wye Connection. The three-phase delta-to-wye connection is often used for distribution service where a four-wire secondary distribution circuit is desired. The delta primary connection permits single-phase loads to be connected between the secondary neutral and the three-phase line leads. The three-phase star voltage is $\sqrt{3}$ times the single-phase (line to neutral) voltage. Fig. 5-35 shows three single-phase transformers connected in delta on the primary side and in wye on the secondary side, having a neutral or fourth wire as indicated by the dotted line. The voltage-vector diagram of a delta-wye connection is shown in Fig. 5-36.

The delta-to-wye arrangement, with the neutral of the wye connection available, is sometimes used for supplying power to a synchronous converter delivering direct current for an Edison three-wire circuit. The third wire of the three-wire direct current circuit is obtained by connecting the third wire of the

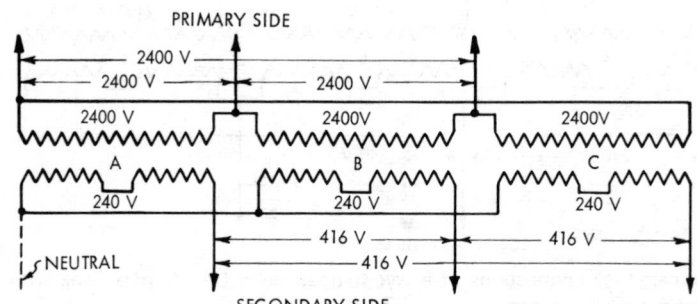

Fig. 5-35. Three single-phase transformers connected delta-wye.

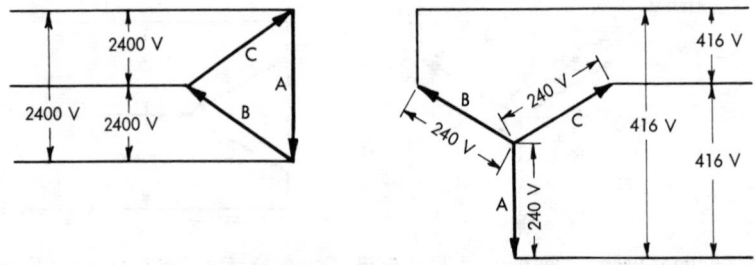

Fig. 5-36. Voltage-vector diagram of Fig. 5-35.

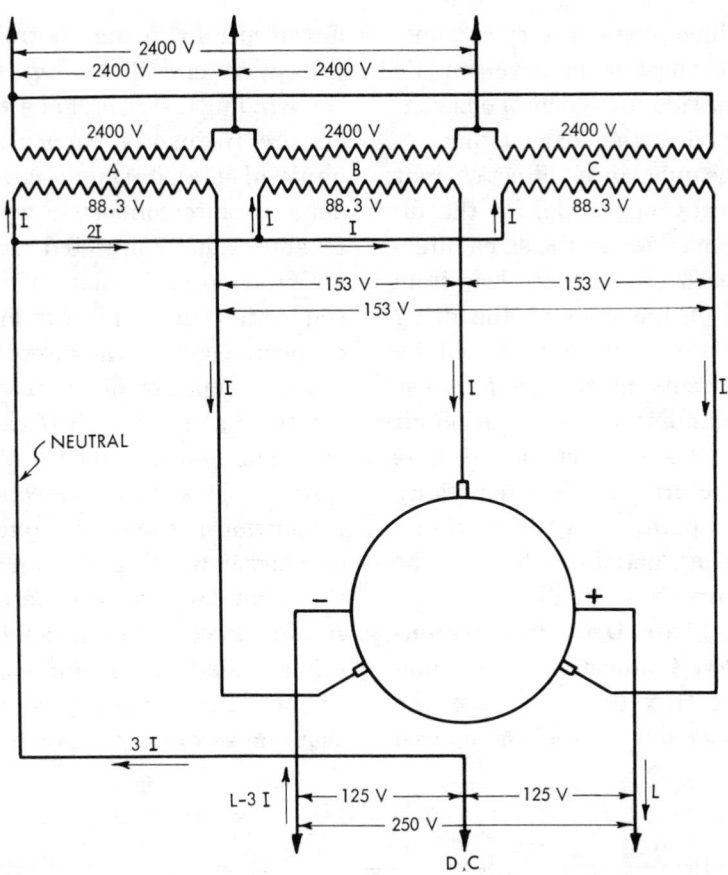

Fig. 5-37. Three single-phase transformers with a voltage ratio of 2400 volts to 88.3 volts connected delta to wye with the secondary connected to a rotary converter for 250/125 volt, three-wire direct current.

system directly to the neutral of the wye-connected secondary of a delta-to-wye bank of transformers. Such a connection is shown in Fig. 5-37. The unbalanced direct current *3I* (Fig. 5-37) divides at the neutral point of the wye connection and one third flows through each of the transformer secondary windings. This tends to increase the magnetic density of the transformer and increased exciting current and iron loss is produced. Therefore it is recommended that such a connection be used only on relatively small capacities of rotary converters where the maximum unbalanced direct current flowing through each transformer is not in excess of 10 percent of the single-phase current rating.

In a three-phase core type transformer connected in wye on its secondary side for synchronous converter service, the magnetizing action of the unbalanced direct current is practically negligible, for the direct current flows in the same direction in each of the three legs from one end of the core to the other end and the fluxes established by these currents must therefore use the surrounding media, such as air or oil and transformer tank or case for their return path. The reluctance of such a path is very high, therefore the magnetizing effect of the direct currents is small.

Three-Phase Delta-to-Interconnected-Wye Connection. To eliminate the flux distortion in the transformer due to the unbalanced direct current in the neutral of the three-wire circuit flowing through the windings, synchronous converters are frequently connected to a bank of transformers connected delta to interconnected wye. Two separate interconnected windings are used on each phase of the wye connection and the windings are connected so that the direct current flows in opposite directions around the two halves of each phase winding, thus neutralizing the flux distortion. This interconnected wye connection is sometimes referred to as a zigzag winding. Fig. 5-38 shows three single-phase transformers *A*, *B*, and *C* connected in delta on the primary side and in interconnected wye on the secondary side. Note that the secondary winding of each

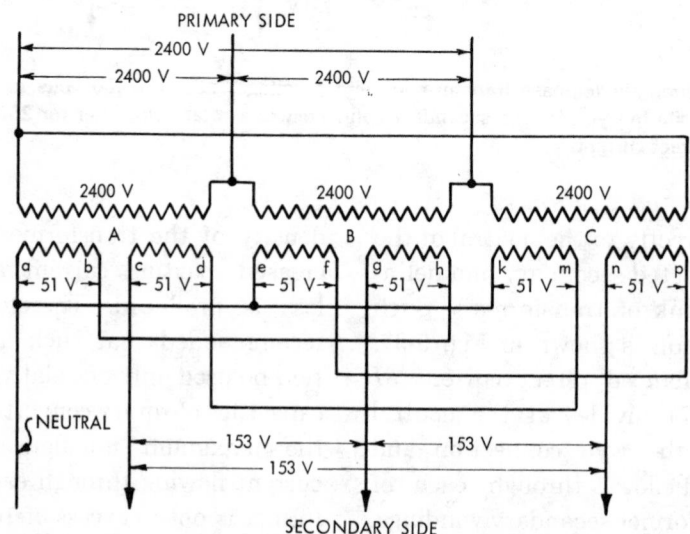

Fig. 5-38. Three single-phase transformers connected delta to interconnected wye.

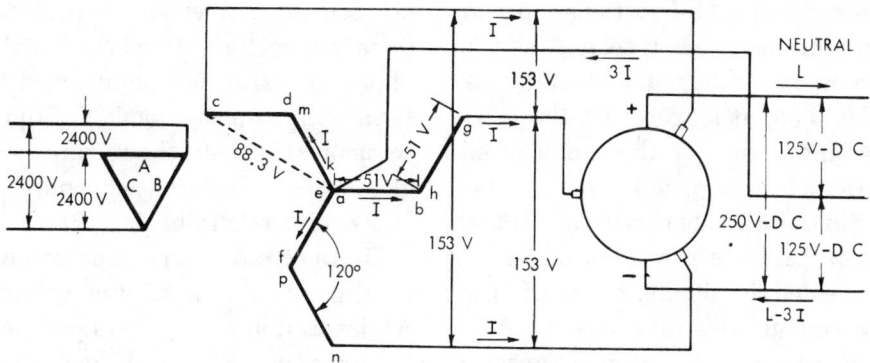

Fig. 5-39. Voltage-vector diagram of transformers of Fig. 5-38 connected to a three-ring synchronous converter, showing the neutralizing action of the unbalanced direct current flowing in the neutral of a 250/125-volt, three-wire, direct-current system.

transformer is composed of two equal windings and the two halves of each winding are connected in different legs of the interconnected wye, as shown in Fig. 5-39. Each leg of the interconnected wye comprises two half sections connected in opposition, which half sections are 120° out of phase with each other. Each leg voltage is, therefore, the vector difference of these two half-section voltages and is numerically equal to $\sqrt{3}$ times the voltage of one half section. The voltage across the terminals of the interconnected-wye connection is the vector difference of two leg voltages 120° out of phase with each other. This difference numerically equals $\sqrt{3}$ times the voltage of each leg. Since each leg voltage is $\sqrt{3}$ times the voltage of each half section, the terminal voltage of the interconnected-wye connection must be

three times the voltage of each half section winding. This is shown in the voltage vector diagram (Fig. 5-39), in which the primary side of the transformers is connected in delta to a 2,400-volt supply, and the secondary is connected in interconnected wye to a 250/125-volt, three-wire synchronous converter.

Assuming that the direct current in the positive line of the 250-volt direct-current circuit is L amperes and the current in the negative line is L-$3I$ amperes, there is, then, an unbalanced current of $3I$ amperes in the neutral. This unbalanced current flows into the neutral of the interconnected-wye windings and divides equally through each of the three legs. Note that the direction of this current flow in the half section a-b is opposite in direction to the flow of the current in the half section c-d, as indicated by the arrows.

251

Since the two half sections *a-b* and *c-d* form the two half sections of the secondary winding of Transformer *A* (Fig. 5-38) it is seen that the magnetizing action of this unbalanced current is neutralized.

Since the two halves of each transformer secondary winding are each connected in different phases, the low-voltage side operates at only 86.6 percent of its normal capacity if operated in straight wye connecation. Therefore transformers arranged for interconnected-wye operation are somewhat larger than those for straight wye connection. The actual ratio of kilovolt-amperes of transformer capacity to kilovolt-amperes transformed for the interconnected-wye system is 1.075 to 1.

Three-Phase Delta-to-Delta Connection. Three-phase transformers are seldom connected in delta-to-delta connection. However, single-phase transformers connected to form a three-phase bank are quite common. Fig. 5-40 shows three single-phase transformers connected in a delta-to-delta bank.

Delta-connected transformers must be wound for full-line voltage. An inspection of Fig. 5-40 and vector diagram, Fig. 5-41, will show that the voltage of each transformer is equal to the three-phase line voltage. The current in each of the transformers is only 57.7 percent of the line current. The windings therefore have a greater number of turns than for wye connection of the same line voltage, while the cross section of the turns is only 57.7 percent of that for the wye-connected transformer.

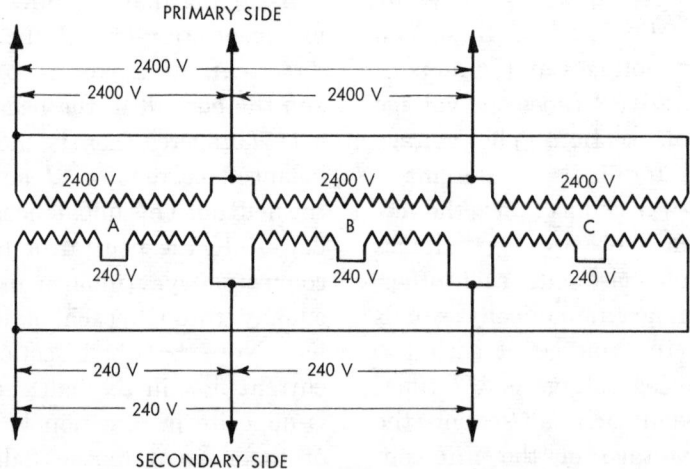

Fig. 5-40. Three single-phase transformers connected delta to delta.

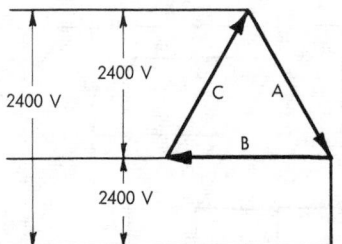 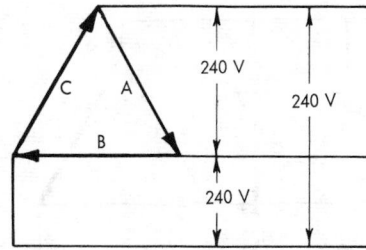

Fig. 5-41. Voltage-vector diagram of Fig. 5-40.

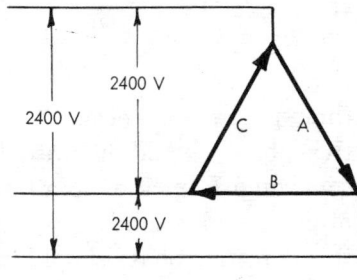 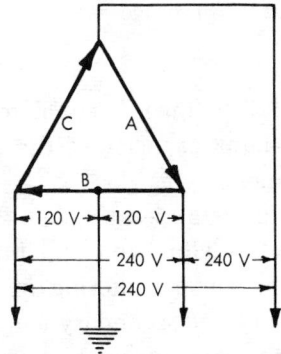

Fig. 5-42. Three single-phase transformers connected delta-delta with neutral point of one phase available for single-phase, three-wire service.

If the secondary is wound in two sections, as shown in Fig. 5-40, the mid-point of one of the windings may be grounded for 240/120-volt, single-phase, three-wire service for lighting, Fig. 5-42. Occasionally the single-phase unit which is to be used for this lighting load is made larger than the other two units so that the maximum kilovolt-amperes available in the other two units for three-phase loading may be utilized. If the mid-points are available on the other two phases, 240/120-volt,

single-phase, three-wire service may also be taken from these circuits; the middle wire of each of these two phases, however, must not be grounded as this amounts to a short circuit. A voltage equal to one-half of the line voltage exists between these mid-points, and three-phase loads at half voltage may therefore be obtained as shown in Fig. 5-43. Simultaneous loading at full voltage and half voltage is therefore possible. The maximum load that can be obtained at half voltage, with no

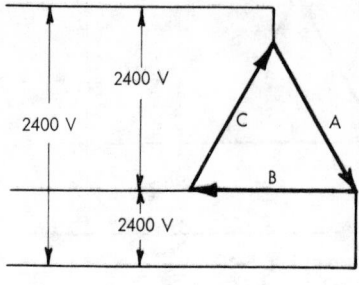 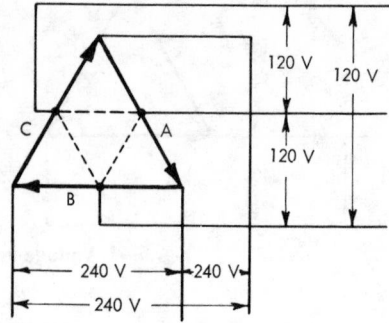

Fig. 5-43. Three single-phase transformers connected delta-delta with mid-points available for three-phase at half-voltage.

load at full voltage, is equal to one half the bank capacity of the three transformers.

Transformers operating in a three-phase delta-delta bank will have a circulating current flowing in both the primary and secondary windings if their ratios of transformation are different and will fail to divide the load properly in each of the windings if their respective impedances and the ratios of reactances to resistances are not equal.

Three-Phase Open-Delta Connection. If one of the single-phase transformers or one winding of a three-phase shell-type transformer in a delta-delta system should become defective, it is still possible to maintain operation by use of the open-delta connection, as shown in Fig. 5-44. (The vector diagram is shown in Fig. 5-45.) The defective unit is entirely removed from the circuit on both the primary and secondary sides. In the case of the

three-phase transformer, the defective phase should likewise be disconnected from the circuit and the primary and secondary windings of this defective phase should be short circuited to prevent stray fluxes from the other phases inducing voltage in the faulty windings.

The capacity of the two units or two windings of the three-phase unit

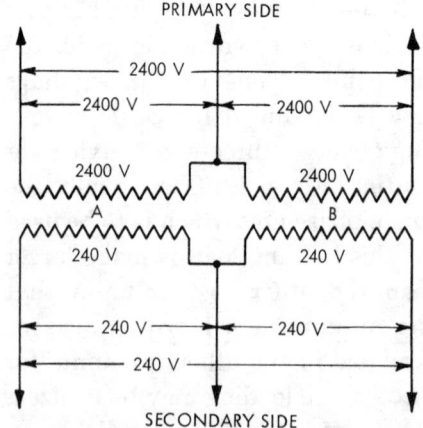

Fig. 5-44. Two single-phase transformers connected in open delta.

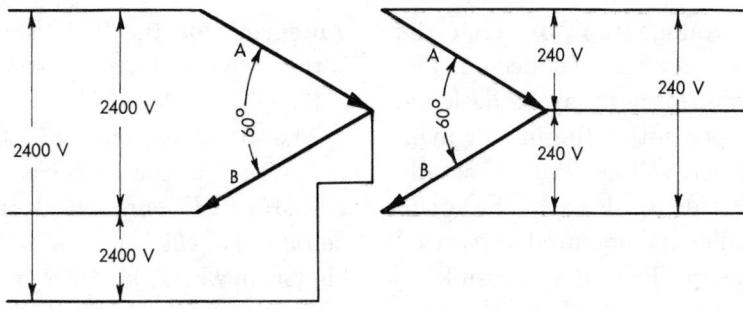

Fig. 5-45. Voltage-vector diagram of Fig. 5-44.

connected in open delta is 57.7 percent of the normal three-phase bank capacity.

The open-delta connection can be used to supply a three-phase load which is presently light but is expected to increase in the future. This type of installation requires only two single-phase transformers which keeps the initial investment low but also provides for future increase of load.

As an example, assume that we are providing a transformer bank for a new industrial building that will have an initial three-phase load of 150 kVA with an anticipated future load of 300 kVA. If two, single-phase, 100 kVA transformers were installed and connected open-delta as in Fig. 5-44, the present three-phase capacity of the transformer bank would be 57.7 percent of the normal three-phase bank capacity or 57.7 percent of 300 kVA which would provide a present capacity of 173 kVA. As the load increased to the 173 kVA open-delta capacity a third 100 kVA

transformer could be installed which would increase the capacity of the delta connected bank to 300 kVA.

The open-delta connection is also used when the secondary circuits are to supply both light and power loads. The grounded or neutral conductor of the lighting circuit is derived from a center tap of the 240-volt secondary winding thus providing a 120/240-volt single-phase three-wire lighting system. When the open-delta supplies a large single-phase load and a small three-phase load, the two transformers would be of different kVA ratings, the one across the lighting load being the larger. This application is also economical in that only one small single-phase transformer need be added to a large single-phase transformer to supply a limited three-phase power load.

Three-Phase T-to-T Connection. As with the open-delta connection, the T-T connection requires only two single-phase transformers. One of the units is called the main transformer and is provided with a 50 percent voltage tap to which the teaser transformer is connected. The teaser transformer may be designed for 86.6 percent of the line or main transformer voltage, but generally is made identical with the main transformer and operated at reduced flux density. Thus it is possible to operate two identical single-phase transformers connected T-T as well as open delta should one of the transformers in a delta-delta bank become inoperative. The only re-

quirement for the T-T connection is the 50 percent tap, Figs. 5-46 and 5-47.

The bank capacity of the T-T connection is the same as for two transformers connected in open delta. The efficiency is somewhat higher, however, because the teaser transformer, if identical units are used, is operating at reduced flux density and therefore will have lower iron loss. This increase in efficiency, due to lower iron loss, should be considered when a bank of transformers is likely to be operated for several years.

The T-T connection can be used for operating a synchronous converter provided a neutral is brought out of the teaser winding at a point equal to one third of the total winding measured from the end connected to the main transformer. The magnetizing action of the unbalanced direct current flowing through the transformer windings is neutralized in each of the two transformers, and there will be no increase in iron

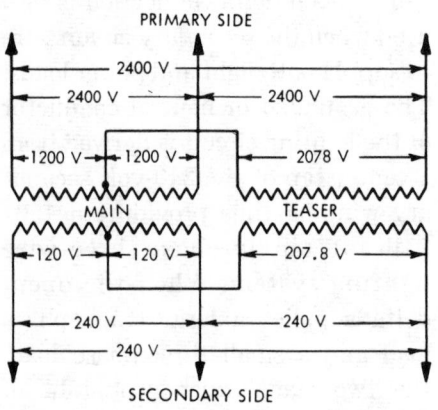

Fig. 5-46. Two single-phase transformers connected T-T.

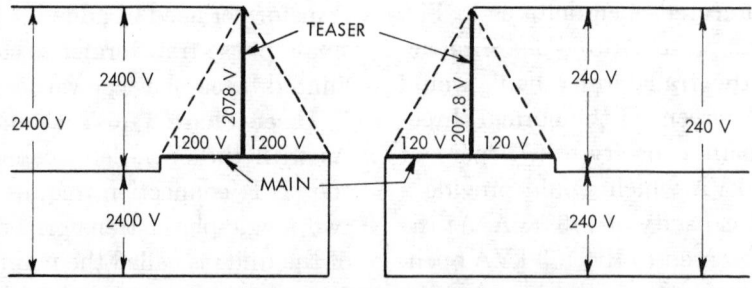

Fig. 5-47. Voltage-vector diagram of Fig. 5-46.

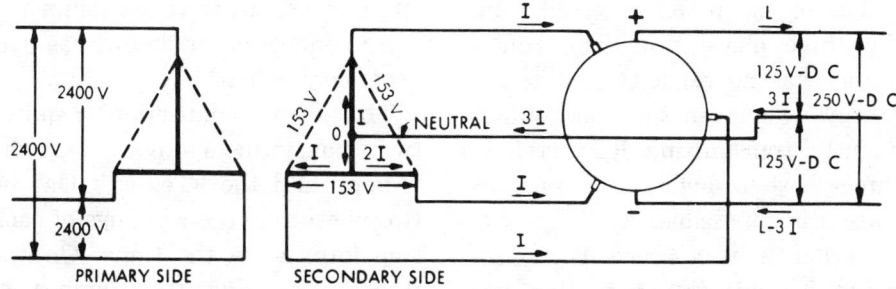

Fig. 5-48. Two single-phase transformers, each having a voltage ratio of 2400 to 153 volts connected in T-T to supply power to a three-phase rotary converter for 250/125 volt three-wire direct current.

loss due to this direct current flow. This is shown in Fig. 5-48, where *3I* represents the unbalanced current in the neutral of the three-wire, direct-current system. At point *O* this current divides, two thirds, or *2I* amperes, flowing towards the main transformer and one third, or *I* amperes, flowing out of the line end of the teaser transformer and returning to the converter circuit. At the junction of the main and teaser windings the *2I* amperes divide and *I* amperes flow through each of the two halves of the main transformer. Note that in the main transformer the direct currents (*I* amperes) are flowing in opposite directions therefore the fluxes established by them neutralize each other. In the teaser winding *2I* amperes flow through one half as many turns and in the opposite direction to that of the *I* amperes. Therefore the magnetomotive forces of the direct currents flowing in the teaser winding also neutralize each other.

Phase Transformation

Three-Phase to Single-Phase

It is practically impossible to deliver single-phase current from a three-phase source of supply by means of transformer action alone and have balanced conditions. This is true because single-phase power is pulsating, that is, it passes from a maximum value through zero and back to a maximum for every half cycle. Three-phase power is continuous in nature, that is, it is delivered at a constant rate. Therefore it would be necessary for the system to store up energy during the interval of time when the power delivered to the single-phase side is

257

less than the power received from the three-phase side. The transformer has no capacity for storing energy at a given time and subsequently transmitting it, therefore, three-phase to single-phase transformation is impossible.

Various schemes have been proposed for this transformation but none of them gives better results than connecting a single-phase transformer across one of the three phases. In a wye-connected source of supply this gives an equal current in two of the phases and zero current in the third. With a delta-connected source, two of the phases have the same current and the third a current twice as large as that in the other two.

Three-Phase to Two-Phase

Three-phase to two-phase or two-phase to three-phase transformation may be accomplished in several different ways, but the most commonly used connection is known as the Scott connection.

The Scott connection requires two transformers known as the "Main" and the "Teaser." On the two-phase side the windings of each transformer are the same. On the three-phase side the number of turns in the teaser winding is 86.6 percent of the number of turns in the main winding. On the three-phase side a 50 percent tap is brought out of the main winding. For the sake of interchangeability and replacement should one of the transformers become inoperative, it is customary to make both transformers identical with respect to the three-phase side, that is, the three-phase windings are each wound with the same number of turns and 50 percent and 86.6 percent taps are placed in each of the windings. When operating as a main winding the

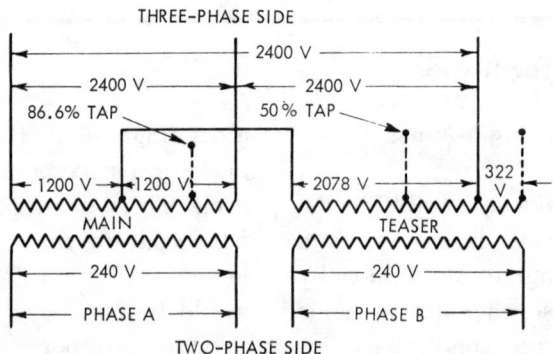

Fig. 5-49. Scott connection for three-phase to two-phase transformation with identical single-phase transformers.

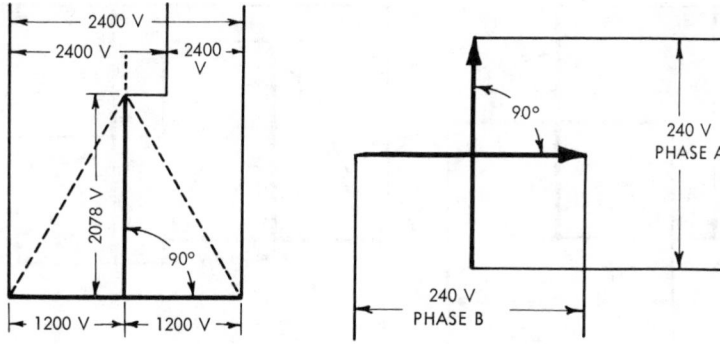

Fig. 5-50. Voltage-vector diagram of Fig. 5-49.

86.6 percent tap is not in use, and when operating as a teaser the 50 percent tap and that portion of the winding beyond the 86.6 percent tap are not in use. The unused tap connections are shown by dotted lines in Fig. 5-49. Fig. 5-50 shows the voltage conditions of a three-phase to two-phase transformation in addition to a voltage transformation from 2,400 volts to 240 volts.

The two halves of the three-phase winding on either side of the 50 percent tap should be distributed over the entire winding length of the magnetic circuit to prevent flux distortion and poor regulation.

In emergency cases where a transformer with an 86.6 percent tap is not available, the T-connection, in which the two transformers are identical with respect to their voltage ratio, may be used. This is an unsymmetrical connection and, when used, no attempt should be made to use it in parallel with a true

three-phase circuit or balanced two-phase circuit.

If this connection is resorted to in transforming from two-phase to three-phase, the three phases will not be exactly 120° apart, and two of the three-phase voltages will be 12 per-cent greater than the third. When transforming from three-phase to two-phase with the T-connection, the two-phase voltages will be in quadrature but the one phase will have a voltage 15 percent greater than the other.

This unsymmetrical condition due to emergency connections may be lessened if a transformer having reduced voltage taps is available. In such case this transformer should be used as the teaser and a three-phase connection should be made on that tap which most nearly approaches 86.6 percent of the total winding on the three-phase side. Fig. 5-51 shows the voltage-vector diagram of two single-phase transform-

259

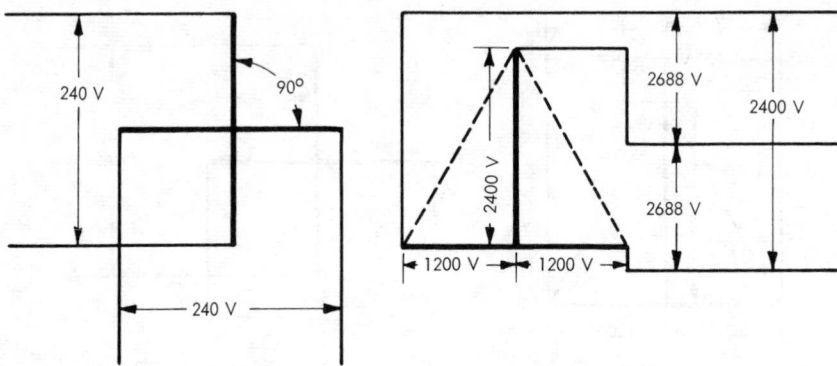

Fig. 5-51. Voltage-vector diagram of two single-phase transformers having a voltage ratio of 240 to 2,400 volts connected in T for two-phase to three-phase transformation.

ers with a voltage ratio of 1 to 10 (240 volts to 2,400 volts) connected in T for emergency operation from two-phase to three-phase. Note that the T-connection is exactly the same as the Scott connection shown in Fig. 5-49 with the exception that the teaser transformer is connected across 100 percent of the winding instead of on the 86.6 percent tap.

In the Scott connected transfor-

mation the two-phase windings are electrically independent. There are, however, a number of connections in which the windings are electrically interconnected on the two-phase side.

A three-phase to two-phase transformation using three single-phase transformers, with the two-phase side interconnected, is shown in Fig. 5-52. The three-phase side is shown

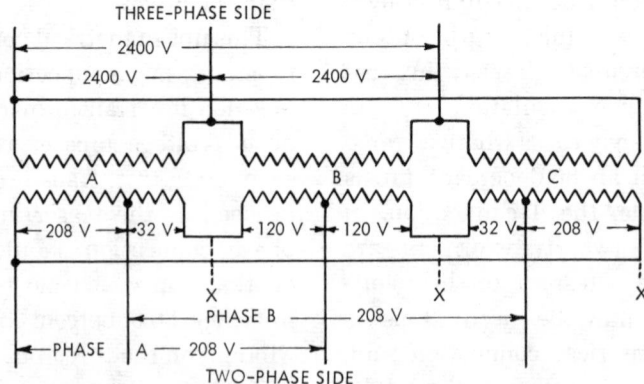

Fig. 5-52. Three single-phase transformers connected for three-phase to two-phase transformation with the two-phase side interconnected.

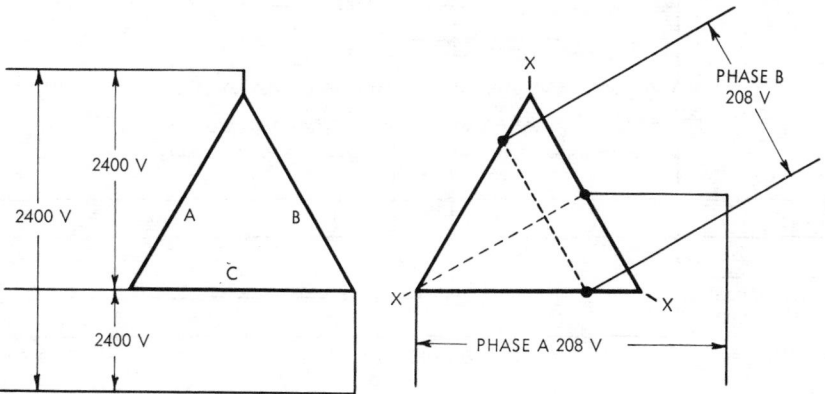

Fig. 5-53. Voltage-vector diagram of Fig. 5-52.

connected in delta. With the proper voltage transformation the three-phase side may be connected in wye. It is to be noted that this connection requires a 50 percent tap in one winding and an 86.6 percent tap on each of the other two windings on the two-phase side. The voltage-vector diagram of the three transformers of Fig. 5-52 is shown in Fig. 5-53.

Instead of using three single-phase transformers, as shown in Fig. 5-52, this transformation may be obtained with a three-phase magnetic circuit. In this case each of the transformers of Fig. 5-52 would represent the primary and secondary windings of one of the three legs of the three-phase transformer.

Three-Phase to Two-Phase and Three-Phase

By connecting to the corners of the delta on the two-phase side, as indicated by the dotted lines *X-X-X* in Fig. 5-52, three-phase power may be drawn from the transformers simultaneously with the two-phase power. However, the voltage of the three-phase circuit will be higher than that of the two-phase circuit. Referring again to Fig. 5-53, if the two-phase voltage is 208 volts, the three-phase voltage will be 240 volts. To obtain the same two-phase and three-phase voltages on the secondary side, two single-phase transformers employing the T- and Scott connections are used. This is shown in Figs. 5-54 and 5-55. The T-connection is used for the three-phase transformation and the Scott connection for the three-phase to two-phase transformation.

Two-Phase to Six-Phase

When operating a six-phase synchronous converter from a two-phase circuit, two single-phase

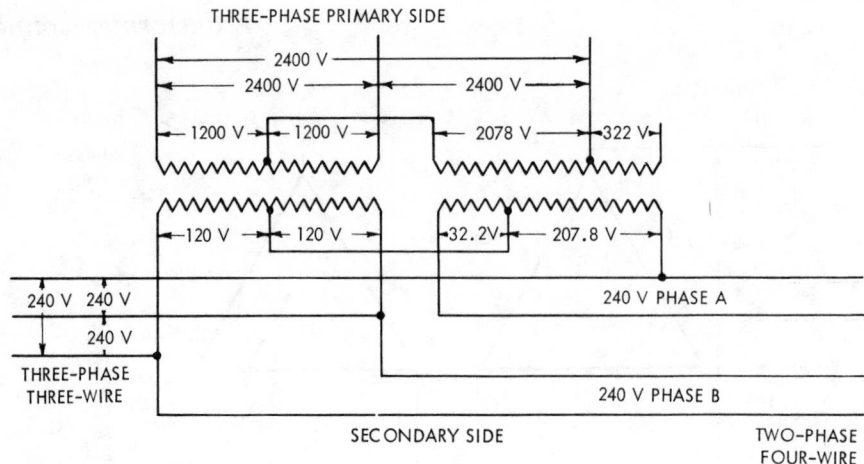

Fig. 5-54. Two single-phase transformers connected for three-phase to three-phase and two-phase transformation.

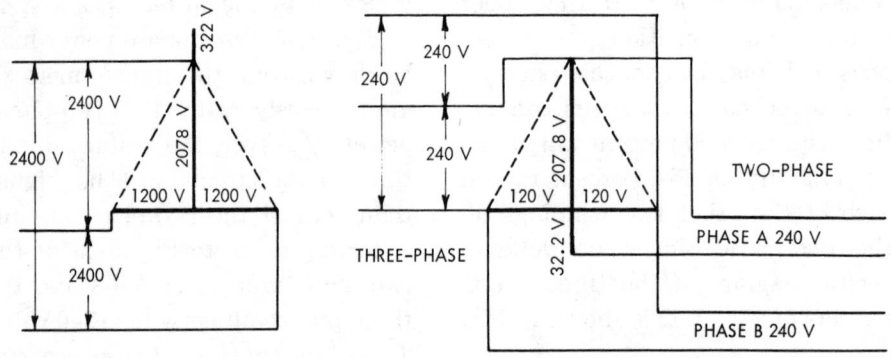

Fig. 5-55. Voltage-vector diagram of Fig. 5-54.

transformers connected in double Scott are employed. This connection requires very special units of the same impedance, each having two low-voltage windings, connected so that their voltages are 180° from each other in phase relation in order to give the six-phase system. Usually the synchronous converter is started from the alternating-current side as a three-phase converter. This further complicates the trans-

former windings. Additional complications are introduced if a neutral must be provided. When all of these conditions must be met in the design of the transformers, they can no longer be made identical as in the case of the single Scott connection.

The two-phase to six-phase connection using the double Scott arrangement is shown in Figs. 5-56 and 5-57. The two-phase primary

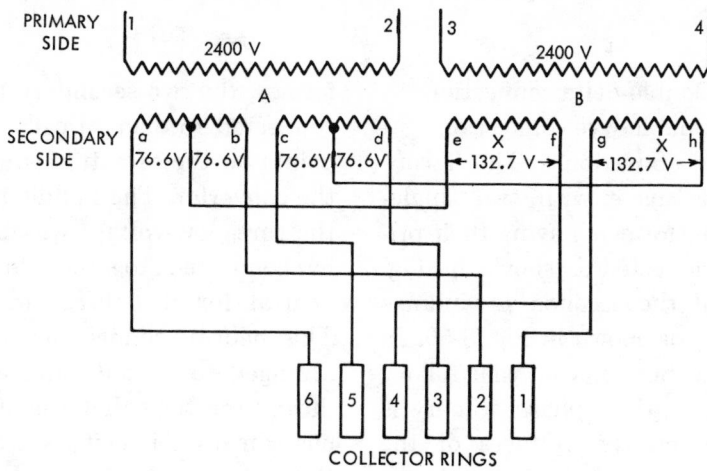

Fig. 5-56. Two single-phase transformers connected for 2,400 volts two-phase to six-phase for synchronous converter which supplies 250 volts DC.

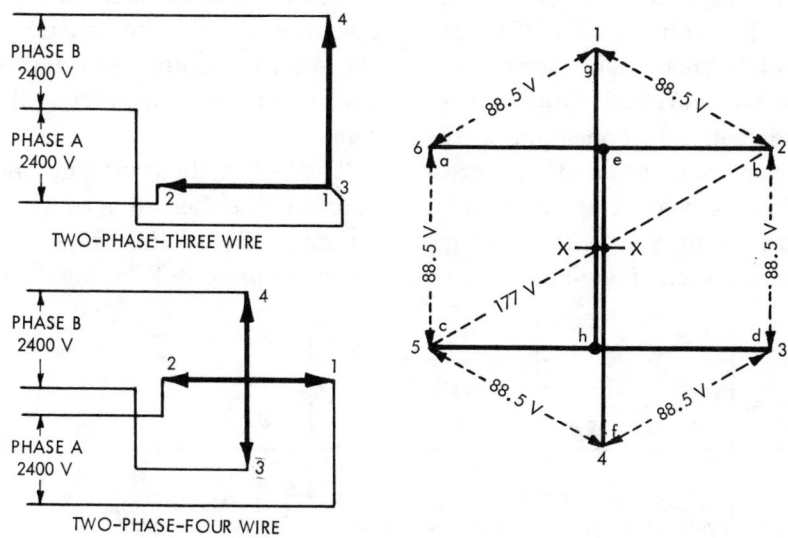

Fig. 5-57. Voltage-vector diagram of Fig. 5-44.

windings may be connected for either three-wire or four-wire service as indicated by the vector connections, Fig. 5-57. The numbers 1 to 6 represent the six successive collector rings of the synchronous converter.

Three-Phase to Six-Phase

For transforming from three-phase to six-phase for synchronous converter service four connections are quite common, namely

1. T-connection
2. Diametrical connection

3. Double-delta connection
4. Double-wye connection

The T-Connection. The T-connection is made with two single-phase transformers having their primaries connected as shown in Fig. 5-46, and the secondary windings connected as shown in Fig. 5-56. In this connection, and all that follow, a 2,400-volt, three-phase primary is assumed and the voltages of the windings on the secondary side are assumed such as to give the proper voltage across the six rings of a six-phase, 250-volt or 250/125-volt, three-wire synchronous converter.

The Diametrical Connection. The diametrical connection is the most commonly used of all three-phase to six-phase transformations. It requires only one low-voltage winding on each single-phase trans-former; the two secondary leads are connected to diametrically opposite points on the armature windings of the converter. The middle points of the three low-voltage windings may be connected together to form a neutral for the three-wire circuit. This neutral connection should be arranged so that it can be opened during the time that the synchronous converter is being started from the alternating-current side. Should one of the transformers become inoperative it is possible to continue operation at reduced capacity with the other two units connected across their respective converter diameters.

The high-voltage windings may be connected either in wye or delta, although the delta-connected primary is preferred because of the

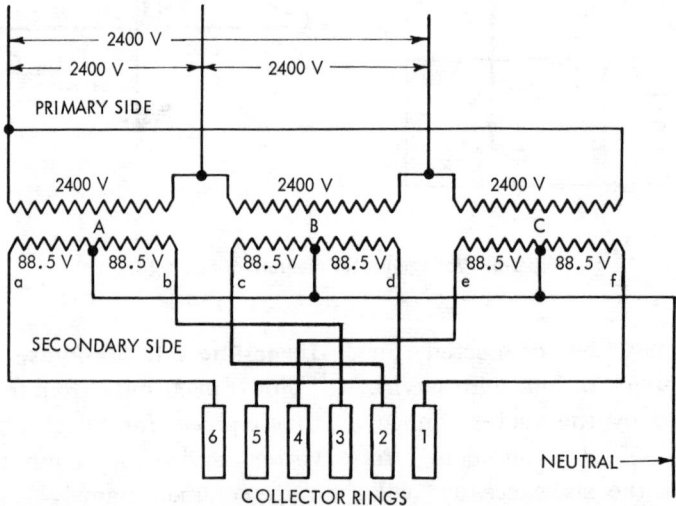

Fig. 5-58. Diametrical connection of three single-phase transformers for transforming 2,400-volt, three-phase to six-phase for 250/125-volt, three-wire synchronous converter.

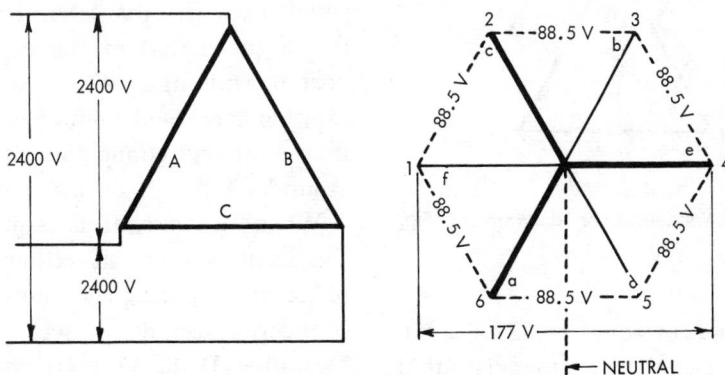

Fig. 5-59. Voltage-vector diagram of Fig. 5-58.

triple-frequency harmonics of voltage which are introduced in a wye-wye connection. Fig. 5-58 shows the diametrical connection of three single-phase transformers having the high-voltage side connected in delta. The voltage-vector diagram is given in Fig. 5-59.

With a six-phase diametrical connection with common neutral, one half the output can be taken from the low-voltage side for operating three-phase without change of diametrical voltage, that is, a three-phase 250/125-volt converter may be operated from one half of the winding, as shown by the heavy vectors of Fig. 5-59.

If full three-phase output should be desired, the low-voltage coils can be connected in delta, in which case the diametrical voltage is increased 15.4 percent as is evident from Figs. 5-60 and 5-61, which show the connection of the low-voltage windings of 5-58 for full three-phase output. Assuming the same ratio of voltage transformation and the same high-voltage as shown in Fig. 5-58, the direct-current voltage is 15.4 per-

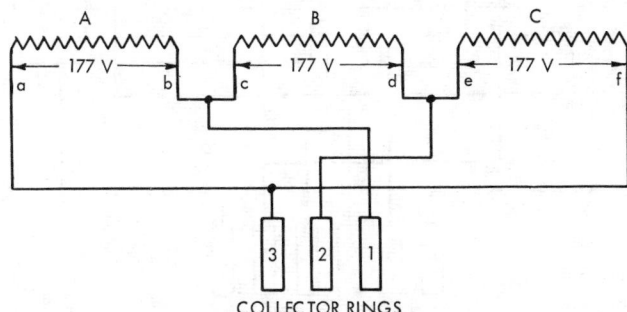

Fig. 5-60. Low-voltage windings of the three single-phase transformers of Fig. 5-58 connected for three-phase, 288.5-volt synchronous converter.

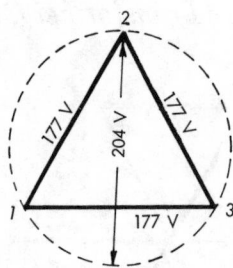

Fig. 5-61. Voltage-vector diagram of Fig. 5-60.

cent higher, or equal to 288.5 volts. The direct-current voltage is equal to the diametrical voltage multiplied by $\sqrt{2}$, regardless of the number of phases. In a six-phase converter the diametrical voltage is equal to the voltage across rings 1–4, 2–5, and 3–6, which are connected to diametrically opposite points of the converter winding.

The full three-phase output may also be obtained by connecting the low-voltage windings in wye, in which case the diametrical voltage is $\sqrt{3}$ times that of the delta-connected windings. When connected for this increased voltage the neutral of the wye connection should be grounded.

When full output is required at the same voltage at either three-phase or six-phase, the double-delta connection usually is used.

Double-Delta Connection. The double-delta connection requires two separate low-voltage windings on each transformer. Both sets of windings are connected in delta, but the one is reversed with respect to the other so that two deltas, displaced 180° from each other, are formed. The high-voltage windings may be connected in either delta or wye,

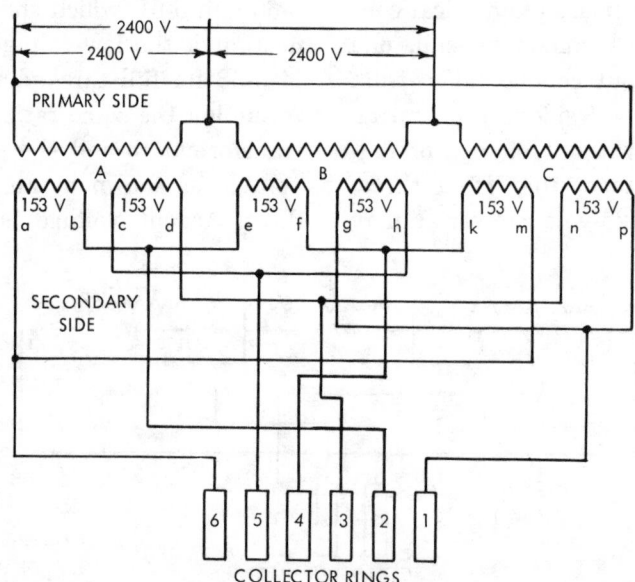

Fig. 5-62. Double-delta connection of three single-phase transformers for transforming from three-phase 2,400 volts to six-phase for 250/125-volt, three-wire synchronous converter.

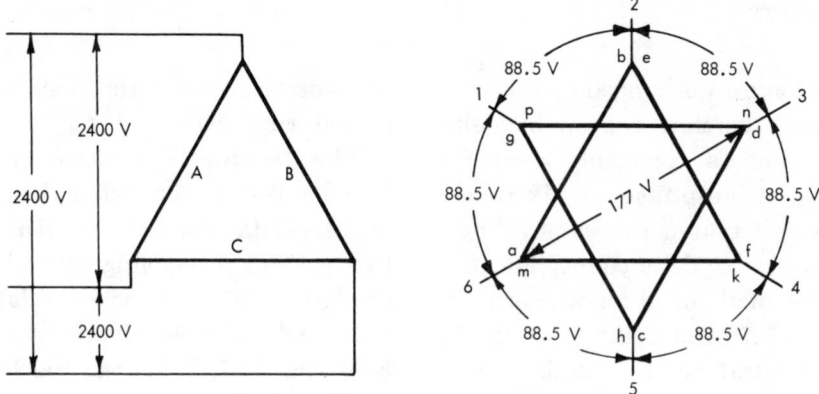

Fig. 5-63. Voltage-vector diagram of Fig. 5-62.

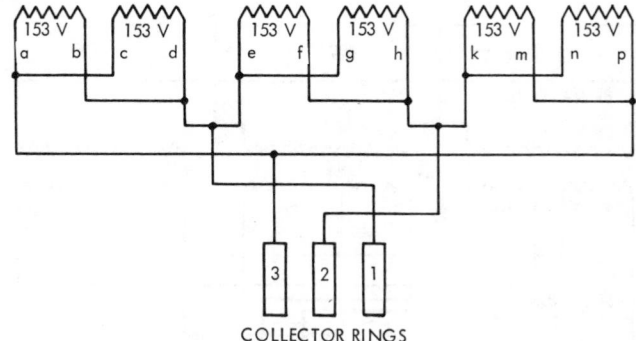

COLLECTOR RINGS

Fig. 5-64. Low-voltage windings of three single-phase transformers connected for full output, three-phase.

although the delta connection is preferred, as this permits continued operation at reduced capacity, with one of the units disconnected from the circuit if it should become inoperative.

The double-delta connection for transforming from three-phase to six-phase, with three single-phase transformers having the high voltage side connected in delta, is shown in Figs. 5-62 and 5-63.

By connecting the two low-voltage windings of each transformer in parallel full output at three-phase

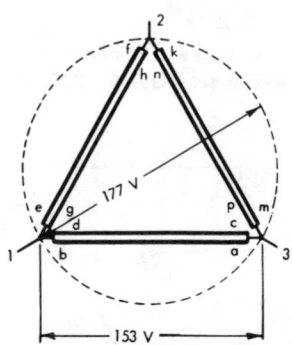

Fig. 5-65. Voltage-vector diagram of Fig. 5-64.

is obtained by connecting the parallel groups in delta, as shown in Figs. 5-64 and 5-65. The diametri-

267

cal voltage is unchanged by this connection, therefore the double-delta connection is commonly used for either three-phase or six-phase transformation at the same voltage.

The double-delta connection cannot be used for three-wire service since no neutral point is available. If a neutral point is desired it is necessary to connect separate auto-transformers across the delta connected windings.

The Double-Wye Connection. The double-wye connection, like the double-delta connection, requires two sets of low-voltage windings, displaced 180° in phase relation from each other as shown in Figs. 5-66 and 5-67. Likewise, the high-voltage windings may be connected

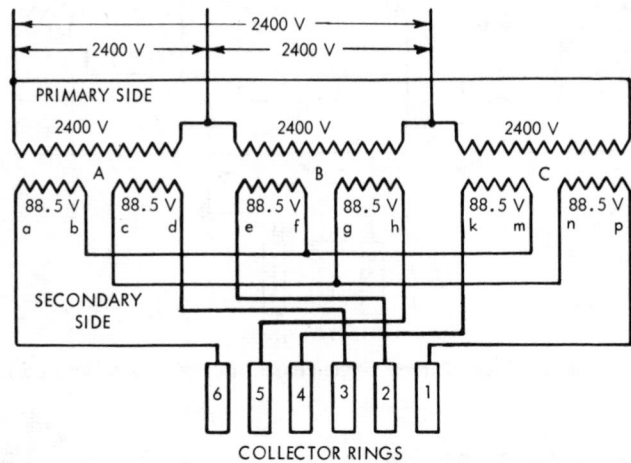

Fig. 5-66. Double-wye connection of three single-phase transformers for transforming from three-phase to six-phase for 250/125 volt, three-wire synchronous converter.

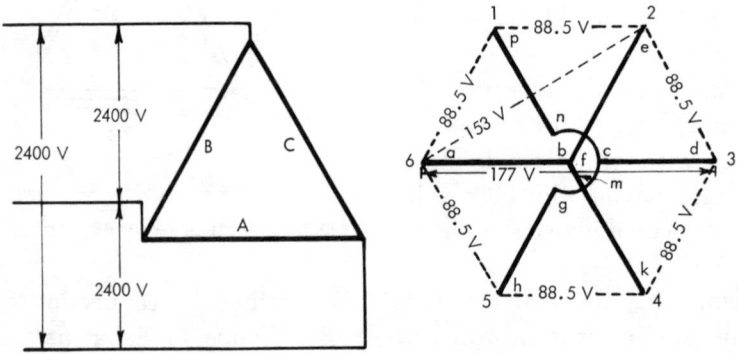

Fig. 5-67. Voltage-vector diagram for Fig. 5-66.

either in delta or wye, although the delta connection is preferable. Fig. 5-66 shows the high-voltage side connected in delta.

For additional single-phase, three-phase, and six-phase connections showing polarity, relative phase rotation, angular displacement and proper lead markings (in accordance with standard practice as recommended by the American National Standards Institute) see Figs. 5-77 to 5-87 inclusive.

Polarity: Three-Phase Transformers

In polyphase transformers, polarity alone is inadequate to represent a definite relation between the high-voltage and low-voltage windings. In addition to the lead markings, voltage-vector diagrams are required to show the angular phase displacement between the high-voltage and low-voltage windings and the time order of phase sequence.

As a rule, all phases of a three-phase transformer have the same relative polarity (expressed in terms of single-phase polarity); that is, if the polarity of one phase is subtractive, the polarity of the other two phases also will be subtractive; if the polarity of one phase is additive, the polarity of the other two phases will be additive.

The three-phase polarity, however, is dependent upon the interphase connections of the respective outlet leads to the full-phase windings, as well as upon the polarity of the separate phases.

The three high-voltage leads and the three low-voltage leads which connect to the full-phase windings are marked H_1, H_2, H_3 and X_1, X_2, and X_3 respectively. The markings are so applied that with the phase sequence of voltage on the high-voltage side in the time order H_1, H_2, H_3, on the low-voltage side it is in the time order of X_1, X_2, X_3. Referring to the low-voltage vector diagram of three single-phase transformers connected in a delta-to-delta three-phase bank, as shown in Fig. 5-68, the phase rotation or phase sequence is in a clockwise direction for both the high-voltage and low-voltage sides, see Fig. 5-69. This phase rotation is only relative. The actual phase rotation is dependent upon and equal to the phase rotation of the supply voltage. If, however, a three-phase motor were first connected across the three leads of the high-voltage side and then transferred directly to similarly

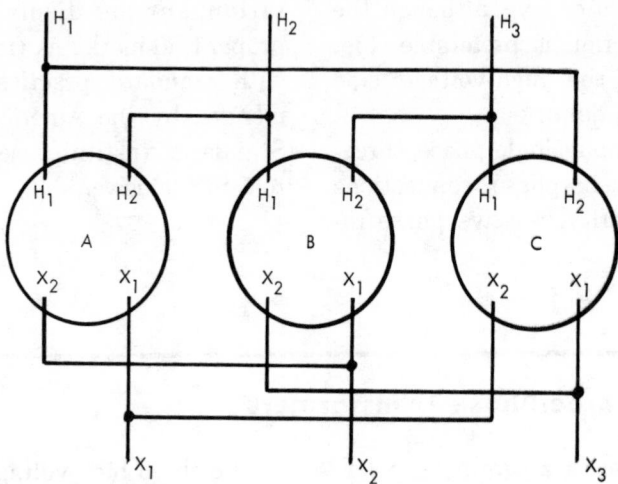

Fig. 5-68. Three single-phase transformers of additive polarity, connected for three-phase, delta-to-delta operation when 0° angular displacement is desired.

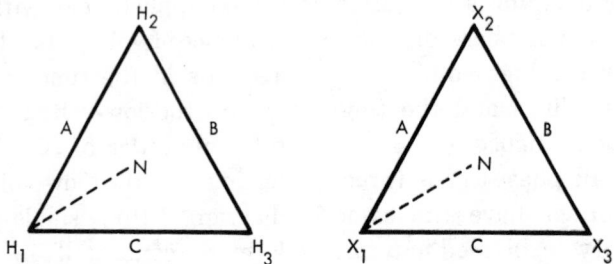

Fig. 5-69. Voltage-vector diagram of Fig. 5-68, showing clockwise phase rotation.

numbered leads of the low-voltage side, that is, transferring its leads from H_1 to X_1, from H_2 to X_2, and from H_3 to X_3, the motor would continue to rotate in the same direction.

The angular displacement between the high-voltage and low-voltage windings as defined by the A.N.S.I. is the angle between the

lines H_1–N and X_1–N where N is the neutral point of the voltage-vector diagram. This angle for the connection of Fig. 5-68 is 0 degrees, as is evident from an inspection of the voltage-vector diagram, Fig. 5-69. Although Fig. 5-68 shows three single-phase transformers, these three transformers may be considered as the windings of each of the three legs of a three-phase transformer. The connections would be identical.

Inasmuch as the three-phase transformer or three single-phase transformers may be connected in any of four combinations the angular displacement will vary, depending upon the combination used.

Transformers connected delta-to-delta or wye-to-wye may have an angular displacement of 0° or 180°. Figs. 5-70 and 5-71 show three single-phase transformers of additive

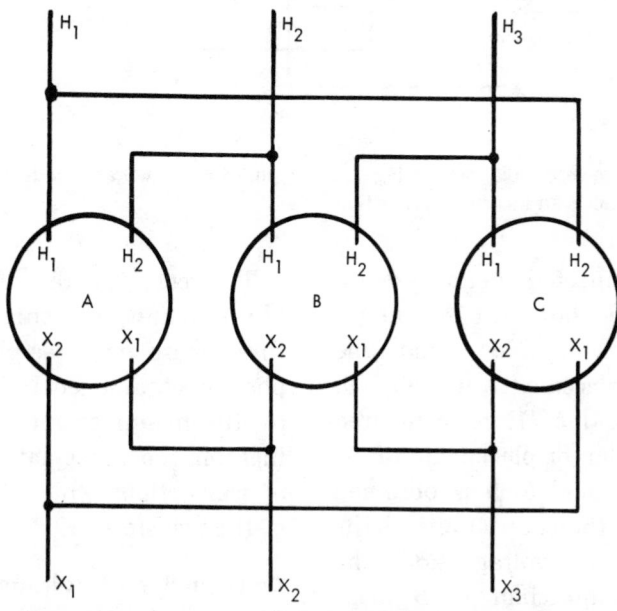

Fig. 5-70. Three single-phase transformers of additive polarity connected for three-phase, delta-to-delta operation when 180° angular displacement is desired.

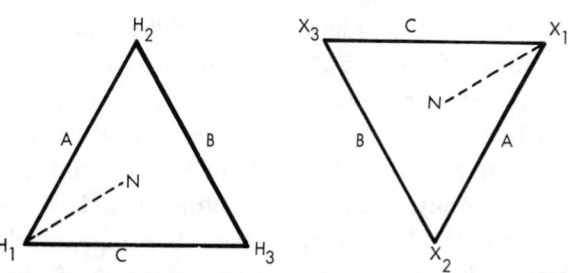

Fig. 5-71. Voltage-vector diagram of Fig. 5-70, showing clockwise phase rotation.

271

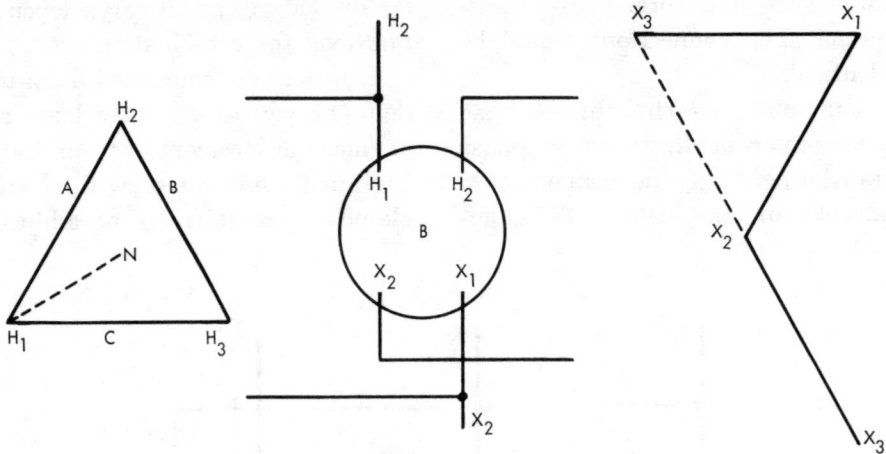

Fig. 5-72. Voltage-vector diagram of Fig. 5-70 if phase B low-voltage leads X_1 and X_2 were interchanged. This is an incorrect connection.

polarity, connected delta-to-delta, as in Fig. 5-68, but with 180° angular displacement. Note that the angular displacement of 180° in Figs. 5-70 and 5-71, as compared to the angular displacement of 0° in Figs. 5-68 and 5-69 is obtained by reversing the connections of the delta on the low-voltage side. The result of not interchanging B phase is shown in Fig. 5-72.

Wye-to-delta connected transformers, with the leads properly marked, will always have an angular displacement of 30°, as is evident from an examination of Figs. 5-26 and 5-28. Therefore all three-phase to three-phase or three-phase to six-phase connections can be grouped in one of five different groups, having an angular displacement of 0°, 30°, or 180°, as shown in Figs. 5-73 and 5-74.

The vector diagram of three-phase transformers for three-phase to three-phase or three-phase to six-phase operation generally is included by the manufacturer in the markings on the nameplate or diagram of connections which forms a part of the transformer.

Construction of Voltage-Vector Diagrams for Three-Phase Transformers

The polarity of each phase of a three-phase transformer may be determined in the same manner as already described for single-phase transformers. The voltage-vector diagram can then be constructed. The construction of a voltage-vector diagram can best be explained by the following example: Assume that the polarities of three single-phase transformers or the windings of each

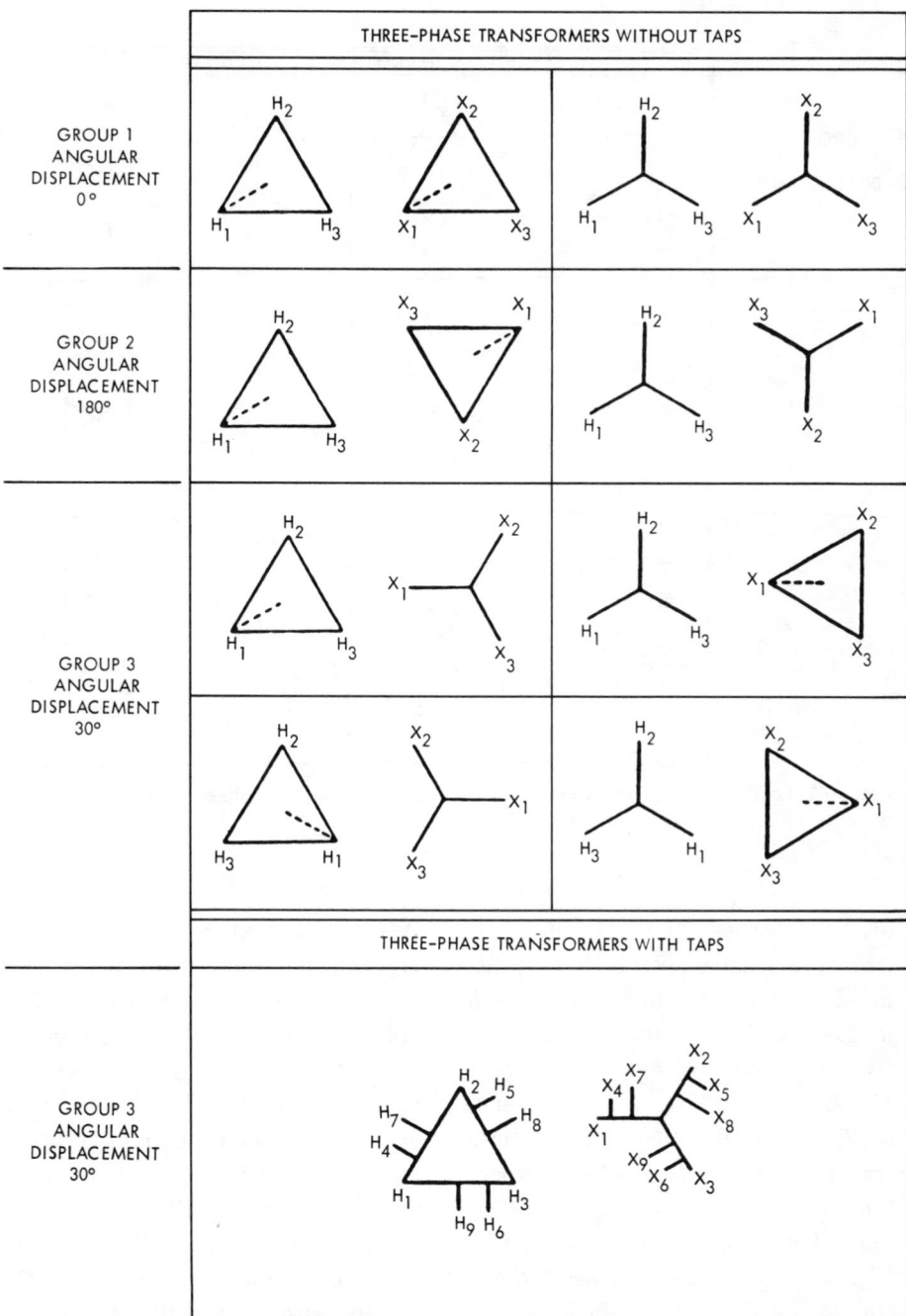

Fig. 5-73. Lead markings and voltage-vector diagrams. Usual three-phase transformer connections.

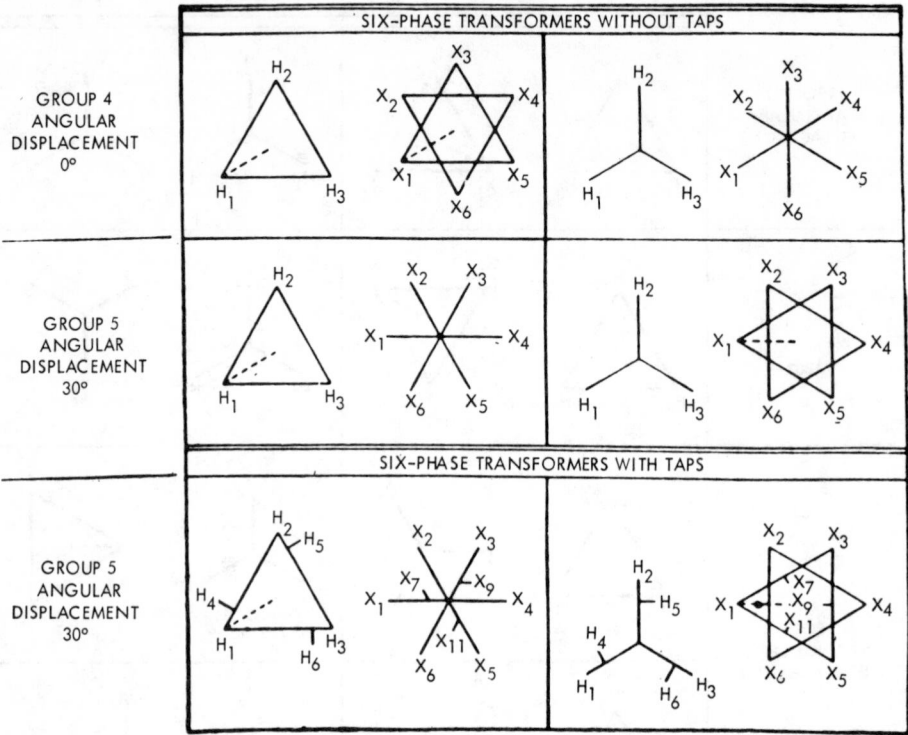

Fig. 5-74. Lead markings, and voltage-vector diagrams. Usual six-phase transformer connections.

of the three legs of the three-phase transformer have been tested as outlined above for the connection shown in Fig. 5-70 and found to be additive. Then the voltage vectors H_1–H_2, H_2–H_3, and H_3–H_1, representing the induced voltages of the three transformers A, B, and C respectively, can be drawn 120° out of phase with each other, as shown in Fig. 5-71, to form a closed delta connection. Since the polarity is additive, the low-voltage vector of A, that is, the full-phase voltage vector

X_1–X_2 is opposite in direction to H_1–H_2; likewise the full-phase voltage vector X_2–X_3, which is the low-voltage vector of phase B, is opposite to that of H_2–H_3, and the voltage vector X_3–X_1, which is the low-voltage vector of phase C, is opposite to H_3–H_1. Therefore the low-voltage vector diagram of Fig. 5-71 is constructed by drawing X_1–X_2 opposite in direction to H_1–H_2, and X_2–X_3 opposite in direction to H_2–H_3, followed by X_3–X_1 opposite in direction to H_3–H_1. This forms a closed delta and is therefore drawn correctly. If the delta connections forming the three-phase connection on the low-voltage side were improperly made, for example, if the X_1–X_2 leads of phase B were interchanged in forming the delta, the voltage vector diagram would not close, as shown by the heavy lines of Fig. 5-72. In this case an unbalanced voltage equal to twice the voltage of one phase would exist in the delta thus formed and an enormous current would flow through this short-circuited connection.

Test Procedure for Checking the Voltage-Vector Diagrams

A committee of the American National Standards Institute has prepared a "Test Code for Transformers" and much of the following is quoted directly or adapted from this test code.

The polarity of each phase may be determined in the same manner as previously explained for single-phase transformers. These voltage-vector diagrams can be constructed and checked with Fig. 5-75 if the transformer is connected for three-phase operation, or with Fig. 5-76 if the transformer is connected for six-phase operation. For example, assume that the three-phase connections are delta-delta with 180° angular displacement as per Group 2 of Fig. 5-73. A three-phase voltage of low value would be applied to leads H_1, H_2, and H_3. With leads H_1 and X_1 connected, voltage readings should be

taken between H_2 and X_2, H_3 and X_3, H_1 and H_2, H_2 and X_3. Then for the proper connections and lead markings the voltage relations should be: H_2-X_2 equals H_3-X_3; H_2-X_2 is greater than H_1-H_2; H_2-X_3 is less than H_2-X_2. The voltage-vector diagram constructed to scale from above voltage measurements should coincide with the diagram as shown for the delta-delta connected transformer in Fig. 5-75. This test is a check for both angular displacement and time-order of phase sequence. This method of measurement becomes quite difficult when the high voltage is very large compared to the low voltage and, like the single-phase test, is practically limited to transformers having a ratio of transformation of 30 to 1 or less.

If the internal connections of the three-phase transformer are unknown and inaccessible, the polarity test may be omitted. The angular displacement and time-order of phase sequence, however, can be determined from test although it may be necessary to take a greater number of readings. The procedure is the same as described above. Voltage measurements should be made as indicated for each of the angular displacement groups, as shown in Fig. 5-75 or Fig. 5-76 and the voltage-vector diagram constructed for check measurement. If none of these diagrams check with those shown it is evident that the lead markings on the low-voltage side are incorrect. The lettering of two of the low-voltage leads should then be interchanged and the voltage measurements repeated, after which a new voltage-vector diagram can be constructed for check measurement. This procedure must be continued until a measurement is found which agrees with the diagram for check measurement as given in Fig. 5-75 or Fig. 5-76.

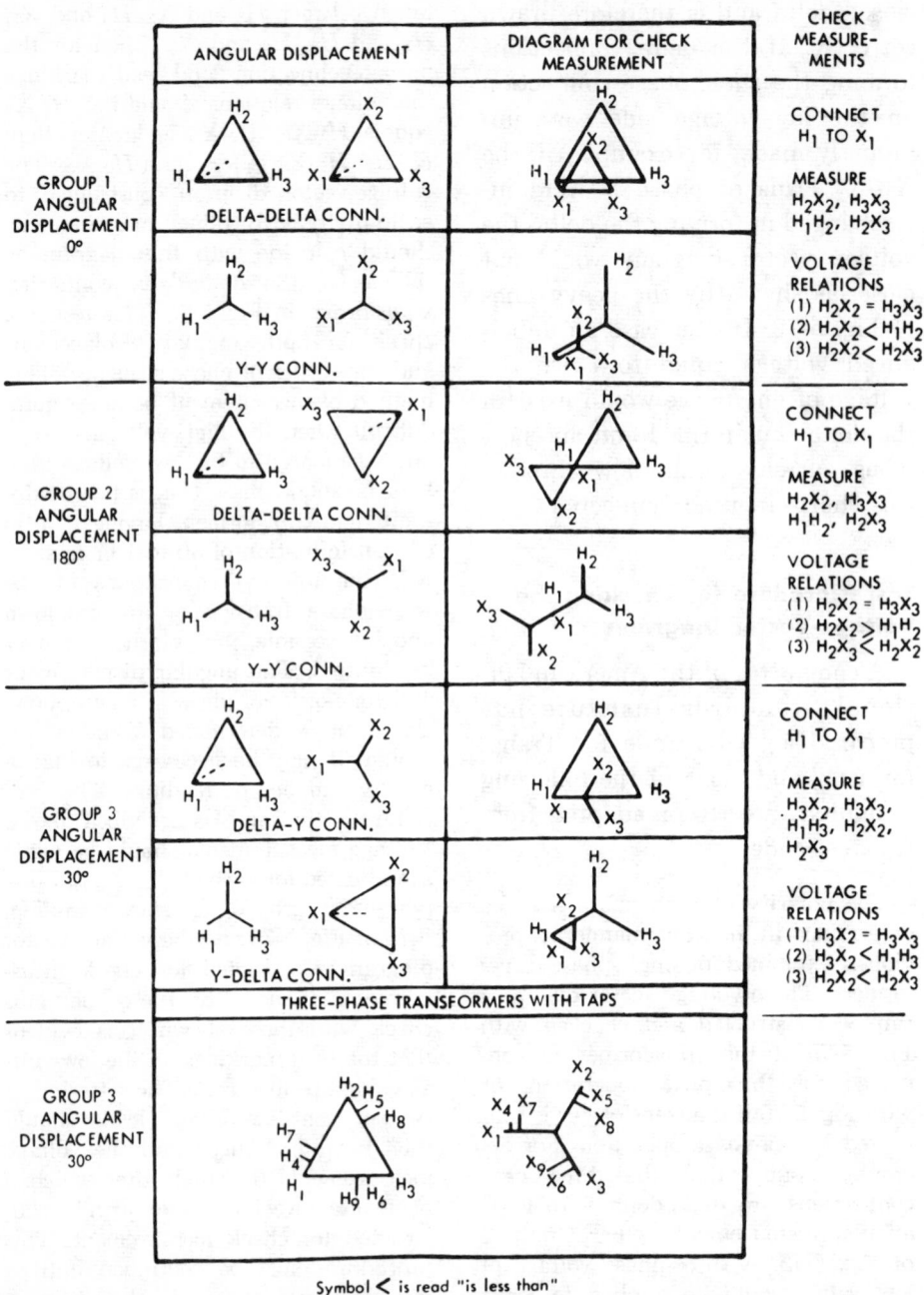

	ANGULAR DISPLACEMENT	DIAGRAM FOR CHECK MEASUREMENT	CHECK MEASURE-MENTS

GROUP 1 ANGULAR DISPLACEMENT 0°

DELTA-DELTA CONN.

Y-Y CONN.

CONNECT H_1 TO X_1

MEASURE H_2X_2, H_3X_3 H_1H_2, H_2X_3

VOLTAGE RELATIONS
(1) $H_2X_2 = H_3X_3$
(2) $H_2X_2 < H_1H_2$
(3) $H_2X_2 < H_2X_3$

GROUP 2 ANGULAR DISPLACEMENT 180°

DELTA-DELTA CONN.

Y-Y CONN.

CONNECT H_1 TO X_1

MEASURE H_2X_2, H_3X_3 H_1H_2, H_2X_3

VOLTAGE RELATIONS
(1) $H_2X_2 = H_3X_3$
(2) $H_2X_2 > H_1H_2$
(3) $H_2X_3 < H_2X_2$

GROUP 3 ANGULAR DISPLACEMENT 30°

DELTA-Y CONN.

Y-DELTA CONN.

CONNECT H_1 TO X_1

MEASURE H_3X_2, H_3X_3, H_1H_3, H_2X_2, H_2X_3

VOLTAGE RELATIONS
(1) $H_3X_2 = H_3X_3$
(2) $H_3X_2 < H_1H_3$
(3) $H_2X_2 < H_2X_3$

THREE-PHASE TRANSFORMERS WITH TAPS

GROUP 3 ANGULAR DISPLACEMENT 30°

Symbol $<$ is read "is less than"
Symbol $>$ is read "is greater than"

Fig. 5-75. Transformer lead markings and voltage-vector diagrams for three-phase transformer connections.

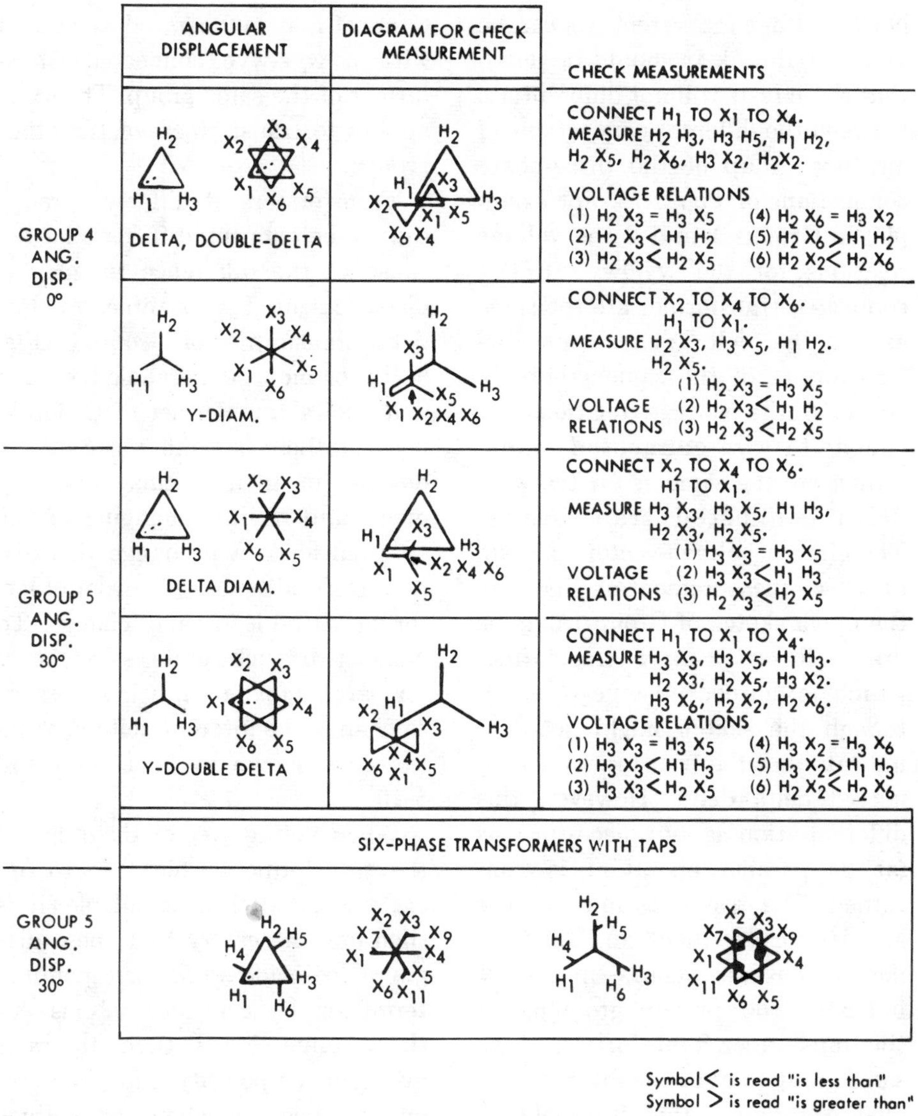

	ANGULAR DISPLACEMENT	DIAGRAM FOR CHECK MEASUREMENT	CHECK MEASUREMENTS
GROUP 4 ANG. DISP. 0°	DELTA, DOUBLE-DELTA		CONNECT H_1 TO X_1 TO X_4. MEASURE $H_2 H_3$, $H_3 H_5$, $H_1 H_2$, $H_2 X_5$, $H_2 X_6$, $H_3 X_2$, $H_2 X_2$. VOLTAGE RELATIONS (1) $H_2 X_3 = H_3 X_5$ (4) $H_2 X_6 = H_3 X_2$ (2) $H_2 X_3 < H_1 H_2$ (5) $H_2 X_6 > H_1 H_2$ (3) $H_2 X_3 < H_2 X_5$ (6) $H_2 X_2 < H_2 X_6$
	Y-DIAM.		CONNECT X_2 TO X_4 TO X_6. H_1 TO X_1. MEASURE $H_2 X_3$, $H_3 X_5$, $H_1 H_2$. $H_2 X_5$. (1) $H_2 X_3 = H_3 X_5$ VOLTAGE (2) $H_2 X_3 < H_1 H_2$ RELATIONS (3) $H_2 X_3 < H_2 X_5$
GROUP 5 ANG. DISP. 30°	DELTA DIAM.		CONNECT X_2 TO X_4 TO X_6. H_1 TO X_1. MEASURE $H_3 X_3$, $H_3 X_5$, $H_1 H_3$, $H_2 X_3$, $H_2 X_5$. (1) $H_3 X_3 = H_3 X_5$ VOLTAGE (2) $H_3 X_3 < H_1 H_3$ RELATIONS (3) $H_2 X_3 < H_2 X_5$
	Y-DOUBLE DELTA		CONNECT H_1 TO X_1 TO X_4. MEASURE $H_3 X_3$, $H_3 X_5$, $H_1 H_3$. $H_2 X_3$, $H_2 X_5$, $H_3 X_2$. $H_3 X_6$, $H_2 X_2$, $H_2 X_6$. VOLTAGE RELATIONS (1) $H_3 X_3 = H_3 X_5$ (4) $H_3 X_2 = H_3 X_6$ (2) $H_3 X_3 < H_1 H_3$ (5) $H_3 X_2 > H_1 H_3$ (3) $H_3 X_3 < H_2 X_5$ (6) $H_2 X_2 < H_2 X_6$
GROUP 5 ANG. DISP. 30°	SIX-PHASE TRANSFORMERS WITH TAPS		

Symbol $<$ is read "is less than"
Symbol $>$ is read "is greater than"

Fig. 5-76. Transformer lead markings and voltage-vector diagrams for six-phase transformer connections.

It is evident that this is a laborious method, but after a few voltage measurements are taken and voltage-vector diagrams constructed to scale from these measurements it becomes apparent what the correct lead lettering should be to obtain the proper diagram, and the number of voltage measurements can be greatly reduced. It should be noted that these tests will not differentiate between the different connections of any one group of the three-phase connections of Fig. 5-75. For example, in Groups 1 and 2 the voltage relations for the wye-wye (Y-Y) connected transformer are the same as for the delta-delta connected transformer of the same group; in Group 3 the voltage relations for the delta-wye connected transformer are the same as for the wye-delta connected transformer. Therefore a voltage-vector diagram for check measurement as shown in the upper sketch of Group 1 may be drawn correctly for a transformer which is connected wye-wye, although the sketch might seem to indicate that the transformer is connected delta-delta. However, this differentiation is not necessary as far as parallel operation is concerned. The test does indicate the angular displacement and time order of phase sequence,—in sort, it indicates the proper grouping of the transformer. Transformers of the same group may be operated in parallel regardless of their internal connections provided similarly marked leads are connected and provided their ratios, voltages, resistances, and reactances are such as to permit parallel operation. For example, if all these factors are the same, a delta-delta connected transformer of

Group 1 may be operated in parallel with a wye-wye connected transformer of the same group. The same applies to transformers of the other groups.

Transformers of different groups cannot be connected in parallel because of the difference in angular displacement. For example, a delta-delta transformer of Group 1 (Fig. 5-75) cannot be paralleled with a delta-delta transformer of Group 2, because there is a difference of 180 degrees in their angular displacement, and no interchanging of the external leads will change this displacement although the order of lettering of the leads may change. To make two such units satisfactory for parallel operation it is necessary to change the internal delta connections, as shown in Figs. 5-68 and 5-70.

When voltage-vector diagrams of the transformers which are to operate in parallel are available, it is then only necessary that these diagrams coincide and corresponding terminals be connected. It is entirely unnecessary then to raise questions of polarity and phase sequence, because when the voltage diagrams coincide, leads which are to be connected will have the same potential, this being the basic requirement for paralleling; whereas polarity, phase sequence, and angular displacement are merely means to arrive at this condition.

Transformer Connections

Figs. 5-77 to 5-87 inclusive show the proper connections and lead lettering of single-phase transformers of the same or different polarity for each of the five groups shown in Figs. 5-73 and 5-74. Any one connection of any one figure may be paralleled with any one connection of another figure by connecting similarly marked leads together, provided the two figures are of the same group and provided also that their ratios, voltages, resistances, and reactances are such as to permit parallel operation.

For example, any bank connection of Fig. 5-78 may be paralleled with any bank connection of Fig. 5-79 by connecting similarly marked leads together, provided the other conditions as given above are such as to permit this parallel operation.

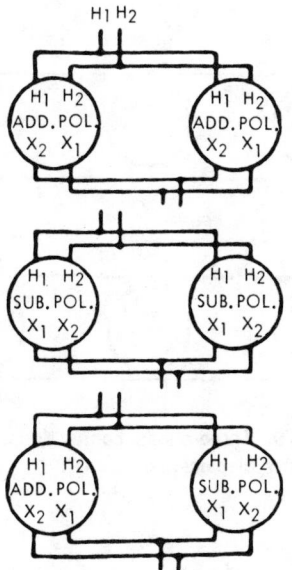

Fig. 5-77. Single-phase connections.

Parallel Operation

Parallel operation of transformers involves two or more transformers connected to carry a common load.

When a given transformer is insufficient in capacity to deliver a particular load it may either be taken

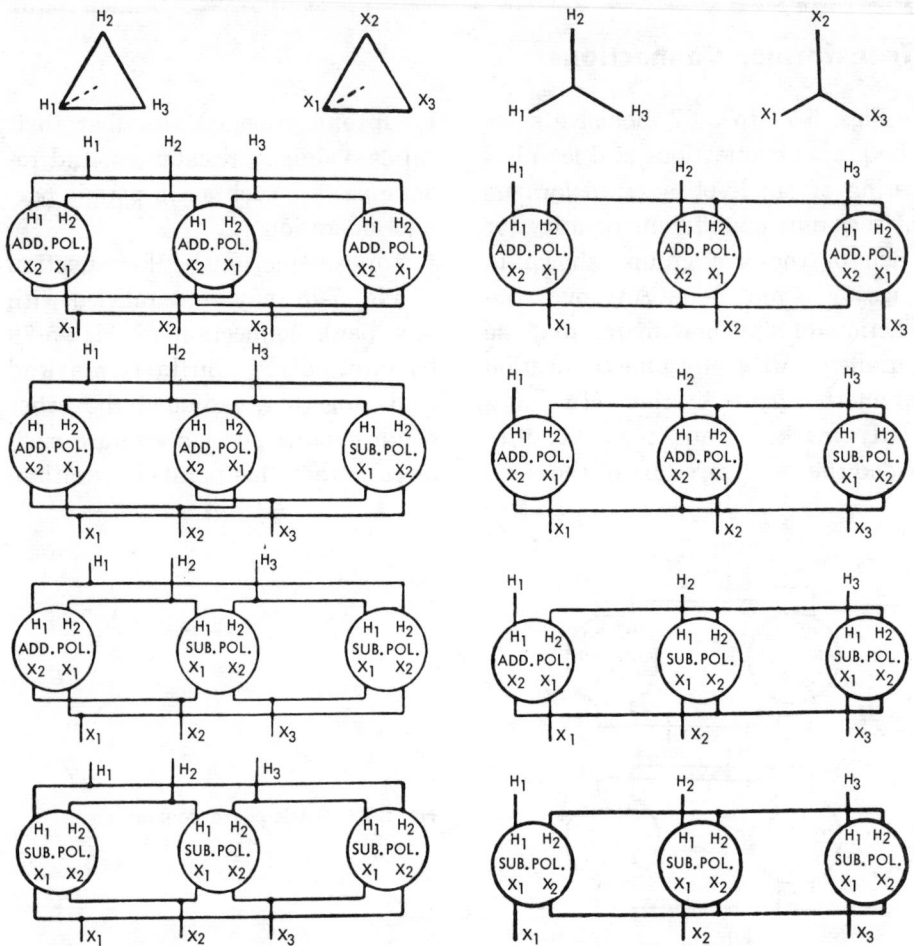

Fig. 5-78. Three-phase connections. Group 1, angular displacement = 0°

Fig. 5-79. Three-phase connections. Group 1, angular displacement = 0°.

out of the circuit and replaced with a larger unit or an additional unit may be added to the circuit by connecting its primary side to the same source of supply and its secondary side to the same load circuit. The second unit is then operating in parallel with the first unit. In mak-

ing this connection it is imperative that similarly marked leads of both transformers be connected to the same sides of each circuit. Generally, the paralleling connections consist in joining common primary and secondary leads and then connecting the bank thus formed to the

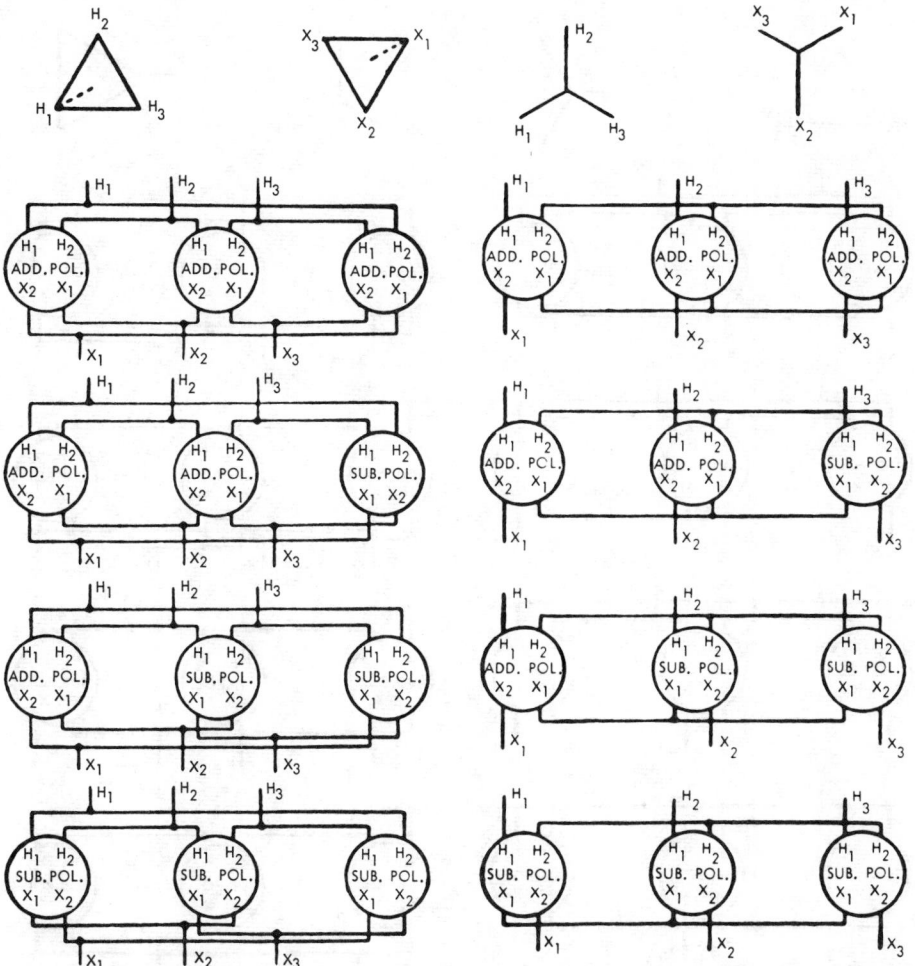

Fig. 5-80. Three-phase connections. Group 2, angular displacement = 180°.

Fig. 5-81. Three-phase connections. Group 2, angular displacement = 180°.

primary and secondary circuits. Fig. 5-77 shows three separate banks each consisting of two single-phase transformers connected in parallel. Note the distinction between the parallel connection as shown in Fig. 5-77 and the parallel connection mentioned in the preceding paragraph. In both cases, the two transformers are connected in parallel because they are energized from the same source of supply and are connected to the same load circuit. In the first instance, however, the

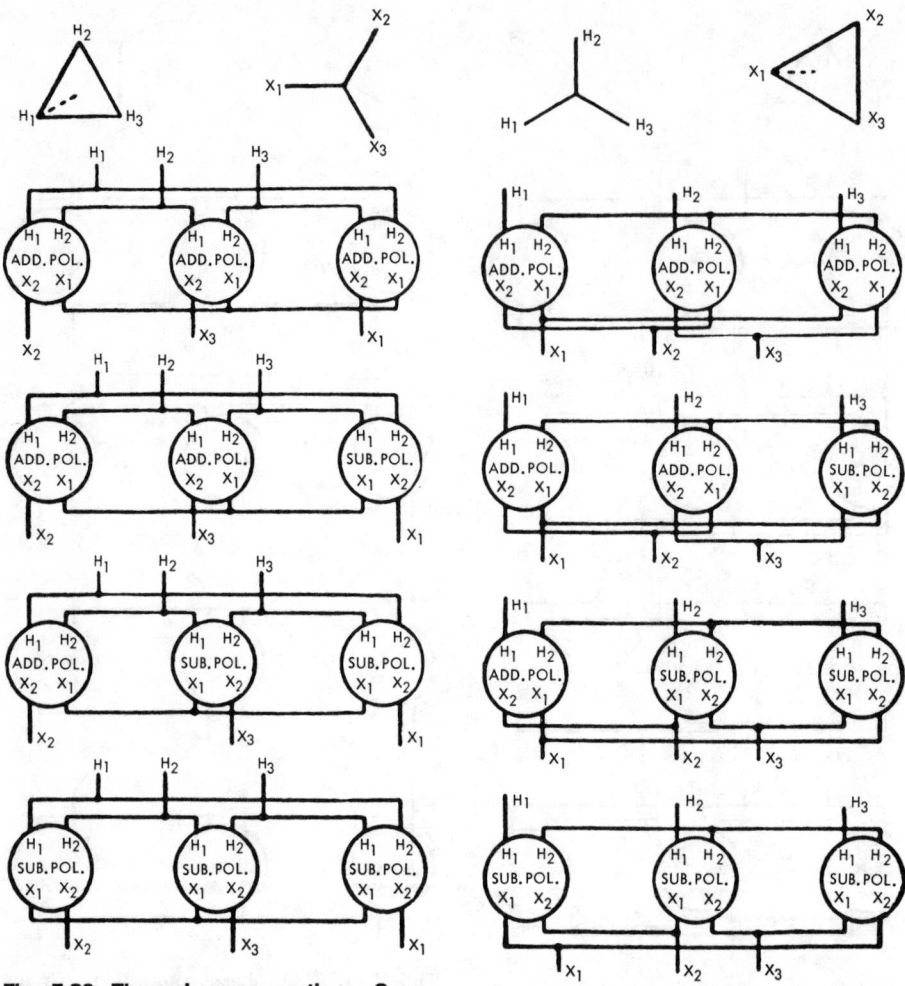

Fig. 5-82. Three-phase connections. Group 3, angular displacement = 30°.

Fig. 5-83. Three-phase connections. Group 3, angular displacement = 30°.

transformers are each connected directly to the supply and load circuit, whereas the two units of Fig. 5-77 are first connected together and then connected to the respective supply and load circuits. In both cases the resistance and reactance

of the leads from the transformer to the common points of connection must be included in the resistance and reactance of the transformer itself. The division of load between the two transformers connected in parallel by either of the two meth-

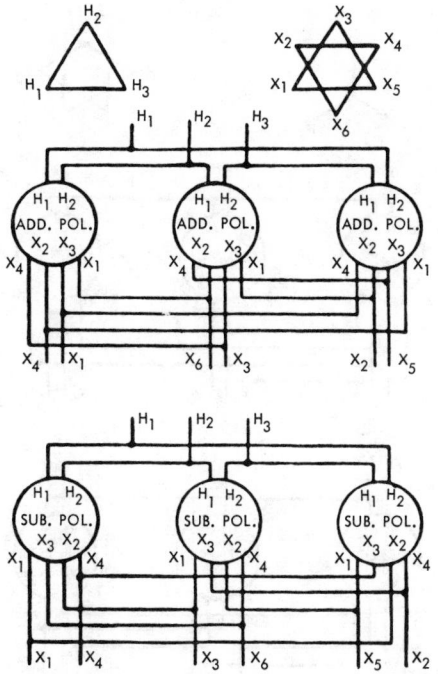

Fig. 5-84. Three-phase to six-phase connections. Group 4, angular displacement = 0°.

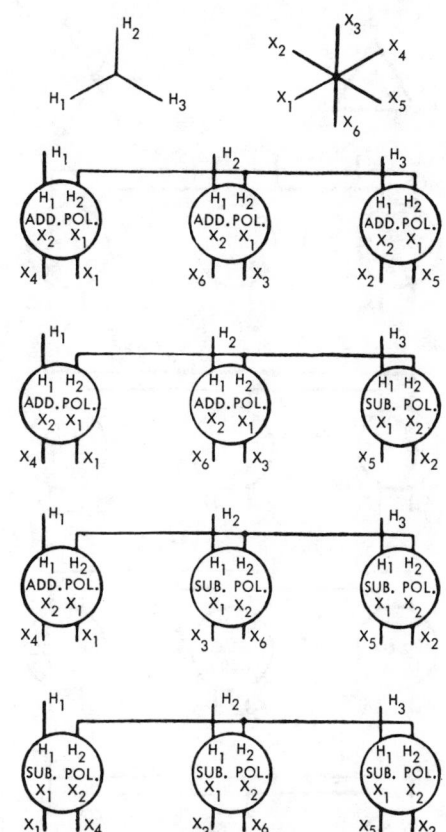

Fig. 5-85. Three-phase to six-phase connections. Group 4, angular displacement = 0°.

ods therefore may be quite different as will be seen later in considering what constitutes satisfactory parallel operation and what factors affect the division of the load.

Satisfactory parallel operation means that the transformers connected in parallel share the common load approximately in proportion to their ratings. The most satisfactory condition is obtained when the units divide the common load exactly in proportion to their ratings. For example, if two 50 kVA (kilovolt-ampere) transformers are

connected in parallel with a third transformer of 100 kVA rating, and the common load supplied by this parallel combination is 150 kVA, the two 50 kVA units should each supply 37½ kVA, and the 100 kVA should supply 75 kVA. If the common load is 200 kVA, which is the maximum bank capacity of the three transformers obtainable without exceeding the sum of the indi-

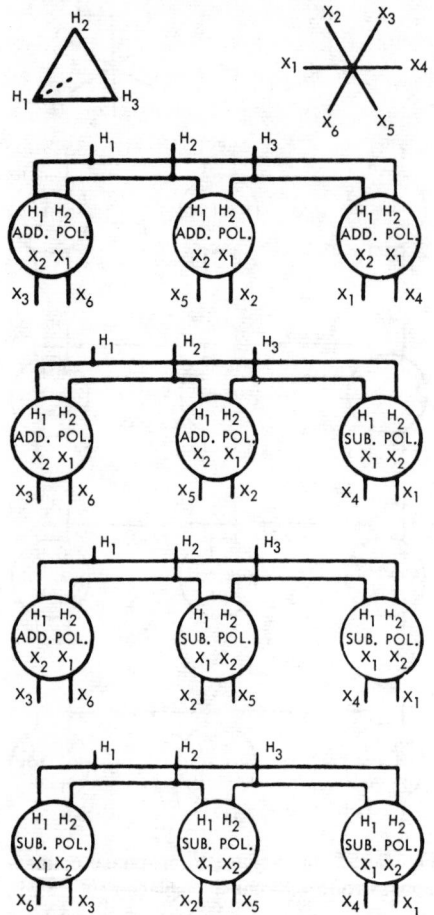

Fig. 5-86. Three-phase to six-phase connections. Group 5, angular displacement = 30°.

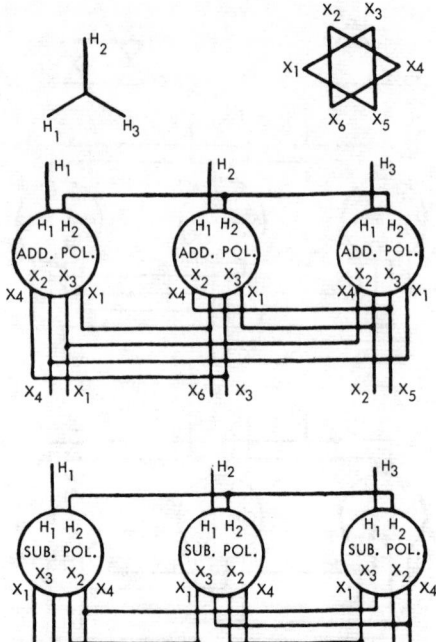

Fig. 5-87. Three-phase to six-phase connections. Group 5, angular displacement = 30°.

Requirements for Satisfactory Parallel Operation

vidual ratings, the two transformers rated at 50 kVA should each supply 50 kVA, and the 100 kVA unit should supply 100 kVA to the common load. In both instances the division of load between the three transformer units is proportional to their normal ratings.

In order that two or more transformers may operate satisfactorily in parallel it is necessary that they have,

(a) the same polarity, or, in case of three-phase transformers, the same angular displacement and same phase sequence

(b) the same voltage ratio

(c) the same percent impedance. Furthermore, the ratio of the

284

resistance component to the reactance component of each impedance should be the same.

Condition A. In a single-phase circuit the secondary windings of two transformers connected in parallel are either in the same time-phase relationship, the two voltages having 0 degrees displacement with respect to each other, or they are in opposite time-phase relationship, the two voltages having 180 degrees displacement with respect to each other. The parallel connection of the two windings constitutes a closed series circuit. If these two voltages are of the same time-phase relationship the induced voltages from either common connection to the other are in the same direction (therefore opposing each other) and there is no tendency for any current to flow in this series circuit. Satisfactory paralleling conditions have been established if there is no current flow in this series circuit, with no load connected across the terminals of this parallel connection. The two windings so connected are therefore of the same polarity. If, however, these two voltages are in time-phase opposition the induced voltages from either common connection to the other are in opposite directions, therefore adding to one another in the series circuit, and a large current will flow in the series circuit which, in effect, is equal to a short

circuit on both transformers. The two windings are, in this case, of opposite polarity and the parallel connection thus formed is unsatisfactory. The polarity designation must not be confused with the terms additive polarity and subtractive polarity as applied to the individual units. These latter terms merely indicate the relative directions of the induced voltages and are given to assist in making the proper paralleling connections. If the leads are properly marked with respect to the definitions as given for additive polarity and subtractive polarity it is only necessary to connect similarly marked leads in common to obtain the proper paralleling connection. For example, referring to Fig. 5-77, the upper connection shows two transformers of additive polarity connected in parallel; the middle connection shows two transformers of subtractive polarity connected in parallel, whereas the lower connection shows a transformer of additive polarity connected in parallel with a transformer of subtractive polarity. In each of these connections the X_1 leads of each transformer are connected and the two X_2 leads are likewise connected. The induced voltages in each of these connections from X_1 to X_2 are the same through each of the transformers inasmuch as the lead marking is determined by the direction of the induced voltages. In each of these three parallel connections the two transformers are of the same polarity.

If when making a parallel connection the lead marking is unknown, it is recommended that the two leads located in similar positions with respect to the bank be first connected together and a fuse of low-current rating be inserted between the remaining two leads. If the windings are of the same polarity the voltage across the fuse will be zero and no current will flow. If, however, the windings are of opposite polarity the voltage across the fuse will be equal to twice the voltage of either winding and an equivalent short-circuit current will tend to flow and the fuse will be blown. To obtain satisfactory operation in this latter case it is then necessary to interchange the connections of one of the units. This is equivalent to reversing the direction of the induced voltage of the one unit, thereby making them both of the same polarity.

In paralleling three-phase transformers of the same voltage ratio, polarity may be neglected entirely. For satisfactory paralleling conditions the angular displacement and phase rotation between the two units to be paralleled must be the same. For example, a wye-wye connected transformer having an angular displacement of 0 degrees cannot be connected in parallel with a wye-delta connected transformer having an angular displacement of 30 degrees. (See Fig. 5-73.) The secondary voltages of these two connections would be 30° out of phase with each other, as indicated in Figs. 5-26 and 5-28. This can more readily be seen if Fig. 5-28 is superimposed on Fig. 5-26. Since the voltages are out of phase with each other, if connected in parallel, a voltage will exist within the parallel connection tending to send current through each of the two transformers, thereby resulting in unsatisfactory parallel operation.

Similarly, if the phase rotation of the secondaries of the two three-phase transformers are in opposite directions it is apparent that a voltage exists within the parallel connection, which tends to send current through each of the two transformers, thereby resulting in unsatisfactory parallel operation.

To simplify the paralleling of three-phase transformers without having to test for polarity, angular displacement, and phase rotation, the lead markings have been standardized and the various three-phase connections have been placed in three different groups, depending upon their angular displacement. These groups are shown in Fig. 5-73. All transformers of any one group may be connected in parallel with each other. A transformer of one group, however, cannot be connected in parallel with a transformer of another group because of the difference in angular displacement. Neither is it possible to parallel transformers of Group 2 (Fig. 5-73) with those of Group 1 by interchanging the external leads. To change the angular displacement from 0 degrees to 180 degrees or vice versa, it is necessary to change the internal connections of the coils forming the wye or delta connection.

Condition B. If the voltage ratio of two transformers connected in parallel is not the same, the difference in voltage between the two windings will cause a current to flow within the parallel circuit at all times. If we again consider the parallel connection comprising two windings connected in series, the voltage difference will send a circulating current through these two windings. The amount of circulating current will be limited only by the sum of the impedances of the two transformers. If the impedance voltage of the two transformers is each 5 percent this means that 5 percent of normal voltage impressed on one winding will circulate full-load current when the other winding is short circuited. When the two transformers are connected in parallel, the total impedance of the two windings in series is 10 percent. In order to limit the circulating current to 10 percent of normal value, the voltage producing this current flow could

not exceed 1 percent of the voltage of the winding in terms of which it is expressed. For example, if the secondary voltage of one of the foregoing transformers is 120 volts the voltage of the other unit would have to be within 1.2 volts (1 percent of 120 volts) above or below 120 volts to limit the circulating current to a value of 10 percent of the normal full-load rating, assuming both units are of the same capacity rating. This circulating current which is at 90° phase angle to the voltage will flow at no load as well as at full load. Therefore when an external load is placed upon the two units connected in parallel this circulating current adds vectorially to the load current in the one winding and subtracts vectorially from the load current in the other. The amount of circulating current can be determined from the following formula:

$$\text{Percent } I_c = \frac{100 \times \%e}{\%IZ_1 + \%IZ_2} \quad (1)$$

Where

Percent I_c = circulating current in percent of the normal load current of one transformer.

Percent e = difference in voltage between the two windings expressed as a percent of the normal voltage.

Percent IZ_1 = the percent impedance of unit No. 1

Percent IZ_2 = the percent impedance of unit No. 2

In the formula given above it is

assumed that the capacity of both units is the same. If the capacities are different, the percent impedances in the formula should be based upon the same kVA rating, and the percent circulating current is then a percent of the normal load current at this same kVA rating. For example, assume that a 50 kVA transformer having a secondary voltage of 100 volts and an impedance of 4 percent is placed in parallel with a 100 kVA transformer having a secondary voltage of 102 volts, and an impedance of 4 percent. The impedance of the 100 kVA transformer at 50 kVA is 2 percent. Then

$$e = 102 - 100 = 2 \text{ volts}$$

$$\text{Percent } e = \frac{2 \times 100}{100} =$$

2 percent of the 100-volt winding

$$\% IZ_1 = 4 \qquad \% IZ_2 = 2$$

$$\text{Then percent } I_c = \frac{100 \times 2}{4 + 2}$$
$$= 33\frac{1}{3} \text{ percent}$$

circulating current expressed as a percent of the normal load current of the 50 kVA-100 volt transformer.

The normal load current is 50,000 divided by 100 = 500 amperes. Therefore the circulating current is $33\frac{1}{3}$ percent of 500 or 166.7 amperes.

Therefore it is evident that only a small difference between the ratios of transformers connected in parallel will produce a relatively large circulating current. For satisfactory parallel operation the circulating current should not exceed 10 percent of the normal load current.

Two transformers of different voltage ratios can be connected in parallel by the use of a balance coil connected as shown in Fig. 5-88. The balance coil is a single-winding transformer having a tap located at some point between its two ends. The position of this tap is determined by the desired division of load on the two transformers connected in parallel. The balance coil receives its excitation from the voltage difference of the two windings connected in parallel; therefore the current which would normally flow in the two windings due to this voltage difference is reduced to an amount equal to the magnetizing current of the balance coil. The tap is so located that when the desired division of current through each of the windings is obtained, the ampere-turns on one side of the tap are equal to the ampere-turns on the other side. Since the ampere-turns on the two sides are in opposition, as in an autotransformer, they neutralize each other with respect to the magnetizing effect of the magnetic circuit. A three-phase balance coil may similarly be used for paralleling three-phase transformers.

Condition C. The division of the external load between trans-

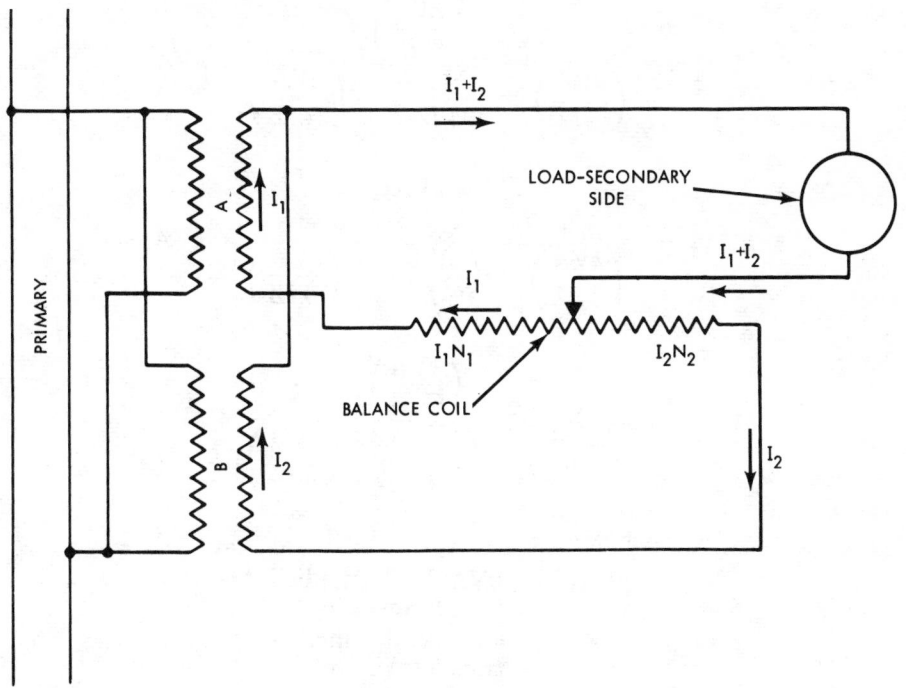

Fig. 5-88. Diagram of connections of two transformers of unequal voltage ratios connected in parallel by the use of a balance coil to obtain the proper division of load.

formers connected in parallel is inversely proportional to their ohmic impedances. The load will divide between the several transformer units until the terminal voltages of all of them have the same value. The unit which tends to maintain a higher voltage, that is, has a lower impedance, will take more current until its terminal voltage falls to a value equal to the terminal voltage of the other unit, or the currents through the several units in parallel will be such that the impedance drops through the several units are the same. Therefore the loads on the transformers connected in parallel will vary inversely as the impedances of the transformers expressed in ohms. If the ratings of the two transformers are the same, the loads will divide inversely as the impedances of the transformers expressed in percent of normal voltage.

The division of load between any number of single-phase transformers operating in parallel may be found by the use of the following formulas:

289

$$kVA_1 = \frac{\left(\dfrac{kVA}{\%IZ}\right)_1}{\left(\dfrac{kVA}{\%IZ}\right)_1 + \left(\dfrac{kVA}{\%IZ}\right)_2 + \left(\dfrac{kVA}{\%IZ}\right)_3} \times kVA_L$$

$$kVA_2 = \frac{\left(\dfrac{kVA}{\%IZ}\right)_2}{\left(\dfrac{kVA}{\%IZ}\right)_1 + \left(\dfrac{kVA}{\%IZ}\right)_2 + \left(\dfrac{kVA}{\%IZ}\right)_3} \times kVA_L$$

$$kVA_3 = \frac{\left(\dfrac{kVA}{\%IZ}\right)_3}{\left(\dfrac{kVA}{\%IZ}\right)_1 + \left(\dfrac{kVA}{\%IZ}\right)_2 + \left(\dfrac{kVA}{\%IZ}\right)_3} \times kVA_L \qquad (2)$$

where:

kVA_1 = kVA load supplied by transformer No. 1.

kVA_2 = kVA load supplied by transformer No. 2.

kVA_3 = kVA load supplied by transformer No. 3.

kVA_L = total kVA of the connected load.

$\left(\dfrac{kVA}{\%IZ}\right)_1$ = kVA rating of transformer No. 1 divided by its percent impedance.

$\left(\dfrac{kVA}{\%IZ}\right)_2$ = kVA rating of transformer No. 2 divided by its percent impedance.

$\left(\dfrac{kVA}{\%IZ}\right)_3$ = kVA rating of transformer No. 3 divided by its percent impedance.

For example, assume it is desired to parallel a 50 kVA-2,400- to 240-volt transformer having an impedance of 4 percent with a 75 kVA-2,400- to 240-volt transformer having an impedance of 6 percent in order to supply a load of 125 kVA.

The portion of the total load supplied by the 50 kVA transformer is:

$$kVA_1 = \frac{\dfrac{50}{4}}{\dfrac{50}{4} + \dfrac{75}{6}} \times 125$$

$$= 62.5 \; kVA$$

The portion of the total load supplied by the 75 kVA transformer is:

$$kVA_2 = \frac{\dfrac{75}{6}}{\dfrac{50}{4} + \dfrac{75}{6}} \times 125$$

$$= 62.5 \, kVA$$

The 50 kVA transformer is overloaded, whereas the 75 kVA unit is operating below its capacity. If the load on the 50 kVA unit is to be limited to 50 kVA, the external load of the bank must be reduced in the ratio of 62.5 to 50 or from 125 to 100 kVA. The load on the 75 kVA unit would be reduced by the same ratio and would therefore be 50 kVA also.

The formulas given above are correct only when the ratio of resistance to reactance of each of the transformers is the same, in which case the currents in each of the transformers are in phase. If these ratios are not the same, the currents in the transformers are not in phase and the sum of the winding currents will be greater than the line current. The phase angle of the current in the transformer is equal to the angle whose cosine is the percent reactance divided by the percent impedance. If the difference in phase angle between two transformers connected in parallel is less than 15 degrees, the line current may be assumed to be the arithmetical sum of the winding currents as found

from the above formulas without appreciable error.

For satisfactory parallel operation the percent impedances of all transformers connected in parallel should be the same. If this is the case, neglecting the error introduced if their resistances and reactances are not of the same ratio, the load will divide in the individual units in the ratio of their capacities, and the total permissible load will equal the sum of the capacities of the individual units connected in parallel.

To obtain such a division of load with units of unlike impedance, it is possible, but not practical or economical, to install an impedance coil in the form of a reactor in series with the transformer having the lower impedance. Although this method is seldom used today we will explore it for theoretical purposes. The reactance of this reactor should be such that when placed in

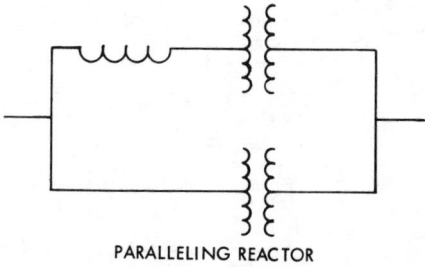

PARALLELING REACTOR

Fig. 5-89. Parallel connection of two transformers of unequal impedance showing the use of a current limiting reactor for increasing the percent reactance of transformer A to that of transformer B for proper division of load.

series with the low impedance unit, the percent impedance is raised to a value equal to that of the other transformer in the paralleling circuit. Fig. 5-89 shows the use of a dry-type reactor connected in series with the low-voltage secondary side of a distribution transformer to raise the percent impedance of transformer A to that of transformer B, thus assuring that the total load will be divided between the two transformers in proportion between their kVA ratings. In the particular connection shown, the transformers are connected for 2,400- to 120-volt service, and the increased impedance required in the circuit of transformer A is obtained by connecting a reactor in series with the low voltage leads as shown in Chapter 7, Fig. 7-1. A reactor of this type or of similar construction is suitable for voltages up to 600 volts, Fig. 7-2. For voltages above 34,500 volts it becomes more economical to use the oil-insulated type of reactor as shown in Fig. 7-6.

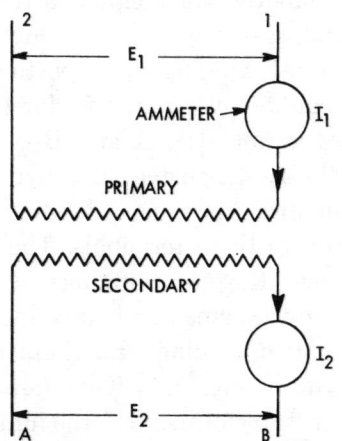

Special Transformers

Chapter

6

Autotransformers

When speaking of a transformer we generally have in mind two independent circuits; the primary circuit which receives the energy and the secondary circuit which delivers that energy to the load circuit. Such a transformer is shown in Fig. 2-2 and again in Fig. 6-1. In the latter, the load current of I_2 amperes at a voltage of E_2 volts on the secondary winding is supplied by the primary winding having a current of I_1 amperes when connected across a source of voltage of E_1 volts. The power transformed is equal to $I_2 E_2$ volt-amperes which in turn is equal to the power input of $I_1 E_1$ volt-amperes if we neglect the losses in the transformer. If we connect these two windings in series to form one single continuous winding as shown in Fig. 6-2, we have an autotransformer. If a voltage is impressed across the total of these two windings in series, current may be taken from the transformer between the point where the two windings are connected together and either of the

Fig. 6-1. Primary and secondary windings of a two-winding transformer.

Transformers

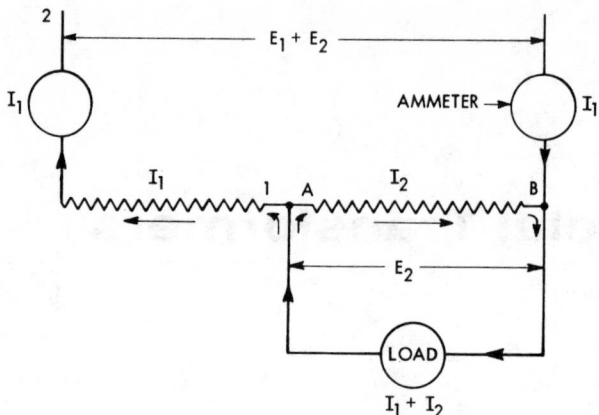

Fig. 6-2. Primary and secondary of Fig. 6-1 connected as an autotransformer.

two ends of the entire winding. Any transformer in which the primary and secondary circuits have a portion of the winding in common constitutes an autotransformer.

It will be seen from a comparison of Figs. 6-1 and 6-2 that the electrical characteristics of the two-winding transformer change when connected as an autotransformer. For example, the primary voltage rating of the transformer as connected in Fig. 6-2 is the sum of the voltages of the separate windings connected as in Fig. 6-1; with the same currents flowing in the respective windings as in Fig. 6-1, the external load current is now the sum of these two winding currents. The power input of this transformer as an autotransformer is $I_1 (E_1 + E_2)$ volt-amperes, which in turn is equal to the power output of $(I_1 + I_2) E_2$ volt-amperes. The output therefore has been in-

creased by $I_1 E_2$ volt-amperes because of this connection.

Principle of Operation

When the two windings of a transformer are connected together to form an autotransformer, as shown in Fig. 6-2, and a voltage is impressed on terminals *2-B*, a counter electromotive force equal to that of the impressed voltage is induced in the total winding. The portion of the counter electromotive force between terminals *1-A* and *B* is that which sends current through the secondary side when a load is connected to these terminals. The currents in the two windings of Fig. 6-2 are the same as when connected as a two-winding transformer as shown in Fig. 6-1. Therefore the power transformed by the unit in either case is equal to $I_2 E_2$ volt-amperes. The remainder of the load, equal to I_1E_2 volt-amperes, is not transformed but is conducted directly from the primary side to the secondary load. In Fig. 6-2 we might consider the portion of the winding from *2* to *1-A* as the primary, and the portion from *1-A* to *B* as the secondary. The volt-amperes of the primary winding are equal to the volt-amperes of the secondary winding, just as in the case of a two-winding transformer. Because part of the secondary load in the auto-transformer connection is simply transferred rather than transformed, it is seen that the output when connected as an autotransformer is greater than when connected as a two-winding transformer.

Output Rating

Since a single winding in an auto-transformer serves for both the primary and the secondary circuits, the material in the autotransformer will be less for a given output than that in a two-winding transformer, or with the same amount of material in both, the output of the autotransformer will be greater.

For example, assume that the transformer in Fig. 6-1 is rated at 10 kilovolt-amperes (10,000 volt-amperes) with a primary voltage, E_1, of 1000 volts and a secondary voltage, E_2, of 100 volts. The primary current, I_1, will then be 10 amperes and the secondary current, I_2, will be 100 amperes.

When connected as an autotransformer as in Fig. 6-2 the primary voltage

E_1+E_2 will be 1100 volts and the secondary voltage E_2 will still be 100 volts. If we assume the same currents in the winding, then $I_1 = 10$ amperes and $I_2 = 100$ amperes. The secondary load current will be $I_1+I_2 = 110$ amperes and the total output will be $110 \times 100 = 11000$ volt-amperes or 11 kilovolt-amperes. The ratio of transformation as a two-winding transformer is

$$\frac{E_1}{E_2} = \frac{1000}{100} = 10 \text{ to } 1;$$

the ratio of transformation as an auto-transformer is

$$\frac{E_1 + E_2}{E_2} = \frac{1100}{100} = 11 \text{ to } 1.$$

It will be noted that the ratio of transformation has been increased 10 percent. Likewise the output has increased from 10 to 11 kilovolt-amperes, or 10 percent. The rating of this transformer when operating as an autotransformer will therefore be 11 kilovolt-amperes.

The output rating of the auto-transformer may therefore be expressed in terms of the ratio of its voltage transformation and the rating of the same winding when operated as a two-winding transformer. This expression is as follows:

$$\text{Auto rating} = \text{two-winding transformer rating} \times \frac{1}{1 - \dfrac{1}{R}} \qquad (1)$$

where R is the ratio of the high voltage to the low voltage of the auto-transformer.

In the foregoing example R equals $1100 \div 100 = 11$. Therefore the autotransformer rating $= 10000 \times$

$$\frac{1}{1 - \dfrac{1}{11}} = 10000 \times \frac{11}{10} = 11{,}000 \text{ or}$$

11 kilovolt-amperes.

The expression as given in the above equation (1) may be stated in another form to give the kilovolt-amperage of transformer parts required to make an autotransformer of a specified output rating and voltage ratio. This form is:

kVA of transformer parts $=$

autotransformer rating $\times \left(1 - \dfrac{1}{R} \right)$

$$(2)$$

Using the same example as given above, the autotransformer rating is 11 kilovolt-amperes; the ratio of transformation is 11 to 1 or R equals 11. The kVA of transformer parts $=$

$$11 \times \left(1 - \frac{1}{11} \right) = 11 \times \frac{10}{11}$$
$$= 10 \text{ kVA}.$$

By the use of the autotransformer connection, therefore, an output of 11 kilovolt-amperes can be obtained from the same parts that would be required to give an output of only 10 kilovolt-amperes when connected as a two-winding transformer.

An examination of equation (2) will show that as R decreases, the parts necessary to produce the auto-transformer capacity become smaller.

The quantity $\left(1 - \dfrac{1}{R} \right)$ approaches

unity when the voltage ratio becomes very large. In this latter case the kilovolt-amperage of transformer parts is practically equal to the rating of the transformer and it would be no advantage to connect the unit as an autotransformer. As R becomes smaller and approaches 1 in value, the quantity $\left(1 - \dfrac{1}{R} \right)$ approaches zero. This means that as the secondary voltage of the auto-transformer approaches in value the primary voltage, the parts required to build such a unit of a given kilovolt-ampere rating becomes smaller and smaller. This reduction in size as the ratio of transformation decreases is one of the main reasons for the use of an autotransformer in place of a two-winding transformer.

For example, suppose it is desired to step up the voltage of a line from 220 volts to 250 volts by means of an auto-transformer. Furthermore, let it be assumed that the load to be stepped up to this higher voltage is equal to 150 kilovolt-amperes. What would be the size of such an autotransformer compared to a two-winding transformer having a voltage ratio of 220 volts to 250 volts? If this change in voltage were obtained with a two-winding transformer it would of course require a transformer having a rating of 150 kilovolt-amperes. The ratio of transformation of the high voltage to low voltage is $250 \div 220 = R$.

Then $\dfrac{1}{R} = \dfrac{220}{250} = .88$ and $\left(1 - \dfrac{1}{R} \right)$

$$= 1 - .88 = .12$$

From equation (2) the kilovolt-amperes of the transformer parts required to transform 150 kVA from 220 to 250 volts is therefore equal to $150 \times .12 = 18$ kVA. This means that an autotransformer required to transform 150 kVA from 220 volts to 250 volts would be no larger than an 18 kVA two-winding transformer having primary and secondary windings of 220 volts and 30 volts respectively.

As the ratio increases, the economic advantage of the autotransformer over the two-winding transformer grows less; therefore, autotransformers are rarely made with a ratio greater than 4:1 or 5:1. Not only is the saving in material in a high ratio autotransformer comparatively small, but the fact that the high- and low-voltage windings are electrically connected requires, if the circuit is ungrounded, that both have the same insulation to ground. This would add considerably to the cost of an autotransformer of high ratio; consequently, a two-winding transformer would be better for such applications.

Limitations of Autotransformers

The autotransformer has two very important limitations. The first is impedance. The impedance of an autotransformer is much less than that of the two-winding type transformer. For instance, if the latter type has an impedance of about 3 percent or rated primary voltage, as an autotransformer, its impedance, Z, is equal to 3 percent times the difference between primary and secondary voltages over the primary voltage because the characteristics are based on the size of transformer parts.

By reducing the impedance, a much higher current will flow on a short-circuit fault. As transformers are not built to withstand this type of short-circuit, means must be found to make the transformer self-protecting should such a fault occur. This may be done by providing an external impedance, which may be accomplished by other transformers in the circuit, by reactors or other devices or, in some cases, in the lines themselves.

The other important limitation of the autotransformer is that the two circuits as previously mentioned are electrically and magnetically connected. Thus, it is possible that the section of the winding common to the primary and secondary may become open. In such a case, anyone coming in contact with the secondary is subject to the high voltage, or the primary may become grounded and establish a high voltage between one of the low-voltage wires and ground.

Application of Autotransformers

The autotransformer is used in starting rotating apparatus such as synchronous and induction motors. The synchronous motor is started at about one-third of normal voltage and the induction motor at about two-thirds of running voltage. Since the voltages of such machines are relatively low, the fact that the primary and secondary voltages of the autotransformer are not insulated from each other is not objectionable, especially so since the starting autotransformer is connected to the line only during the starting cycle. Since this starting cycle is of short duration, usually only one minute, the size of the autotransformer is further decreased, for the materials can be worked at much higher densities without exceeding a safe temperature rise. Such starting autotransformers are usually connected open-delta or wye. The open-delta connection is used for smaller motors and the wye connection for larger machines. The wye connection gives a better balanced condition of voltage, whereas the open-delta connection provides a simpler means of voltage control, especially when more than one starting voltage is required.

Transformers connected open-delta require the changing of taps in two coils only, whereas those connected in wye must be changed in three windings. Autotransformers for starting duty may be single-phase units connected in open-delta or wye, or they may be three-phase units. The three-phase unit consists of only two windings on a three-phase magnetic circuit. Three-phase units are the preferred type because they require less space than three single-phase units of same bank capacity.

Autotransformers for power purposes are used when the values of primary and secondary voltages are not far apart such as 115 kV and 69 kV and where greatest economy results. Such autotransformers are usually connected in wye with the neutral grounded, as shown in Fig. 6-3. Such transformers are usually provided with a tertiary winding connected in delta to provide a path for the flow of the third harmonic of exciting current. The value of this current is usually very small and the

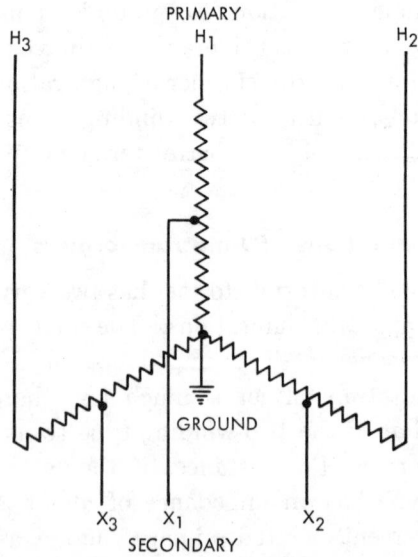

Fig. 6-3. Wye-connected autotransformer with grounded neutral.

delta winding added for that purpose is therefore but a small percentage of the capacity of the main autotransformer winding. The delta winding is sometimes increased in size and used for supplying power to a synchronous condenser for power-factor correction as previously mentioned.

While a three-phase transformation may be obtained with three single-phase autotransformers or one three-phase autotransformer connected in delta, Fig. 6-4, they are seldom used because they are very inefficient. Furthermore, a voltage transformation greater than 2 to 1 cannot be obtained. The delta connection also causes a phase shifting of the secondary voltage as compared with the wye connection. Such a shift in phase of secondary is shown by the dotted lines in Fig. 6-4. This phase displacement will also vary for each different ratio of transformation. Therefore it is impossible to parallel wye-connected and delta-connected autotransformers.

Autotransformers For Miscellaneous Uses

The autotransformer has many other applications of which only two will be mentioned: (1) As a preventive autotransformer in bridging across successive taps in a power transformer in which the voltage ratio is changed while the transformer is carrying load. Such an autotransformer is shown in Figs. 4-6 and 4-9. (2) The National Electrical Code permits the use of autotransformers (or equipment such as buck and boost transformers connected as autotransformers) on existing installations without requiring a common primary and secondary grounded conductor if the transformation is 208 to 240 volts or 240 volts to 208 volts as in Fig. 6-5.

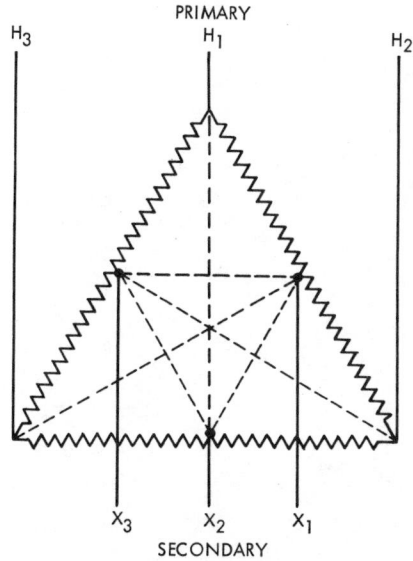

Fig. 6-4. Delta-connected autotransformer.

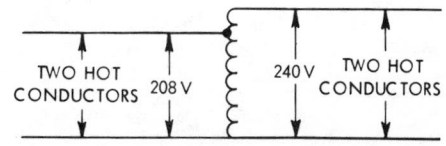

Fig. 6-5. Autotransformer used to convert 208 to 240 volts or 240 volts to 208 volts.

Another wide application of the autotransformer in common use is the variable transformer such as General Radio's Variac.

A variable transformer is basically an adjustable autotransformer consisting of a single layer of wire wound on a toroidal core and a carbon brush which traverses this winding. The brush track is made by removing a portion of the insulation from each turn of the winding, thus forming a series of commutator elements. The basic principle is that of a tap changing transformer. The brush is always in contact with one or more wires and continuously taps off any desired fraction of the winding voltage. It is possible, therefore, to remove the contact under load without interrupting the circuit.

Current-Limiting Reactors

Inductance

Whenever a current is flowing in a conductor, a magnetic field is established about the conductor at right angles to it. The strength of this field is directly proportional to the quantity of electricity flowing. As the current increases the magnetic field increases; as the current decreases the magnetic field also decreases. This magnetic field, or lines of force, commonly referred to as flux, may be considered as originating from the center of the current-carrying conductor and expanding or contracting in ever increasing or decreasing circles, as the current increases or decreases. With the lines of force cutting the conductor, a counter electromotive force is induced in it. When one volt is induced in a circuit by a uniform change of the current at a rate of one ampere per second, the circuit is said to have an inductance of one henry.

Reactance

The measure of the induced voltage due to the inductance of a circuit is called "Inductive Reactance," and it is directly proportional to the rate of change of current or the frequency of the circuit. It can be shown that the inductive reactance X, in ohms is equal to $2\pi fL$, where f is the frequency of the circuit in cycles per second, and L is the inductance of the circuit in henrys; the induced voltage is $2\pi fLI$ where I is the rms (root-mean-square), or effective current in amperes. The reactive voltage may therefore be expressed as equal to $2\pi f L I$, or IX. It is generally desirable that this reactive voltage be kept to a low value because of its adverse effect upon voltage regulation and the power factor of the system. For example, referring to Fig. 2-7, if the transformer had little or no leakage reactance, the reactive voltage vectors $I'X_p$ and I_sX_s would be very small, or equal to zero. Assuming the same induced voltages E_p and E_s on the primary and secondary sides respectively, the impressed voltage E' required to give this induced voltage of E_p volts would be greatly reduced, and, at the same time, the secondary terminal voltage E'', at a load current of I_s amperes, would be increased with a resultant improvement in

voltage regulation. Similarly, if the load circuit had little or no reactance the load current I_s would be nearly in phase with the terminal voltage E'' with a resultant improvement in the power factor of the circuit.

However, if an alternating-current circuit could be made without reactance, the short-circuit current would be limited only by the resistance of the circuit. This would result in large currents flowing during a short circuit, especially if there is a large amount of power available for feeding current into the short circuit. The conductor in the circuit must be braced to withstand the stresses set up by the magnetic forces of such large currents, and circuit breakers must be capable of interrupting these currents. There-

Fig. 7-1. Single-phase aluminum wound, current limiting reactor, dry-type (rated to 16 kVA, 600 volts and below). (Westinghouse Electric Corp.)

fore it is common practice to insert in the circuit enough additional reactance so that the short-circuit current is limited to some predetermined value which can be handled safely and economically. This additional reactance is provided in devices known as current-limiting reactors. Such a reactor (dry-type) is shown in Figs. 7-1 and 7-2.

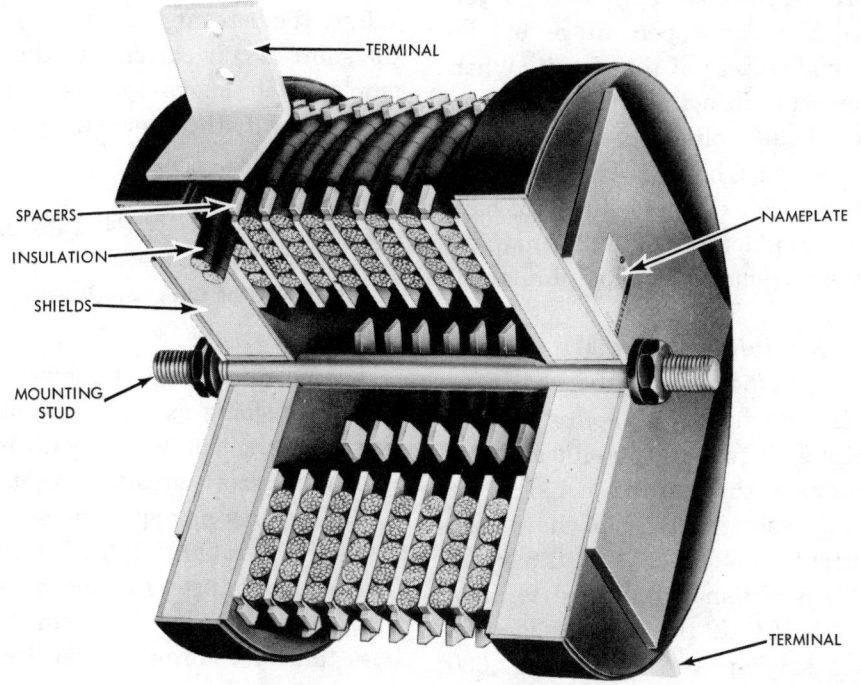

Fig. 7-2. Cutaway of single-phase aluminum wound, current limiting reactor, dry type (rated to 16 kVA, 600 volts and below):

1. **TERMINALS.** Silver-plated aluminum.
2. **SPACERS.** Polyester-glass radial spacers, provide optimum insulation and mechanical strength.
3. **SHIELDS.** End shields are made of a Micarta ring with a fiberboard bottom, backed by a steel plate and filled with a special cement. The shields serve to confine the stray flux, common to reactor fields and straighten flux-leakage paths in the windings which lowers winding eddy-current loss.

4. **MOUNTING STUD.** A high-resistance, nonmagnetic mounting stud (one inch in diameter, eight threads per inch) is used. Stud must be mounted horizontally to insure proper cooling of reactor.
5. **INSULATION.** Fiberboard insulating washers confine special cement and increase electrical creepage distance. Strong, moisture-proof, high-surge strength Class B insulation is used throughout.
6. **NAMEPLATE.**
(Westinghouse Electric Corporation)

Rating of Current-Limiting Reactor

A current-limiting reactor may be rated in terms of ohms reactance at a given frequency and current-carrying capacity or by expressing the voltage drop across the coil when carrying the given current at a given frequency as a percentage of the normal voltage of the circuit. When connected in a three-phase system, the circuit voltage used should be the voltage between the line and neutral. For example, a reactor having an inductance of .002 henry will have a reactance of .754 ohm at 60 hertz.

$$X = 2\pi f L = 2 \times \pi \times 60 \times .002$$
$$= .754 \text{ ohm}$$

The voltage drop E_x across this reactor with, for example, 400 amperes flowing in the circuit is equal to the ohms reactance multiplied by the current, or $400 \times .754 = 301.6$ volts. This is obtained from the formula $E_x = 2\pi f L I$. If this reactor is used in a single-phase circuit of 11,000 volts, the percent reactance would be

$$\frac{301.6 \times 100}{11000} = 2.742 \text{ percent}$$

If this reactor is used in a 11,000 volt, three-phase circuit in which the voltage to neutral is $11,000 \div \sqrt{3}$, or 6350 volts, the percent reactance with 400 amperes flowing in the line would be

$$\frac{301.6 \times 100}{6350} = 4.749 \text{ percent.}$$

The short-circuit current through this reactor with full voltage maintained at its terminals can be calculated either from the ohms or the percent reactance. Dividing the circuit voltage by the ohms reactance, or dividing the normal current times 100 by the percent reactance equals the short-circuit current. In the example given above for the three-phase circuit, the short-circuit current would be

$$\frac{6350}{.754} \text{ or } \frac{400 \times 100}{4.749} = 8422 \text{ amperes}$$

Construction of Current-Limiting Reactors

The current range of the reactor in the foregoing example is from a normal current of 400 amperes to a maximum short-circuit current of 8,422 amperes or approximately 21 times its normal full-load rating. The current range of a current-limiting reactor may vary from 4 to $33\frac{1}{3}$ times its normal full-load current rating. Throughout this entire range of current variation the ohms reactance of the reactor must be constant otherwise the short-circuit current would likewise vary.

If an iron core is used for the magnetic circuit in order to obtain the necessary inductance at normal current rating, the magnetic induction in the core increases as the voltage builds up across the reactor. As the current and voltage increase, the saturation point in the core is reached and each succeeding increase in current must be increasingly larger to produce the same increase in voltage. This means that the current increases more rapidly than the voltage, which is equivalent to a corresponding decrease in ohms reactance. In order to maintain a constant value of ohms reactance it would be necessary to work the magnetic circuit of the reactor under normal operating conditions at so low an induction point that the saturation point is not reached until after the maximum short-circuit current has been obtained. For current-limiting purposes it is not economical to use a magnetic circuit of iron with air gaps as is done in the case of a paralleling reactor in which the range of current seldom exceeds two and one half times its normal full-load current rating. The magnetic circuit of a current-limiting reactor is therefore an air core, giving a voltage drop across the reactor at all times directly proportional to the current flowing through it. For a more detailed function of the magnetic circuit, refer to "Function of the Magnetic Circuit," Chapter 3, and especially to Fig. 3-2. It may more easily be understood for this particular application if in Fig. 3-2 the ordinate which is marked "Flux Density" is changed to read "Reactance Volts." In making this change it must be understood, however, that the vertical scale of reactance volts for each of the three curves shown in Fig. 3-2 would be different; the underlying principle, however, is the same for each.

The current-carrying circuit of the current-limiting reactor is circular in shape, consisting of several turns of copper cable insulated with cotton, asbestos, or spun glass tapes. The turns are wound into discoidal layers. The horizontal spacers between layers provide extra layer insulation. These spacers have a hole through which vertical tie rods are passed for bolting the entire structure between concrete disks at the top and bottom. These concrete disks have inserts cast in them for attaching the insulating "feet" and terminal supports. Cable ends are brazed into the terminal supports, and all connections not bolted are either welded or brazed. This form of construction, shown in Figs. 7-1 and 7-2, gives maximum strength to resist the mechanical stresses set up by the magnetic field during short-circuit conditions.

Types of Current-Limiting Reactors

Dry-Type. The dry-type current-limiting reactor, as described in the previous paragraph, is dependent entirely upon the surrounding air for carrying away the heat developed in the winding. Therefore adequate space must be provided for a free circulation of air. Furthermore, since the reactor uses the air as a return path for the flux, care must be taken to keep the reactor at some distance from all magnetic materials such as steel channels, S beams, plates and other steel structures. The minimum distance usually is specified by the manufacturer. If the reactor is to be mounted in a metal cell, care must be taken to break up electrical and magnetic circuits adjacent to the reactor in order to eliminate any paths in which the flux from the reactor might induce a voltage sufficient to cause excessive eddy currents to flow, Figs. 7-3, 7-4 and 7-5.

The general field of application of the dry-type reactor is that of indoor use for all voltages up to approximately 34,500 volts and outdoor use for voltages to 25,000 volts.

Fig. 7-3. Vertical stack of three single-phase reactors to make a three-phase unit. (Westinghouse Electric Corp.)

Fig. 7-4. Single-phase current limiting reactor, dry-type (for circuits of 1200-34,500 volts, 5 to 6000 kVA). (Westinghouse Electric Corp.)

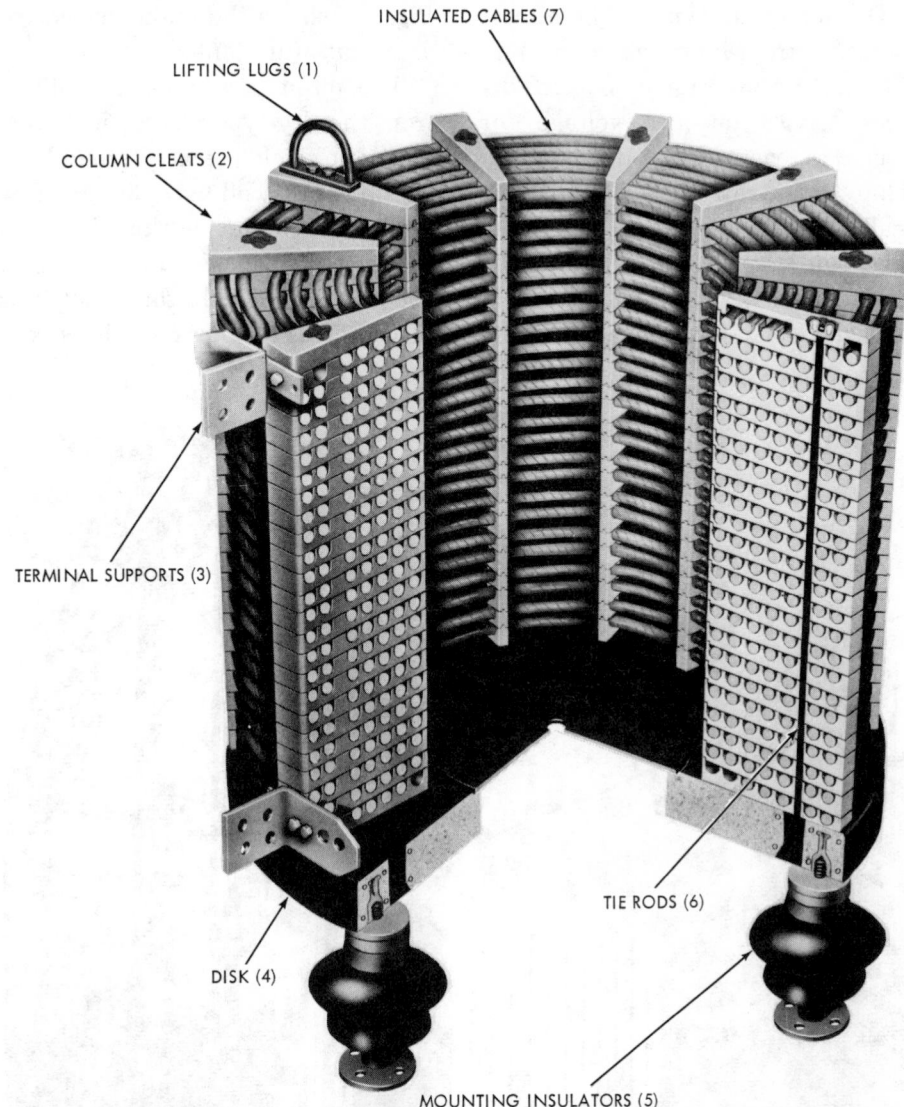

INSULATED CABLES (7)

LIFTING LUGS (1)

COLUMN CLEATS (2)

TERMINAL SUPPORTS (3)

TIE RODS (6)

DISK (4)

MOUNTING INSULATORS (5)

Fig. 7-5. Cutaway of single-phase current limiting reactor:

1. **LIFTING LUGS.**

2. **COLUMN CLEATS.** Molded from polyester-glass resin, they provide columns having high dielectric value and resiliency to thermal and mechanical shock.

3. **TERMINAL SUPPORTS.**

4. **DISK.** Cleat columns are mounted on a concrete or asbestos lumber disk.

5. **MOUNTING INSULATORS.**

6. **TIE RODS.** Each column of cleats supporting the cable is solidly bonded together by a special high-strength glass reinforced epoxy-resin tie rod.

7. **INSULATED CABLES.** Most dry type reactors are now wound with aluminum cable. (Westinghouse Electric Corp.)

Oil-immersed Type. The oil-immersed current-limiting reactor, Fig. 7-6, may be used for indoor or outdoor service on any voltage for which a reactor may be required. The current-carrying coils are normally wound either in cylindrical layers or in pancake coils on voltages of 34.5 kV and below and in pancake coils for above 34.5 kV.

Fig. 7-7 shows the oil-immersed reactor out of its tank.

The oil-immersed reactor differs from the dry-type reactor in that a magnetic shield is placed around the outside of the coil to confine the flux within the limits of this shield so that no flux will flow in the wall of the tank in which the reactor is placed. This magnetic shield is con-

Fig. 7-6. A bank of oil-immersed current limiting reactors, 11,136 kVA, type OA, 138 kV high voltage, single-phase, 60 hertz, 800 amps installed on the power systems of a large western utility. (Westinghouse Electric Corp.)

Fig. 7-7. Oil-immersed reactor out of its tank. (Westinghouse Electric Corp.)

structed of laminated high-silicon sheet steel which is welded to the tank wall at each end, as shown in Fig. 7-8. The heavy steel angles at each end as well as the hoops about the body of this iron shield have insulated joints to avoid the formation of short-circuited turns in which the flux would otherwise set up a large circulating current. The magnetic shield reduces the reluctance

Fig. 7-8. Magnetic shield of oil-immersed reactor as mounted in its tank. (Westinghouse Electric Corp.)

of the magnetic path, causing the normal-current reactance to be 15 to 20 percent higher (approximately) than at the time of short circuit. Sufficient iron is used for normal operation only. During short-circuit the iron saturates and the reactor performs substantially the same as an air-core reactor. Since the rating of the reactor is based upon the short-circuit reactance the increase in reactance at normal current flow is not objectionable.

The oil-immersed reactor has a high factor of safety against flashover not obtained in any other way. This feature makes it desirable for outdoor service and for high-voltage application. The high thermal capacity of the oil and the ease of cooling compared to that of a dry-type reactor make it desirable for large units or where space is limited.

The general appearance of the oil-immersed current-limiting reactor (Fig. 7-6) is similar to that of an oil-insulated transformer. In fact, many times a standard oil-insulated transformer case is used to hold these reactors.

Applications of Current-Limiting Reactors

A current-limiting reactor, being primarily a protective device, should be connected as near as possible to the service or apparatus it is to protect. A few of the applications are listed as follows:

1. Alternator reactors are those which are connected in the leads between alternators and station bus bars. When several large generators

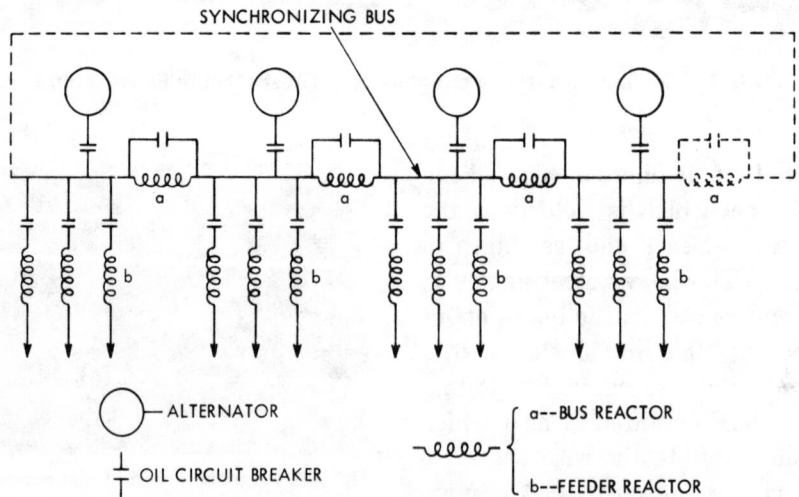

Fig. 7-9. Continuous straight sectionalized bus, with bus reactors.

are paralleled at one station it is often desirable to use generator reactors to prevent excessive damage to a winding due to the current rushes from the other machines in case of fault in the generator winding.

2. Bus reactors are those connected between different bus sections, as *a* in Fig. 7.9. In order to reduce the expense of generator reactors and also to eliminate the voltage drop under normal operating conditions, generators are often tied together to a continuous straight sectionalized bus, Fig. 7-9, and each generator supplies its own group of feeder circuits. If another bus reactor is connected to Fig. 7-9 as shown by the dotted line, it becomes a ring bus without generator reactors.

3. Tie-line reactors are used in series with the tie lines between two generating stations or between two substations which are interconnected. It is customary to divide the total tie-line reactance into two equal parts and place each half as near as convenience permits to the station bus at each end, since a fault is most likely to occur in the line between the two stations.

4. Feeder reactors are used to localize voltage drop on the feeder where trouble occurs. They are used extensively, since most short circuits occur in feeder circuits. Each feeder is connected to the bus through a

reactor, *b* as shown in Fig. 7-9. Such feeder reactors afford great protection to the system and make possible the use of smaller and cheaper feeder circuit-breakers and fewer bus supports. The feeder reactor also prevents opening of the main breaker in case of trouble, and thus prevents cutting other feeders out of service due to a fault in one of the feeders.

A thorough study of present and future conditions of the particular system in which current-limiting reactors are to be used is necessary to determine the most economical amount of reactance to be used and to decide whether this reactance should be located in the feeders, busses, or alternator leads. In the calculation of the possible short-circuit conditions, the inherent reactance of the generators must be taken into consideration since the reactance of the generator usually is sufficient to offset materially the total reactance of the various branches of the system.

Saturable-core Reactor

It is often desirable to control the value of the inductive reactance in an alternating current circuit. This may be accomplished by using a device referred to as a saturable-core reactor, Fig. 7-10. Its performance is similar to a variable resistor in a DC circuit, except that the value of inductive reactance (or impedance)

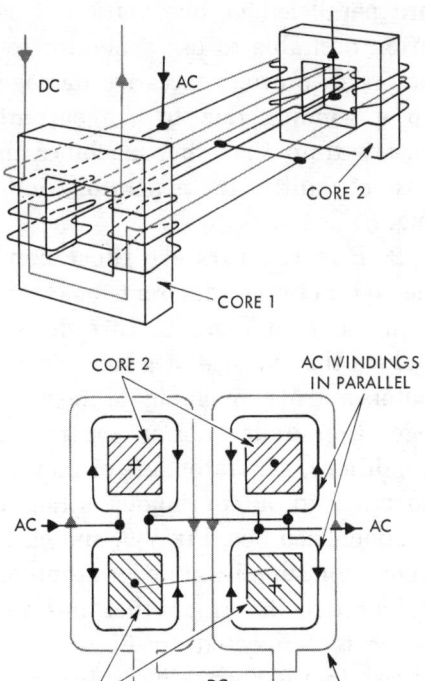

Fig. 7-11. Construction of a saturable core reactor. (Westinghouse Electric Corp.)

Fig. 7-10. Saturable core reactor. Single-phase, 37½ to 200 kVA, 5000 volts and below. (Westinghouse Electric Corp.)

is changed by controlling the magnetic saturation in an iron core by circulating a direct current in a sep-

arate winding, Fig. 7-11. With no DC current flowing in the separate winding the device will function as a simple iron core reactor. As the DC current is increased the core becomes saturated with DC flux thereby reducing the inductive reactance as far as AC is concerned. A complete range of control of the AC circuit can be obtained in a smooth stepless manner. By using a saturable-core reactor large amounts of AC power can be controlled by a

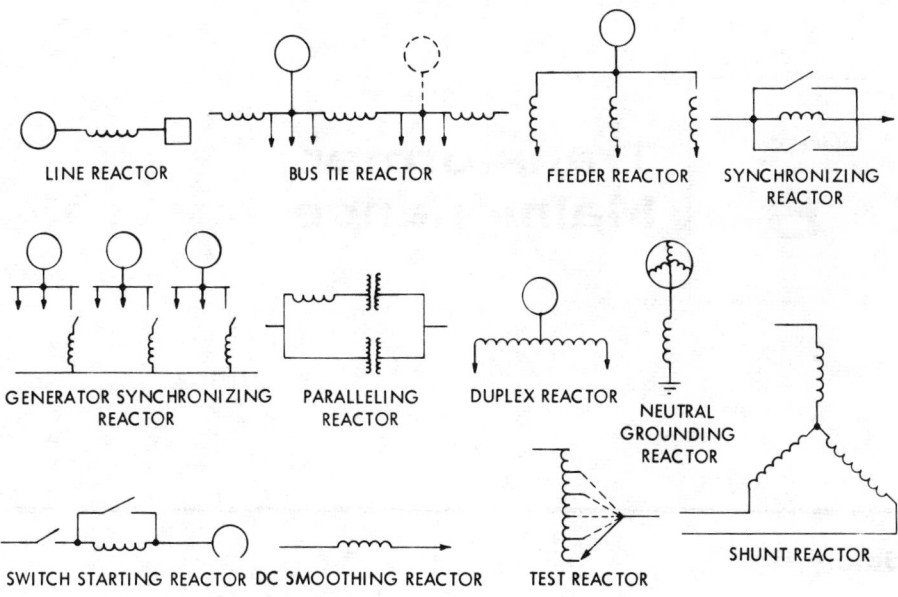

LINE REACTOR BUS TIE REACTOR FEEDER REACTOR SYNCHRONIZING REACTOR

GENERATOR SYNCHRONIZING REACTOR PARALLELING REACTOR DUPLEX REACTOR NEUTRAL GROUNDING REACTOR

SWITCH STARTING REACTOR DC SMOOTHING REACTOR TEST REACTOR SHUNT REACTOR

Fig. 7-12. Reactor classification by use. (Westinghouse Electric Corp.)

small amount of DC power, usually 1 percent or less.

Fig. 7-12 shows diagrams of a few of the practical applications of reactors, some of which we have studied in this text.

Transformer Maintenance

Safety First

The basic safety rules that apply to all electrical equipment also apply to transformer installations except that with transformers additional hazards are involved and additional safety precautions must be taken.* The primary rule is to disconnect the transformer from all sources of electrical energy and ground the windings. To effectively disconnect the transformer, both the primary and the secondary must be disconnected. The purpose of disconnecting the secondary is to eliminate any possibility of a secondary system feedback. After the transformer primary and secondary windings

have been disconnected, the disconnecting means must be locked in the open position. If fuses or fuse cutouts are used, the fuses should be removed to a special inaccessible area so they cannot be accidentally reinstalled. When the transformer has been disconnected the primary and secondary windings should be grounded to discharge any capacitance energy that has been stored in the transformer. This ground should not be removed until all work has been completed. The permanently installed transformer tank ground should be checked before starting work on the tank and cooling coils.

If internal work or an internal inspection is necessary to liquid-immersed transformers, special precautions must be taken to relieve the internal tank pressure before any

---

*An excellent bulletin entitled "Transformer Accident Prevention Bulletin," which contains information relating to transformer care and maintenance, is published by S. D. Meyers, Inc., P.O. Box 3575, Akron, Ohio 44310.

attempt is made to remove the tank top or manhole cover. This is accomplished by partially operating a valve or plug above the liquid level. If an inert gas is used, the transformer tank must be purged and a continuous source of clean air circulated in the tank.

Special precautions must be taken when working with transformers using askarel as an insulating liquid. Direct contact with askarel should be avoided as it has an irritating effect on the skin, especially the eyes, nose and lips. Severe skin eruptions may develop by continuous absorption of askarel through the pores of the skin. An immediate application of castor oil is recommended for treatment. Transformers containing askarel should not be opened when the transformer is hot as the fumes are toxic. If it is absolutely necessary to open a hot transformer, it should be done only in a well-ventilated area, and direct exposure to the fumes should be avoided. It should be mandatory procedure that when a man enters a transformer tank a second man be on duty outside the tank.

Special care must be taken that no tools or other equipment are dropped or left inside the transformer tank when the work has been completed. It is recommended that a list of all tools or other articles that are used to make repairs are listed before the job is started so that an inventory can be taken when the job is completed to be sure nothing was dropped or left inside the transformer tank.

Do not take a chance. Play safe.

National Safety Council

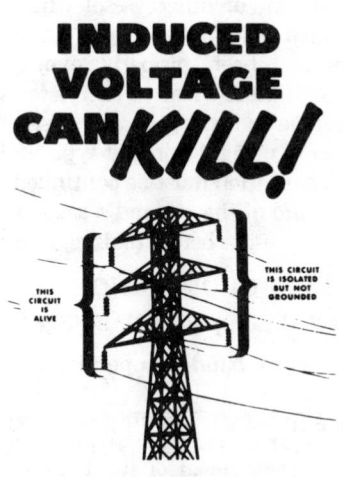

National Safety Council

Do not assume that a transformer is dead (de-energized). Make sure breaker, fuses, or switches are open before doing work. After work is completed, make sure all is clear before energizing.

Suggested Transformer Safety Standards*

CIRCUIT GROUNDING
Transformer Circuit Grounding

Secondary circuits of current transformers and potential transformers on circuits of primary voltage, 460 volts and above, shall be grounded in an approved manner.

CURRENT TRANSFORMERS

Current transformers shall not be worked on while energized, except under the immediate direction of an authorized person who shall remain at the worksite until the job is completed.

EXPLOSION HAZARD
Entering Dangerous Areas or Vessels

Persons shall not enter a transformer or circuit breaker tank, oil storage tank or tank car, or other vessel which has contained flammable liquid until the vessel has been properly expurgated and ventilated. Prior to entry, tests for flammable or toxic vapors or bases shall be made using approved-type indicators. Ventilation must be continued during the life of the job and a second man must be present before making entry.

GROUNDING
Oil-Filled Equipment Handling

During oil-handling operations on oil-filled equipment, such as transformers and circuit breakers, the following precautions shall be observed:

Equipment Bonding: All apparatus tanks, shielded hoses, pumping or filtering equipment, drums, tank cars, trucks and storage tanks shall be solidly bonded through a common interconnecting grounding cable.

Grounding Exposed Conductors: All exposed conductors, such as transformer or circuit breaker bushing, or coil ends of transformers which have bushings removed, either physically or electrically shall be connected to the same ground cable.

(Exceptions: For treatment of insulating oil in equipment under energized

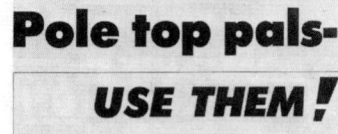

National Safety Council

*From: POWER SYSTEM SAFETY STANDARDS, United States Department of the Interior, Bureau of Reclamation.

conditions, grounding is not required on exposed conductors but equipment bonding shall be made as specified in *Equipment Bonding,* above.)

Oil-Filtering Jobs: When returning to a partially completed oil-filtering job after shutdown for any reason, all switching, grounding, and bonding shall be checked before resuming operations.

POTENTIAL TRANSFORMERS
Clearances on Potential Transformers

In placing clearances on potential

transformers or equipment to which potential transformers are connected, the fuses or switches shall be opened to prevent energizing the equipment cleared from the low-voltage side.

SECONDARY CIRCUITS
Transformer Secondary Circuits

Secondary circuits of current transformers shall not be opened while the primary circuit is energized unless the secondary terminals of the transformer are effectively jumpered by approved methods.

Purpose of Transformer Maintenance

Costly transformer failures and interruption of service can be kept to a bare minimum if a few basic procedures are followed. Probably the most important factor is that the proper transformer is selected for each specific installation. When the proper transformer has been selected it is of utmost importance that the transformer be properly installed in an application for which it is designed.

After the transformer has been properly installed it is necessary to periodically inspect and maintain the installation to prevent costly failures and interruption of service. Because transformers basically have no moving parts it is often erroneously assumed that there is no need for inspection and maintenance. A transformer that has been operating

satisfactorily for many years may, because of neglect, suddenly fail without warning or any outward sign of distress thereby causing expensive repairs or replacement and loss of production.

The proper maintenance of transformers basically consists of: (1) keeping all parts clean and protected from rust, dirt and corrosion, (2) testing the winding insulation and the insulating and cooling liquid, (3) inspecting and testing the protective and indicating devices, (4) inspecting the transformer internally, (5) inspecting the auxiliary equipment such as fans, coolers, lightning arresters and grounds.

Although a transformer with its associated equipment does not have as many moving parts as most electrical equipment, the same basic rules still apply. These basic fundamental rules are: (1) current-carrying components must operate in a *moisture-free* insulating liquid or area, (2) the installation must be kept *clean* from dirt, rust or corrosion, (3) all moving parts must be kept well-lubricated, (4) all enclosures containing insulating and cooling liquids as well as weatherproof enclosures that protect the equipment from the weather must be kept *tight*.

Inspection and Maintenance

The frequency of inspection and maintenance of transformers and their associated equipment varies but basically depends upon their size, type and use. Normally the large power transformer that is costly to replace and usually serves electrical equipment that is critical for plant production, should receive more frequent inspections and maintenance than the small lighting and distribution transformers that serve less critical circuits. Also large power transformers normally have more associated equipment such as cooling fans, forced oil pumps or protective alarms that must be properly maintained to insure a trouble-free installation. It is recommended that personnel responsible for the proper and continuous operation of the transformer installation develop an inspection and maintenance schedule. Once a schedule has been established it should be closely followed and the readings and conditions observed should be recorded in a log or inspection sheet. A detailed record should also be kept, listing that information which is relative to the installation showing the transformer specifications and characteristics, past history, repairs and tests made, spare parts and other relative information.

The following includes most of the equipment that requires periodic inspection and maintenance and should help plant personnel and operators to establish suitable schedules for continuous service and a trouble-free installation.

Load and Voltage

Since the current of a transformer determines the amount of heat generated by the transformer, frequent readings should be taken. It is recommended that hourly readings be taken and recorded if they are not continuously recorded on instrument charts. For further information on the loading of oil-immersed power transformers, reference should be made to "Guide for Loading Oil-Immersed Power Transformers With 65°C Average Temperature Rise", NEMA publication No. TR 98-1964. Excerpts from this publication are found in this book in Chapter 3 under "Loading of Transformers".

Liquid Level

To prevent the transformer from overheating, the proper liquid level must be maintained. This should be checked daily and if there is any loss from evaporation or leakage, it should be replaced immediately and repairs made to prevent further loss.

Temperature

For large power or substation transformer installations, the ambient temperature of the area where the transformer is installed should be observed hourly. For small lighting and distribution transformers feeding less critical circuits, a less frequent observation may be made. This temperature is an important factor in determining the amount of load a transformer can carry since the ambient temperature added to the transformer temperature rise determines the operating temperature. Normally transformers are designed to operate continuously at full load with an average ambient temperature of 30°C (86°F) for cooling air with a maximum ambient not to exceed 40°C (104°F). The liquid temperature of the transformer should be observed hourly. This temperature will normally be 10°C to 15°C less than the hottest-spot winding temperature. The hottest-spot winding temperature should be observed hourly. This is an important preventive maintenance factor as this temperature determines the aging or rate of deterioration of the insulation of the transformer. Some experts use the rule that for each 8°C temperature rise above the rated temperature of the transformer, the aging rate for Class A insulation doubles. According to this rule if a transformer having Class A insulation were to operate at a hottest-spot temperature of 8°C above its rated temperature the life of the transformer would be cut in half.

The effect that heat has upon a transformer is very important and it is recommended that hourly inspections be made and the observations noted of the ambient temperature, liquid temperature and the hottest-spot winding temperature. If the maximum temperatures are exceeded, the load should be decreased or the cooling should be increased. If, under the same conditions of load and ambient temperatures, a gradual or sudden increase in temperature is observed from previous readings, immediate steps must be taken before serious damage occurs within the transformer.

Air Temperature of Dry-Type Transformers. Heat can also be a problem with dry-type transformers and hourly checks are recommended. The ambient temperature of the surrounding air should be noted and the temperature of the ingoing and outgoing cooling air should be carefully observed. An obstruction to the cooling air or a rise in the ambient temperature could result in serious damage to the transformer, especially if it is operating at rated load.

Water Temperature of Water-Cooled Transformers. Daily checks are recommended to observe the temperature difference between the input and output of water through the cooling coils. The results should be examined and recorded so that the efficiency of the cooling system can be determined. Semi-annual checks are recommended to determine the water pressure and flow through the cooling coils. If the water pressure remains constant but the flow decreases, there could be an obstruction such as scale in the cooling coil. The water pumps, driving motors and intake screens should also be examined to be sure they are operating properly. The temperature of the input cooling water should not exceed 25°C with the temperature difference between the input and output cooling water through the cooling coils not exceeding 10°C. If the specified rate of flow cannot be achieved with the proper water pressure and all pumps operating properly, the scale or sediment that has formed on the inside walls of the cooling system will have to be removed. If proper heat transfer is not achieved after the cooling coils have been repaired and are functioning properly, the trouble is probably due to oil sludge deposits on the external surface of the cooling coils. This sludge must be removed to effect the proper heat transfer. Special care should be taken so as not to expose the water in the cooling coils to a temperature that could cause the water to freeze thereby causing severe damage to the transformer. If a water-cooled transformer is taken out of service and the transformer is exposed to below freezing temperatures, the water must be drained from the cooling coils.

Oil Temperatures of Forced-Oil-Cooled Transformers. The temperature of the oil entering and leaving the heat exchanger should be carefully checked and recorded daily. All other associated equipment such as pumps and fans should also be checked daily and a determination made that the cooling system is performing efficiently.

Pressure Vacuum Gage (Sealed Tank Transformers)

Most liquid-filled power transformers are equipped with a pressure vacuum gage which should be read daily, and all readings should be recorded, Fig. 3-10. The manufacturer's specified normal operating pressure range should be carefully followed. As the oil temperature changes there should be a change in the pressure gage reading, if the reading does not change it normally would indicate an air leak in the tank above the oil level.

Pressure Gage (Gas-Oil-Sealed Preservation System)

The pressure gage used with the type of transformer that utilizes the inert-gas cushion above the oil level should be read daily and recorded, Fig. 3-27. The manufacturer's specified operating range should be carefully followed. A leak in the system would be indicated by a low pressure reading when the transformer temperature is high assuming the nitrogen cylinder pressure is sufficiently high. The nitrogen cylinder pressure gage should be read daily and recorded. It is important that the pressure of the nitrogen cylinder does not fall below the manufacturer's specifications. A 200 cu-ft cylinder normally will last from six months to one year depending upon the size and service conditions of the transformer. Caution should be taken with this type of transformer that the oxygen content in the nitrogen cushion does not exceed 5 percent as more than this amount

of oxygen will create an explosive mixture. The oxygen content of the mixture may be checked by an oxygen analyzer or indicator. The oxygen contents should be checked weekly during the first month of operation of a new transformer and every six months thereafter.

Fans, Pumps and Their Controls

All fans and pumps should be checked daily for proper operation and the observation recorded. This would include the motors, fans and pumps and associated control equipment. If a second stage fan is involved but not frequently used, the motor, fan and controls should be checked weekly. It is recommended that all auxiliary cooling units and their control equipment be thoroughly cleaned and overhauled semiannually or oftener if prevailing conditions so warrant.

Dehydrating Breathers

The chemicals used in dehydrating breathers to remove moisture from the air entering a transformer as it breathes should be checked and recorded monthly. The two most commonly used chemicals are calcium chloride and silica gel. Silica gel breathers are a more recent development and are replacing the calcium chloride breathers. The advantage of using silica gel is that when silica gel becomes saturated with moisture it does not restrict breathing as does calcium chloride. Silica gel, when dry, is blue in color and the color changes to pale pink as it becomes saturated with moisture. If the chemical is found to contain excessive moisture, it must be replaced or reconditioned. Silica gel can be dried by heating it in an open container at a temperature of between 150°C and 200°C for two hours. Calcium chloride can be dried by heating it in the same manner at a temperature of between 180°C and 200°C until completely dry. All air paths in breathers should be checked monthly to make sure they are free from any restrictions.

Pressure-Relief Devices

When an arc occurs under the liquid inside a transformer, a gas is generated. If this arc is severe, such as that caused by a short circuit, the amount of gas generated could create sufficient pressure to rupture the transformer tank. The rate of rise of gas pressure in a transformer is proportional to the arc power and inversely proportional to the volume of the gas space.

Most liquid-filled transformers are equipped with pressure-relief devices such as relief diaphragms or mechanical-relief devices to relieve excessive pressure before serious damage is done to the transformer tank.

Pressure-Relief Diaphragms

Pressure-relief diaphragms are normally constructed of a thin layer of a suitable material such as micarta, glass, bakelite or thin metal and are usually mounted on the end of a large pipe which is connected to the transformer case above the liquid level. The diaphragm is normally designed to rupture at from 10 to 15 psi. These devices should be checked and the condition recorded quarterly. When inspecting the pressure-relief diaphragms, close attention should be given to gasketed joints for evidence of rust, corrosion and tightness. The condition of the actual diaphragms should be observed to determine if there are any defects or cracks. A defective gasket or a cracked diaphragm should be replaced immediately.

Mechanical-Relief Devices

Mechanical-relief devices are normally furnished as standard equipment with transformers using the inert-gas cushion oil-preservation system and are optional with liquid-filled transformer, Fig. 3-28. These devices should be carefully checked and recorded quarterly. Careful inspection should be made of the condition of the diaphragm and gasketed joints for evidence of leaks. If the device includes a visible or audible alarm, it should be carefully checked to assure that it will function properly if the pressure in the transformer tank reaches a predetermined value.

Gas Detector and Pressure Relays

Gas-detector relays are normally used with the expansion tank (or conservator-type) transformers to detect incipient faults or localized overheating of insulating materials. These relays are located in the pipe between the expansion tank and the transformer and are designed to trap any gas which may rise through the oil. The ability of this relay to measure a small accumulation of gas enables it to detect faults in their incipient stage. In normal practice the accumulator contacts are connected to an alarm but they may also be connected to trip the transformer off the line. These relays and their associated alarms should be checked quarterly and recorded.

Pressure relays are designed to operate on a rapid rise in pressure in the transformer tank which normally would be caused by an arcing fault in the transformer. An equalizing orifice in the relay permits a slow change in pressure due to normal loading and ambient temperature without operating the alarm contacts. These relays and their alarms should be checked quarterly and recorded.

Protective Alarms

It is recommended that a monthly inspection be made and recorded of all transformer alarm and trip-out circuits for proper operation.

The settings and operation of the alarm actuating devices such as oil-flow indicators, sudden pressure relays, pressure-relief devices, oil-flow indicators, liquid and hottest-spot winding temperature indicators and outlet air thermometers should be noted and recorded monthly.

Tap Changers

Most transformer installations include a no-load or load type tap changer which permits changes in the voltage ratio of the transformer. Normally the no-load tap changer is manually operable and must not be operated while the transformer is energized. The load-type tap changer may be manually or automatically operated while the transformer is in operation. The original position of the tap changer should be carefully observed and recorded, and when all work has been completed the tap changer must be returned to this original position before the transformer is put back in service.

The exposed external mechanical parts should be carefully examined annually for wear, proper lubrication, corrosion, dirt, looseness, adjustment and seal leakage. If there have been any indications of unsatisfactory operating conditions, a voltage ratio check should be conducted and an internal inspection made. To conduct an internal inspection the transformer must be opened and the oil level lowered so as to expose the electrical and mechanical parts of the tap-changing assembly. The contacts should be carefully examined and cleaned. The tap-changing operating mechanism such as motors, gears, shafts, bearings and control relays should be carefully checked and adjusted for proper operation. The tap changer should be rotated over its entire range several times and a careful check made to assure that a complete contact surface and tension exists at each position. A further check may be made at each contact position by using the wheatstone bridge to detect a poor contact. The manufacturer's specifications and instructions for the operation and maintenance of this equipment should be carefully followed.

Ground Connections

The noncurrent-carrying metal parts of a transformer installation are required by the National Electrical Code to be "effectively grounded." Conductive materials enclosing electric conductors or equipment, or forming part of such equipment are grounded for the purpose of preventing a voltage above ground on these materials. Circuits and enclosures are grounded to facilitate overcurrent device operation in case of insulation failure or ground faults.

Ground connections should be inspected monthly and recorded. The inspection should include close examination for loose, corroded or broken connections. The connections should be observed for any unusual heating at connection. An annual resistance-to-ground check should be made and recorded.

Lightning Arresters

Lightning arresters should be regularly inspected and cleaned. A monthly inspection should be made and recorded. This inspection should include a visual observation of dirt and foreign deposits on the arrester or any undue mechanical strain from insufficient slack in the line.

An annual inspection should be made and recorded. This inspection should include a complete overhaul and cleaning of arresters. All connections should be clean and tight and all broken or damaged parts should be replaced.

Overcurrent Protective Devices

The prime factor when selecting the value of the overcurrent protection for transformers are the requirements of the NEC (National Electrical Code). The basic provisions found in Article 450 of the 1971 edition of the NEC require that each askarel- and oil-insulated transformer (a bank of 3 single-phase transformers operating as a three-phase unit is considered as one transformer) shall be protected by an individual overcurrent device in the primary connection, rated or set at not more than 250 percent of the rated primary current of the transformer. An exception to this is that a transformer having an overcurrent device in the secondary connection, rated or set at not more than 250 percent of the rated secondary current of the transformer, or a transformer equipped with a coordinated thermal overload protection by the manufacturer, is not required to have an individual overcurrent device in the primary connection provided the primary feeder overcurrent device is rated or set to open at a current value not more than 6 times the rated current of the transformer for transformers having not more than 6 percent impedance, and not more than four times the rated current of the transformer for transformers having more than 6 but not more than 10 percent impedance.

It is obvious that the maximum overcurrent protection permitted by the NEC will not protect a transformer from overheating resulting from a constant overload of up to 250 percent of the full load current. It is the purpose of these overcurrent devices to protect the transformer installation from excessive currents caused by lightning, short circuits or grounds. The transformer would be protected from overheating by other protective devices such as the hottest-spot winding indica-

tor, liquid-temperature indicator or the sudden pressure relay.

The type of overcurrent devices selected would normally consist of circuit breakers, fuse links, fuse cutouts, fused switches, oil circuit reclosers, or circuit breakers controlled by protective relays.

It is recommended that the overcurrent protective devices and their associated equipment be inspected monthly for evidence of loose connections, dust, dirt, corrosion or heating. An annual inspection should be made where all equipment is carefully examined and cleaned. All manual and automatic switches and circuit breakers should be operated several times to assure that all moving parts are free and operable.

Insulating Liquids

The purpose and preservation of insulating and cooling liquids was discussed in Chapter 3. Many experts feel that the proper inspection, testing and maintenance of transformer insulating liquids is the most positive way to keep a transformer in proper operating condition and to extend the life expectancy of the transformer to a maximum value.

Chapter 3 discussed the methods used to keep air and moisture, which are the chief enemies of insulating liquids, from coming in contact with the insulating liquid. Oxygen in the air in contact with insulating oil causes the oil to oxidize when heated which results in the formation of acids and sludge. The acid content of the insulating oil is basically a measure of its deterioration and its tendency to form sludge. The sludge is extremely harmful to a transformer as it settles on the internal parts of the transformer and greatly impedes cooling and reduces the "flashover" value of exposed parts and insulators. A sludge deposit of ⅛ to ¼ inch on the core and coils may increase the temperature from 10° to 15°C. When moisture is present in the insulating oil it will form conducting paths on non-conducting parts, such as bushing shanks and insulated leads which creates a definite hazard to the safe operation of the equipment. Moisture not only reduces the dielectric strength of all insulating liquids but it also may be absorbed by the insulating materials used throughout the transformer. The dielectric strength of an insulating liquid (oil or askarel) is basically the voltage required to break down the liquid in a 0.1 inch gap between two electrodes of standard dimensions under specifically controlled testing conditions.

Askarel has the advantage as a cooling and insulating liquid in that it is neutral and under normal operating conditions it does not oxidize or decompose to produce acid or sludge. If, when askarel is tested, a small acidity content is found, this is not due to deterioration of the askarel but is normally caused by its solvent action on certain organic insulating materials in the transformer.

Many plant electricians generally consider the dielectric test to be the most important but there are other tests which are of considerable value. Since the transformer needs good oil for trouble free operation, periodic tests of oil purity should be made and purification processes carried out promptly, if necessary. The interval between oil tests depends upon such conditions as the load carried and the importance of the unit. Power station transformers and those serving vital circuits in industrial plants should have an oil test at three- to six-month intervals. Other transformers should receive oil tests annually. It is strongly recommended that a properly equipped laboratory be used for the accurate testing of oil, Fig. 8-1.

In small transformers, especially of lower voltage, operating companies periodically drain off a small portion of the oil from the bottom of the tank and replace it with more oil at the top. This removes water which may have collected on the bottom. On larger units, however, it is sometimes desirable to withdraw all the oil and recondition it to remove both moisture and sludge held in suspension. This is done by pumping the oil through a filter press or by passing it through a centrifugal

Fig. 8-1. Deteriorated oil caused the failure of this secondary coil of a rectifier transformer. Through inspection the costly damage could have been prevented. (Kemper Insurance Co.)

machine which separates the moisture and objectionable material from the oil.

Samples of the insulating liquid should be taken after the liquid has been given time to settle, the larger the transformer the more time required. Normally 8 hours is sufficient for a small transformer containing a barrel of liquid while several days are necessary for a large transformer. If samples are taken from a transformer in operation a sufficient amount should be taken to insure that it really represents the insulating liquid that is actually circulating through the transformer. The samples taken for askarel should be taken from both the top and bottom of the tank as askarels are heavier than water. Insulating oil is not heavier than water, therefore the samples should be taken from the bottom of the tank. Again a sufficient amount should be withdrawn to insure that it is truly a representative sample of the liquid in the transformer tank. A sufficient quantity of liquid must be drained for testing. For dielectric tests at least 1 pint and for other tests at least 1 quart.

Special precautions must be taken when taking and containing the samples. The samples should be taken on a clear, dry day, and put into a clear glass or polyethylene container. A glass container has the advantage of being able to observe water if it is present in the liquid. The container must be thoroughly cleaned with gasoline or Stoddard solvent and dried. It is recommended the container be further cleaned with soap and water and be rinsed with distilled water and dried. If a tube or hose is necessary to remove samples from the bottom of a transformer tank, special care should be taken to use one constructed of glass or polyethylene. A tube or hose made of rubber should not be used as rubber normally contains sulphur which will dissolve in the liquid and attack the copper windings.

If the insulating oil or askarels test less than those recommended for safe operation, they should be reconditioned or replaced immediately. In all cases the manufacturer's specifications should be closely followed.

Insulation Tests

If the insulating liquid in a transformer is kept free from moisture and the transformer has not been subjected to excessive voltages, overload or excessive heating, the deterioration of the insulating materials within the transformer will be kept to a minimum. It is recommended that as a part of preventive maintenance that the condition of the insulation between the separate windings and the windings and ground be tested annually. This test is recommended even if the transformer has not been subjected to any of the foregoing factors which would contribute to the deterioration of the insulating materials within the transformer. If the transformer has been exposed to an excessive voltage (such as lightning) or overload, has been overheated, or through normal inspection an excessive moisture content has been found in the insulating liquid, immediate tests should be made to determine the condition of the transformer insulation.

Insulation Resistance

The insulation-resistance values vary with the temperature and the amount of moisture in the insulation. The insulation-resistance values vary inversely with the temperature in a somewhat indefinite manner and it is recommended the person who interprets the measurements have considerable experience and judgment in this field. The insulation-resistance values vary in proportion to the moisture content. The measurement is also affected by the kind and size of the transformer being tested and the value of the test voltage and the length of time the voltage is applied.

Because of the many variables involved when making these tests it is important that the temperature, humidity and other similar factors be recorded at the time the test is made. These factors should be compared when making future tests to determine the trend of the insulating resistance. These tests are normally made with a megger and they measure the resistance of the insulation between the various windings of the transformer and between the various windings and ground.

Dielectric-absorption Testing of Insulation

To further test the condition of the insulation the dielectric-absorption testing method is used. In this test a voltage is applied over a period of time, usually ten minutes, to the equipment under test. At one minute intervals during the test, insulation-resistance readings are taken and recorded on log-log graph paper with insulation resistance and time as the vertical and horizontal scales. The resultant curve on the graph of the insulation-resistance with time gives an indication of the condition of the insulation under test. For proper testing the equipment under test must be grounded both before and between tests.

If the dielectric-absorption curve that has been plotted on the graph at one minute intervals is a straight line that increases with time the insulation is in good condition. If the curve does not rise with time or only rises slightly and levels off, the insulation being tested is not safe and immediate repairs must be made.

The insulation-resistance and the dielectric-absorption tests are only two of many tests that can be made to determine the condition of the insulation. There are many other tests that can be made but they are more complicated and normally require additional testing equipment.

Unless the person performing the test and evaluating the test results has had considerable experience, it is recommended that when the tests indicate a dangerous condition, a qualified transformer consultant be engaged before the transformer is taken out of service and expensive overhauling is started.

Detailed External Inspection

A semi-annual inspection should be made of the entire transformer installation. A detailed list of all of the component parts should be made and each item that is inspected should be checked off to assure maintenance personnel that no parts were overlooked. This inspection should be made to determine that all parts are free from damage, leakage, deterioration or of accumulation of foreign deposits.

The operating area of a transformer installation normally dictates the amount and frequency of maintenance work that must be performed to assure a trouble free installation. Areas where carbon or metallic dusts are present require special precautions as these dusts are conducting and could cause a flashover. Another troublesome area outdoors is where grain or other similar dusts are present. Although these dusts may not be conducting, a combination of the accumulation of these dusts and a high ambient moisture, such as fog, rain or sleet could cause a flashover.

Bushings and potheads should be closely inspected to detect any chips, cracks, dirt or dust accumulations. The porcelain may be cleaned by an oil free solvent, such as an ammonia and water solution or carbon tetrachloride and must be applied with a soft non-abrasive material such as a cloth. Because the fumes from carbon tetrachloride are toxic it can only be used in a well-ventilated area. There are solvents available such as a special silicone that can be applied to porcelain bushings to retard the accumulation of dust and dirt. If the bushing contains an oil or compound, it should be examined for evidence of leakage. If leakage is detected the equipment should be checked for moisture content and the manufacturer's specifications should be closely followed before adding compound or oil.

All metal parts of the transformer should be cleaned and examined for rust or corrosion. These parts should be kept well-painted. All gasket and pipe joints must be examined for leakage and kept in good repair. All electrical connections should be inspected for corrosion or evidence of heating. These connections should be checked for tightness as a poor connection in an electrical system, under load, will cause heat and an eventual breakdown of the system.

Internal Inspection of Core and Coils

It is recommended that for open-type liquid-immersed power transformers that have a removable main cover or manhole cover that the cover be removed and inspected annually. (Be sure to relieve any internal pressure by cracking a valve or pressure relief device before removing cover). The inside of the cover and the bushing supports should be closely examined for any evidence of condensation or rust that would indicate a moisture problem. If there is evidence of moisture or rust immediate repairs to eliminate this problem must be made.

For open-type liquid-immersed power transformers it is recommended that every five years an internal inspection be made of the top of the core and coil assembly. This is done by dropping the liquid level to expose the top of the core. The core is then inspected for sludge deposits, loose connections, condition of the insulation or damaged or defective parts. If there is any indication of trouble, all of the liquid must be removed and the necessary repairs made. The frequency of this type of inspection is determined by the operating conditions and the age and type of the transformer. If the transformer operates at normal loads and is of the inert-gas cushion type and all of the regular maintenance checks have indicated that the transformer is in good operating condition this inspection may be performed at less frequent intervals. If the transformer has been exposed to an overvoltage or overloads or the regular maintenance tests have indicated a problem this inspection should be performed at more frequent intervals.

It is recommended that, unless a person has the necessary experience and qualifications to make judgment decisions on questionable conditions of transformers, an experienced transformer specialist be consulted. It is also recommended that the manufacturer's specifications be closely followed when making tests or repairs to transformers.

Tips on Transformer Maintenance*

1. Transformers should be inspected thoroughly as soon as received from the manufacturer.

2. Distribution transformers should never be energized unless the oil is at the proper level.

3. Transformers should be handled with care, avoiding severe jarring that might damage the internal structure.

4. Transformers should always be inspected by the crew before installation. This means that bushings, access plates and other fittings should be inspected and thoroughly tightened before the transformer is placed in operation.

5. When replenishing oil, care must be taken so that, (1) moisture does not enter the transformer and, (2) nothing falls into the tank.

6. Transformers should never be moved or lifted by the bushings or other attachments. Lifting lugs are provided for this purpose.

7. When a transformer is mounted on a pole, care should be taken to see that the transformer is in a vertical position and securely fastened so that a severe impact against the pole will not jar it loose.

8. When performing maintenance inspections, all bushings, connections, protective devices, gaskets, the paint finish on the tank, and all other exterior fixtures should be inspected for evidences of rust, corrosion or over-heating. The oil level should be checked, and the condition of the oil tested.

9. If an inspection cover is removed from a transformer for inspection, it should be properly replaced. If the gasket is not in perfect condition, it should be replaced.

10. When working with any transformer that is connected with any other transformer, the connections on both the high and low sides should be opened, to avoid danger from feedback voltage.

*Florida Power and Light

Inspection Check List

*This and the following pages from the UNITED STATES DEPARTMENT OF THE INTERIOR, BUREAU OF RECLAMATION, DIVISION OF POWER OPERATIONS.
**Refers to Bureau of Reclamation numbering. Section 6 not included in this book.

TRANSFORMERS AND REGULATORS *

INSPECTION CHECK LIST

ITEM OF INSPECTION	LARGE UNITS ATTENDED	LARGE UNITS UNATTENDED	SMALL UNITS ATTENDED	SMALL UNITS UNATTENDED	PROCEDURE REFERENCE
FOUNDATION, RAILS AND TRUCKS	A	A	A	A	12.01
TANKS AND RADIATORS	D A	W A	W A	M A	12.02
OIL AND WATER PIPING	D A	W A	W A	M A	12.03
VALVES AND PLUGS	D A	W A	W A	M A	12.04
OIL LEVELS, GAGES, AND RELAYS	D A	W A	W A	M A	12.05
BREATHERS AND VENTS	D A	W A	W A	M A	12.06
RELIEF DIAPHRAGM	D A	W A	W A	M A	12.07
WATER-COOLING COILS AND PIPING	A	A	A	A	12.08
FLOW INDICATORS AND RELAYS	D A	A	D A	A	12.09
HEAT EXCHANGERS	A				12.10
OIL PUMPS	D A	SEE SECTION 6**			12.11
COOLING FANS	D A		D A		12.12
TEMPERATURE INDICATORS AND RELAYS	D A	W A	W A	M A	12.13
INERT GAS TANKS	D	W			12.14
GAS REGULATOR, GAGES, AND RELAYS	D A	W A			12.15
GAS PIPING AND VALVES	A	A			12.16
GAS ANALYSIS	Q	Q			12.17
BUSHINGS	W A	W A	W A	M D	12.18
BUSHING CURRENT TRANSFORMERS AND POTENTIAL DEVICE	A	A			12.19
MAIN TERMINAL AND GROUND CONNECTIONS	D A	W A	W A	M A	12.20
CORE AND COILS	NS	NS	NS	NS	12.21
INTERNAL INSPECTION	A	A	A	A	12.22
TERMINAL BOARD AND CONNECTIONS	A	A	A	A	12.23
RATIO ADJUSTER	W A	W A	W A	M A	12.24
TAP CHANGER OR REGULATOR	D A	W A			12.25
MOTOR AND DRIVE	A	A			12.26
AUXILIARY AND LIMIT SWITCHES	A	A			12.27
POSITION INDICATORS	A	A			12.28
OPERATION COUNTER	W A	W A			12.29
OPERATION	A	A			12.30
POWER SUPPLIES AND WIRING	D A	W A			12.31
INSULATION RESISTANCE	A	A	A	A	12.32
OIL DIELECTRIC	A	A	A	A	12.33
OIL ACIDITY	5 yr	5 yr	5 yr	5 yr	12.34
FILTER AND RECLAIM OIL	NS	NS	NS	NS	12.35
FIRE PROTECTION	M A				12.36

D--ROUTINE DAILY INSPECTION
W--ROUTINE WEEKLY INSPECTION
M--ROUTINE MONTHLY INSPECTION
Q--QUARTERLY INSPECTION
A--ANNUAL INSPECTION
NS--NOT SCHEDULED

12.01 FOUNDATION, RAILS, AND TRUCKS	Annual inspection. Check foundation for cracking and settling. A slight shift of the transformers may break bushings or connecting oil or water lines. See that rail stops are firmly in place to hold transformer in position on the rails. Check transfer car and matching of its rails with transformer deck rails at each position. Paint metalwork as needed.
12.02 TANKS AND RADIATORS	Daily, weekly, or monthly inspection. Check for unusual noise and oil and water leaks.
12.03 OIL AND WATER PIPING 12.04 VALVES AND PLUGS	Annual inspection. Clean dirt and oil from radiating surfaces. Repaint as necessary. Stop excessive vibration of radiator tubes. Tighten loose or vibrating parts. Check for unusual internal noises. Inspect oil and water piping, valves, and plugs. Manipulate radiator cutoff valves to see that they are in operating condition, and secure in the open position. See that all oil drain valves which can be operated without wrenches are plugged or locked to prevent unauthorized opening.
12.05 OIL LEVELS, GAGES, AND RELAYS	Daily, weekly, or monthly inspection. Check oil level in main and auxiliary tanks, oil-filled bushings, etc. Changes in oil levels from time to time should be noted, taking into consideration the change in level caused by change in oil temperature. A rise in level in a water-cooled transformer (for a given temperature) indicates that water is leaking from the cooling coils into the oil. Annual inspection. Clean dirty gage glasses and connections into tank. Check oil level indicators and relays for proper operation. Replenish oil if below normal. Drain out and replace bushing oil if dirty or discolored.
12.06 BREATHERS AND VENTS	Daily, weekly, or monthly inspection. See that relief diaphragm has not opened and breathers and vents appear to be normal.
12.07 RELIEF DIAPHRAGM	Annual inspection. See that relief diaphragm is in operating condition and closes tightly. The nonshattering-type diaphragm should be actuated to see that it is not stuck shut from rust or paint. Make sure that the material used in the shattering-type diaphragm is not too thick or tough to be broken by reasonable internal pressure. See that screens and baffles in vents or breathers are not obstructed or broken. If breathers are of the dehydrating type, check chemicals and replace if depleted.
12.08 WATER-COOLING COILS AND PIPING	Annual inspection. Check external supply and drain piping for leaks. Flush out cooling coils or heat exchanger water passages with air and water. Test coils for leaks by applying air pressure to coils and observing for bubbles rising in oil and drop in air pressure with supply valve closed, or use a hydrostatic pressure test. A pressure of about 75 pounds per square inch is recommended. If water scale is present, circulate a solution of 25 percent hydrochloric acid and water through the coils until clean. Then flush out thoroughly. Clean external surfaces of coils.
12.09 FLOW INDICATORS AND RELAYS	Daily inspection. See that proper supply of cooling water is flowing. Annual inspection. Check waterflow indicators and relays for proper operation.
12.10 HEAT EXCHANGERS	Annual inspection. Clean and test water tubes similar to cooling coils. Check for oil and water leaks.

Daily inspection. See that oil-circulating pumps are in operation when required.

Annual inspection. See Section 6*.

12.11
OIL PUMPS

Daily inspection. See that fans are in operation when necessary.

Annual inspection. Check motors and control as per Sections 8 and 14*.

12.12
COOLING FANS

Daily, weekly, or monthly inspection. Check and record transformer temperatures.

Annual inspection. Check calibration of temperature indicators and relays. Check and clean relay contacts and operating mechanism.

12.13
TEMPERATURE
INDICATORS
AND RELAYS

Daily or weekly inspection. Check gas pressure left in tanks and change out at about 25-pound pressure.

12.14
INERT GAS
TANKS

Daily or weekly inspection. See that proper gas pressure is being maintained in transformer.

Annual inspection. Check setting and operation of regulator and relay. See that gages are indicating properly.

12.15
GAS REGULATOR
GAGES AND
RELAYS

Annual inspection. Check for gas leaks by applying liquid soap on all joints, valves, connections, etc., with gas pressure raised to the maximum recommended by the transformer manufacturer.

12.16
GAS PIPING
AND VALVES

Quarterly inspection. Check analyzer for proper operation. Analyze gas. Purge if oxygen content is over 5 percent.

12.17
GAS ANALYSIS

Weekly or monthly inspection. Check for chipped or broken porcelain, excessive dirt film, oil level, and oil or compound leaks.

12.18
BUSHINGS

Annual inspection. Clean porcelain with water, chlorothene, or other suitable cleaner. Repair chipped spots by painting with lacquer such as red glyptal. Inspect gaskets for leaks. Tighten bolts. Check power factor. Check oil sample from bottom of bushing for dielectric strength and presence of water which may be entering at top. Replace or replenish oil if necessary.

Annual inspection. Check tap setting and adjustments at terminal board to see that they agree with diagrams. Check insulation resistance of wiring with devices connected. Check ratio and phase-angle adjustments of potential devices if changes have been made in secondary connections and burden. Tighten connections, including potential device tap, into bushing.

12.19
BUSHING CURRENT
TRANSFORMERS AND
POTENTIAL DEVICES

Daily, weekly, or monthly inspection. Check for presence of foreign material, birds' nests, etc., in or near connecting bus work, loose or heating connections, and loose or broken tank ground connections.

Annual inspection. Tighten all bus and ground connections. Refinish joint contact surfaces if they have been overheating. Inspect ground cable to see that it is not loose or broken.

12.20
MAIN TERMINALS
AND GROUND
CONNECTIONS

*Refers to Bureau of Reclamation numbering. Section 8 not included in this book.

12.21 CORE AND COILS	Not scheduled. If the transformer has been properly maintained and not overheated and barring internal failure, it should not require untanking within its normal life. If sludge has been allowed to form due to over-heating and oxidation of the oil, the transformer should be untanked and the core, coils, oil passages, tank, and water-cooling coils washed down with clean oil under pressure to remove sludge and other accumulations which prevent proper circulation of the oil. Inflammable liquids should not be used in cleaning the core, coils, or inside of tank. Provide sufficient fresh air for workman while working inside of tank. While untanked, check for loose laminations, core bolts, insulating blocks, etc., and other pertinent features on the check list.
12.22 INTERNAL INSPECTION 12.23 TERMINAL BOARD AND CONNECTIONS	Annual inspection. Lower the oil level to at least the top of the core. Inspect for sludge on core and windings. Inspect underside of cover for moisture and rust and clean up. Check connections at terminal board. Tighten all bolted connections, core bolts, etc., within reach.
12.24 RATIO ADJUSTER	Weekly or monthly inspection. Note position of ratio adjuster and that it is adequately locked to prevent unauthorized operation. Annual inspection. Inspect contacts and clean if reachable on internal inspection. If not reachable for visual inspection, check each position with wheatstone bridge across winding to detect poor contact. Work adjuster back and forth over complete range several times.
12.25 TAP CHANGER	Daily or weekly inspection. Note position of tap changer. See that positions on all three transformers are the same and that manual operating device is locked. Annual inspection. Drain oil from contact compartment. Clean and re-finish contact surfaces. Check contact spring pressure. Check contact operating mechanism. Tighten connections and other bolts.
12.26 MOTOR AND DRIVE	Annual inspection. Check motor as in Section 8*. Check and adjust brake. Check gears, shafts, and lubrication.
12.27 AUXILIARY AND LIMIT SWITCHES	Annual inspection. Check condition of contacts and refinish if burned or corroded. Check contact springs, operating rods, and levers. Check closing and opening position with respect to position of main contacts.
12.28 POSITION INDICATOR	Annual inspection. See that positions indicated correspond to position of main contacts. Check remote electrical position indicators for correct operation, obstruction to movement of pointer, etc.
12.29 OPERATION COUNTER	Weekly inspection. Check and record reading of operation counter. Annual inspection. Check operation counter for correct registration.
12.30 OPERATION	Annual inspection. Run tap changer or regulator through several complete cycles by both control relay and manual control, and observe contacts and mechanism for proper operation.

---

*Refers to Bureau of Reclamation numbering. Section 3.19 not included in this book.

Daily or weekly inspection. See that all power, control, and alarm supply circuit switches are closed and fuses in place so that circuits are completed.	12.31 POWER SUPPLIES AND WIRING
Annual inspection. Inspect fuses or circuit breakers on all power, control, and alarm supplies to auxiliary equipment and devices. Check and tighten wiring connections at all terminal points. Inspect wiring for open circuits, short circuits, and damaged insulation. Check insulation resistance of wiring with devices connected.	
Annual test. Check the insulation resistance between each winding and between each winding and ground. Disconnect all external leads at the bushing terminals, except where the connecting leads can be suitably isolated at adjacent disconnecting switches, for this test. A similar test using a capacitance bridge is recommended where such an instrument is available.	12.32 INSULATION RESISTANCE
Annual test. Check the dielectric strength of the insulating oil in the main and auxiliary tanks and oil-filled bushings. (See Power O&M Bulletin No. 11.)	12.33 OIL DIELECTRIC
The acidity of the insulating oil in the main tank should be checked at intervals of not more than 5 years. Transformers operating at high temperatures or showing signs of sludging or dark color of the oil should be checked more frequently. Oil may be checked in the field with a Gerin test kit or samples sent to the Denver laboratory.	12.34 OIL ACIDITY
Not scheduled. The necessity for filtering and/or reclaiming the insulating oil will depend on the results obtained from the oil dielectric and oil acidity tests. It may be more economical to replace the oil in small transformers rather than filter or reclaim it.	12.35 FILTER AND RECLAIM OIL
Monthly. Where a water spray system is used, see that control valves and automatic devices are in operating condition and water supply is available.	12.36 FIRE PROTECTION
Annually. With transformer deenergized, try out water spray system. Observe fog nozzles for proper coverage and spray. See Section 3.19* (See O&M Bulletin No. 23.)	

REFERENCES

How to Maintain Electric Equipment, GET-1125, General Electric Company
Operators Lesson No. 12, Alternating Current Machinery--Transformers and Regulators
Power O&M Bulletin No. 14, Painting of Transformers and Oil Circuit Breakers
Power O&M Bulletin No. 3, Testing Electrical Equipment Insulation
Power O&M Bulletin No. 11, Maintenance of Mineral Insulating Oil
Power O&M Bulletin No. 6, Permissible Overloading of Oil Immersed Transformers and Regulators
Power O&M Bulletin No. 1, Testing and Maintenance of High Voltage Bushings
Power O&M Bulletin No. 23, Fire-fighting, Cause and Prevention

*Refers to Bureau of Reclamation numbering. Section 6 and Sections 8 and 14 not included in this book.

Index

Index

Oil-immersed current-limiting reactor, 308-310
Oil-immersed forced-oil cooling transformer, 159-160
Oil-immersed self-cooling transformer, 155-159
Oil-immersed self-cooling with air-blast transformer, 160-161
Oil-immersed transformer, 180-182
Oil-immersed water-cooling transformer, 158-159
Oil-insulated transformer, 126, 129
Oil-type current transformer, 99
Open-core transformer, 18-19, 21
Open-delta connection, three-phase, 254-255
Operation of autotransformer, 294-295
Operation of transformer, 17-18
   parallel, 279-292
   with load, 77-80
   without load, 75-77
Outdoor transformers, 113-120, 126-129
Output, 3
Output rating of autotransformer, 295-297
Overcurrent protective devices, 182-183, 324-325

P

Pad-mounted transformer, 127-129
Pancake coils, 31-33, 149-151
Panelboards and transformers, 122-123
Parallel operation of transformer, 279-292
Partial service of cooling, 174-175
Permeability, 7-8
Phase-angle control, 216-221
Phase transformation, 257-269
   three-phase to single-phase, 257-258
   three-phase to six-phase, 263-269
   three-phase to two-phase, 258-261
   three-phase to two-phase and three-phase, 261
   two-phase to six-phase, 261-263
Phase, transformers, 136-137
Plates, tank wall, 157
Polarity
   additive, 228-229
   markings, 99

single-phase transformer, 227-232
   subtractive, 228-229
   test of, 231-233
   three-phase transformer, 269-278
Pole-mounted transformers, 126-127
Potential transformer, 100-105
Precautions, current transformer, 99-100
Pressure gage, 321
Pressure relay, 323
Pressure-relief devices, 322
Pressure-relief diaphragms, 322
Pressure vacuum gage, 320-321
Primary
   circuit, 3
   coil, 17
   losses, 23
   winding, effect of number of turns in, 21
   winding taps, 225-227
Protective devices
   cooling, 182-189
   overcurrent, 324-325
Pumps, 321

R

Radiators, tank wall, 158-160
Rated output, 171
Rating
   current-limiting reactor, 304
   standard kilovolt-ampere, 138
   standard voltage, 140
Ratio
   transfer, 22-23
   turns, 22-23
   voltage, 84
Reactance, 83-84, 301-303
Reactive voltage, 151-152
Reactor, 301-333
   alternator, 310-311
   bus, 311
   feeder, 311
   saturable-core, 311-313
   tie-line, 311
Reduction of voltage, 2-3
Regulator, constant-current, 108-110
Relative motion, 12
Relay
   gas, 198

Index

Index

W

Water-cooled transformer, 158-159, 320
Water temperature, 320
Windings, 148-151
    concentric, 148-150
    interleaved, 150-151
    loss, 182
Winding taps, primary, 225-227
Window-type current transformer,
    92-99

Wound core transformer, 34-46
Wound-type current transformer, 91-92
Wye connections, 241-244
Wye-to-delta connections, three-phase,
    246-248
Wye-to-wye connection, three-phase,
    244-246

Z

Zinc, as conductor of magnetic field, 7